Current Topics in
Developmental Biology

Volume 47

Somitogenesis
Part 1

Somitogenesis
Part 1

Edited by

Charles P. Ordahl
Department of Anatomy
Cardiovascular Research Institute
University of California
San Francisco, California

Academic Press
San Diego London Boston New York Sydney Tokyo Toronto

Cover photo: The cover picture shows a lateral view of an RaAchLaacZ transgenic mouse embryo histochemically stained for ß-galatosidase. In this 11.5-day postcoitum embryo, ß-gal$^+$ cells in the myotome indicate that a homologous recombination has occurred during development in the LaacZ transgene of one of the ancestral cells of the founder cells of the myotome. The small number of ß-gal$^+$ cells in each segment indicates a polyclonal origin of these segments as well as of the whole myotome. The contribution of genealogical related cells to several segments distributed along the whole axis of the embryo indicates that a single self-renewing stem cell population generates myotomal segments of the embryo. Finally, as the clone also contributes to contralateral segments from rostral to caudal (not shown), this embryo also demonstrates that the pool of ancestral cells resides permanently in nonbilateral structures of the embryo, that is, in the primitive streak and next to the tail bud. Clonal analysis provides information on the spatial and temporal organization of the pool of founder cells of a structure and on where and when the mode of clonal growth of the embryonic cells changes, often in relation to the expression of developmental genes. In addition to genetic clonal analysis in the mouse (Chapter 2), clonal analysis performed by intracellular injection of chemicals is used in the chick and in the frog for prospective tracing of the fate of cells during development. Image supplied by Jean Françoise Nicolas.

This book is printed on acid-free paper. ∞

Academic Press
A Harcourt Science and Technology Company
525 B Street, Suite 1900, San Diego, California 92101-4495, U.S.A.
http://www.apnet.com

Academic Press
24-28 Oval Road, London NW1 7DX, UK
http://www.hbuk.co.uk/ap/

International Standard Book Number: 0-12-153147-3

PRINTED IN CANADA
99 00 01 02 03 04 FR 9 8 7 6 5 4 3 2 1

Contents

3

Segmentation of the Paraxial Mesoderm and Vertebrate Somitogenesis

Olivier Pourquié

4

Segmentation: A View from the Border

Claudio D. Stern and Daniel Vasiliauskas

5

Genetic Regulation of Somite Formation

Alan Rawls, Jeanne Wilson-Rawls, and Eric N. Olson

6

Hox Genes and the Global Patterning of the Somitic Mesoderm

Ann Campbell Burke

7

The Origin and Morphogenesis of Amphibian Somites

Ray Keller

8

Somitogenesis in Zebrafish

Scott A. Holley and Christiane Nüsslein-Volhard

9

Rostrocaudal Differences within the Somites Confer Segmental Pattern to Trunk Neural Crest Migration

Marianne Bronner-Fraser

Contributors

Numbers in parentheses indicate the pages on which authors' contributions begin.

Marianne Bronner-Fraser (279), Division of Biology and Beckman Institute, California Institute of Technology, Pasadena, California 91125

Ann Campbell Burke (155), Department of Biology, University of North Carolina, Chapel Hill, North Carolina 27599

Anne Camus (1), Embryology Unit, Children's Medical Research Institute, University of Sydney, Westmead, NSW 2145, Australia

Sophie Eloy-Trinquet (33), Unité de Biologie moléculaire du Développement, 75724 Paris Cédex 15, France

Devorah Goldman (1), Embryology Unit, Children's Medical Research Institute, University of Sydney, Westmead, NSW 2145, Australia; and Program in Developmental Biology, University of California, San Francisco, California 94143-0452

Scott A. Holley (247), Max-Planck-Institut für Entwicklungsbiologie, Tübingen, Germany

Ray Keller (183), Department of Biology, University of Virginia, Charlottesville, Virginia 22903

Luc Mathis (33), Unité de Biologie moléculaire du Développement, 75724 Paris Cédex 15, France

Jean-François Nicolas (33), Unité de Biologie moléculaire du Développement, 75724 Paris Cédex 15, France

Christiane Nüsslein-Volhard (247), Max-Planck-Institut für Entwicklungsbiologie, Tübingen, Germany

Eric N. Olson (131), Department of Molecular Biology and Oncology, Hamon Center for Basic Cancer Research, The University of Texas Southwestern Medical Center at Dallas, Dallas, Texas 75235

Olivier Pourquié (81), Laboratoire de Génétique et de Physiologic du Développement, Developmental Biology Institute of Marseille, CNRS-INSERM-Université de la Méditerranée-AP de Marseille, Campus de Luminy, Marseille Cedex 09, France

Alan Rawls (131), Department of Biology, Arizona State University, Tempe, Arizona 85287

Gary C. Schoenwolf (1), Department of Neurobiology and Anatomy, University of Utah School of Medicine, Salt Lake City, Utah 84132

Claudio D. Stern (107), Department of Genetics and Development, College of Physicians and Surgeons of Columbia University, New York, New York 10032

Patrick P. L. Tam (1), Embryology Unit, Children's Medical Research Institute, University of Sydney, Westmead, NSW 2145, Australia

Daniel Vasiliauskas (107), Department of Genetics and Development, College of Physicians and Surgeons of Columbia University, New York, New York 10032

Jeanne Wilson-Rawls (131), Department of Biology, Arizona State University, Tempe, Arizona 85287

Preface

Somitogenesis has fascinated embryologists for centuries. Somites were identified as the primordia of the segmented organization of vertebrates more than two centuries ago, but it was only at the end of the 19th century that embryologists deduced that the somites are embryonic tissue progenitor organs giving rise to cartilage, skeletal muscle, and dermis tissues. Today we recognize somites as intermediates in the overall process of somitogenesis, the development of the paraxial mesoderm from gastrulation and on into the fetal period. During the 20th century, the somite has become an increasingly attractive target for experimentalists in their investigations of the cellular and molecular underpinnings of fundamental processes in vertebrate development: segmentation, tissue morphogenesis, and cell fate determination. There are many reasons for the attractiveness of the somite as an experimental target, for example, its compact size and limited cell number. However, the special fascination of somitogenesis clearly derives from its very regularity—regularity in terms not only of the sequential, reiterative, and predictable nature of its development, but also of the relative constancy of somitogenesis in embryos from different branches of the vertebrate subphylum.

The 18 chapters in these volumes address a wide variety of developmental problems through the study of somitogenesis in the embryos of four vertebrate classes (fish, amphibians, birds, and mammals). Each also presents a different strategic approach to the study of somitogenesis encompassing genetics, molecular biology, cell biology, and experimental embryology, including xenotransplantation between vertebrate classes. That diversity, however, serves an underlying commonality of themes regarding fundamental problems in development, the mechanisms of segmentation, embryo growth and cellular morphogenesis, and the molecular/genetic control of development. Finally, because overlap between chapters has been kept to a minimum, it must be acknowledged that the chapters in these volumes cannot fully represent all of the new and exciting research currently being conducted on somitogenesis. The sequence of chapters is intended to proceed from the more general to the more specific; thus, early chapters deal with gastrulation and segmentation of the paraxial mesoderm, while later chapters deal with the subsequent elaboration of specific differentiated tissues from the somite.

Although the subject was originally intended to be covered in a single volume, two volumes became necessary because of the overall length of the 18 chapters. The first volume deals with the earliest phases of paraxial mesoderm development, including gastrulation and segmentation. The chapters in the second volume deal with the later development of somites, including the elaboration of specific differentiated tissues.

I particularly thank all of the authors, whose efforts have made this two-volume set possible and with whom it has been a pleasure to work. My involvement in organizing this project was only possible with the encouragement and support of my friend and colleague Roger Pedersen and with the patience and understanding of Craig Panner at Academic Press, who diplomatically and successfullly guided the project to completion. Finally, I acknowledge my debt of eternal gratitude to Professor Margaret Hess, who first introduced me to the study of embryology 36 years ago.

Charlie Ordahl

1

Early Events of Somitogenesis in Higher Vertebrates: Allocation of Precursor Cells during Gastrulation and the Organization of a Meristic Pattern in the Paraxial Mesoderm

Patrick P. L. Tam,[1,*] *Devorah Goldman,*[1,2] *Anne Camus,*[1] and *Gary C. Schoenwolf*[3]

[1] Embryology Unit
Children's Medical Research Institute, University of Sydney,
Westmead NSW 2145, Australia

[2] Program in Developmental Biology
University of California, San Francisco, California 94143-0452

[3] Department of Neurobiology and Anatomy
University of Utah School of Medicine
Salt Lake City, Utah 84132

I. Anatomy of the Paraxial Mesoderm

The orientation of the anterior (rostral)–posterior (caudal) embryonic axis of higher vertebrates is delineated first during gastrulation by the posterior localiza-

*Correspondence should be addressed to: Patrick Tam, Embryology Unit, Children's Medical Research Institute, Locked Bag 23, Wentworthville, NSW 2145, Australia.

Current Topics in Developmental Biology, Vol. 47

0070-2153/00 $30.00

tion of the primitive streak and later during neurulation by the formation of the neuraxis and the notochord in the midline. In mouse and chick embryos, formation of the paraxial mesoderm is achieved by the recruitment of cells from the epiblast to the primitive streak, the ingression of these cells through the streak, and the organization of the ingressed cells into the mesodermal layer. Once incorporated into the new germ layer, the paraxial mesoderm condenses into two longitudinal strips of tissue flanking the neural tube. The paraxial mesoderm separates the neural tube from the intermediate mesoderm, which forms the ducts and tubules of the urogenital system, and the more laterally positioned lateral plate mesoderm, which forms the mesoderm of the gut and body wall.

A unique feature of the paraxial mesoderm is its segmentation into blocks of somites, which on differentiation give rise to the dermomyotome and sclerotome. Somites are formed as a rostrocaudal series of segmentally registered pairs during the development of the embryonic axis. However, not the entire paraxial mesoderm is overtly segmented. The mesoderm associated with the brain never forms any discrete blocks of tissue, and the portion of the paraxial mesoderm posterior to the most recently formed somite (known as the segmental plate of avian embryos and the presomitic mesoderm of mouse embryos) also appears unsegmented. For simplicity of terminology, we shall refer to the caudal unsegmented portion of the paraxial mesoderm as the presomitic mesoderm for both vertebrate embryos.

During axis development, new somites are formed when groups of mesenchymal cells in the rostral end of the presomitic mesoderm are organized into epithelial structures (Christ and Ordahl, 1995). Embryonic fragments that contain the presomitic mesoderm continue to form somites *in vitro* (Packard, 1978, 1980; Tam *et al.,* 1982). Moreover, in these explants, somites are formed in their appropriate rostrocaudal sequence. Reversal of the rostrocaudal orientation of the segmental plate by transplanting it into a host embryo also does not alter the preset order of segmentation or the rostrocaudal polarity of the somites (Stern and Keynes, 1987; Keynes and Stern, 1988). New somites are formed in these explants in their appropriate rostrocaudal sequence. Results of these experiments suggest that all of the information required for segmentation is contained within the presomitic mesoderm and the associated tissues included in the embryonic fragment. Most intriguing, is the discovery that a consistent number of somites is always formed during the differentiation of the presomitic mesoderm (10–12 somites in chick and quail; 5–6 somites in mouse), irrespective of its tissue mass which changes during embryonic development (Packard and Meier, 1983; Tam, 1986). The presence of the primitive streak or the tail bud in the embryonic fragment is essential to sustain somite formation. Without these tissues, somite formation ceases after the entire presomitic mesoderm has become segmented. *In toto,* these findings suggest that (1) the presomitic mesoderm contains cells that have been allocated to a fixed number of prospective somites, (2) these prospective somites are scheduled for segmentation in a rostrocaudal sequence, and (3) the formation of additional prospective somites beyond the fixed number requires a continuous recruitment of new cells.

The finding that the presomitic mesoderm can only form a limited (and constant) number of somites by itself suggests that the major source of new somitic cells must be outside the presomitic mesoderm and that somitic cells are continuously recruited throughout development. Recently, fate mapping studies using a transgenic *lacZ* marker, fluorescent dyes, chick/quail transplantation chimeras, or intracellular labels have provided a finer resolution of the localization and the morphogenetic movement of the somitic precursors during gastrulation and early organogenesis. An important issue that arises from the discovery that a constant number of prospective somites is contained within the presomitic mesoderm is whether a prepattern of segmental units exists prior to the epithelialization of the mesenchyme to form somites. Although numerous lines of evidence suggest the presence of a meristic pattern in the presomitic mesoderm, it is not known, if such prepattern is indeed present, whether the meristic units are already endowed with information regarding the cellular composition, physical dimensions, and developmental schedule of segmentation.

In this chapter, we focus on the early events of recruitment, specification, and allocation of cells to the paraxial mesoderm in avian and mouse embryos. We also review the morphological and molecular evidence that point to the existence of a meristic prepattern of segmentation in the presomitic mesoderm.

II. Localization of the Somitic Precursor Cells

A. Experimental Strategy for Fate Mapping

Postimplantation mouse embryos develop into a hollow cylindrical structure consisting of the inner epiblast layer enveloped by the primitive endoderm (Fig. 1). The avian embryo, in contrast, develops as a discoidal structure, with the pregastrula containing an epiblast layer overlying an incomplete hypoblast. Despite these differences in embryonic morphology, similar layers are found in mouse and chick embryos at the onset of gastrulation. For both types of embryos, the epiblast has been shown to be the sole precursor for all embryonic tissues. The primitive endoderm or hypoblast does not contribute significantly to any tissues in the embryo.

The developmental fate of cells in the epiblast has been analyzed by following their movement and differentiation during embryonic development. To track the descendants of the epiblast cells, localized populations of cells have to be marked to distinguish them from other populations in the embryo. Cell labels are introduced to either single cells or a group of cells in the epiblast by microinjection of markers, such as horseradish peroxidase, lectin conjugates, carbocyanine dyes that

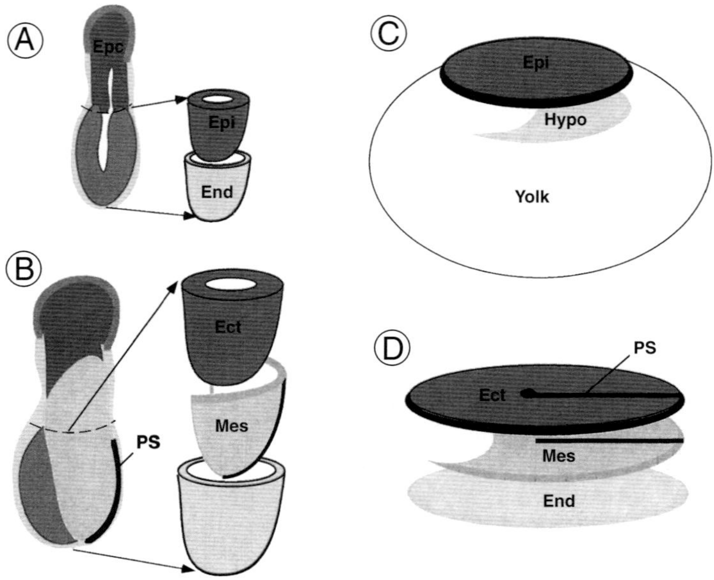

Figure 1 The organization of germ layers of (A, B) mouse and (C, D) chick embryos at the pre-streak and late-streak stages. Anterior–posterior axes of the embryo are from left to right in the figures. (A) The mouse embryo develops as a cylindrical structure containing an extraembryonic region, with an ectoplacental cone (Epc) connecting to the uterine wall, and an embryonic region. Before gastrulation, the embryo consists of an inner epiblast (Epi) and an outer primitive endoderm (End). (B) Cellular ingression at the primitive streak (PS) takes place during gastrulation leading to the formation of mesoderm (Mes) between the ectoderm (Ect) and the endoderm (End). (C) The pre-streak chick embryo is a discoidal structure on the surface of the yolk. It consists of a superficial epiblast (Epi) and a deeper incomplete hypoblast (Hypo) that spreads anteriorly from the posterior margin. (D) The chick embryo also gastrulates by cellular ingression through the primitive streak (PS) to form ectoderm (Ect), mesoderm (Mes), and endoderm (End). In the diagram, the germ layers of the chick embryo which are closely stacked together *in vivo* are shown as separated layers.

bind to the membrane, and fluorescent dyes that are incorporated into the cytoplasm. The progeny of the labeled cells may be tracked as long as the label has not been diluted to undetectable levels during cell proliferation. Alternatively, the fate of cells may be traced by the expression of transgenes encoding β-galactosidase or green fluorescent protein. To exploit these genetic markers, cells expressing the transgene have to be isolated from a donor embryo and then transplanted to a host embryo for the assessment of developmental fate. In avian embryos, the fate of cells may be traced in a similar manner by grafting quail cells or fluorescent cells to host chick embryos, with the grafted cells being uniquely identified using specific antibodies.

In both the microinjection labeling and transplantation approaches, it is necessary to perform micromanipulation on the embryo and to culture the embryo for some time before cell fate can be analyzed. It is, therefore, imperative to ascertain that such experimental manipulation and embryo culture have no adverse effects on the behavior and differentiation of the embryonic cells. In these experiments, cell fate is often assessed by the types of tissues colonized and the histological characteristics of the differentiated cells. Neither criterion is sufficient to distinguish between passive incorporation of the labeled cells into the embryonic tissues and the genuine acquisition of a particular cell fate. With the availability of immunological and riboprobe reagents, it is now possible to assay more critically the differentiation of the cells under study by the coexpression of tissue-specific molecules and the lineage marker (Lawson and Hage, 1994; Tam and Zhou, 1996; Arkell and Beddington, 1997; Darnell and Schoenwolf, 1997; Tam *et al.,* 1997; Yuan and Schoenwolf, 1998).

Results of fate mapping experiments performed in the last decade using more reliable cell markers have enabled the construction of much finer fate maps for the germ layers of the gastrulating avian and mouse embryos (Lawson *et al.,* 1991; Lawson and Pedersen, 1992; Hatada and Stern, 1994; Smith and Schoenwolf, 1997; Tam and Behringer, 1997). The maps show the geographical localization of the precursor cells in the germ layers and allow a prediction of the fate of cells during normal development.

B. The Paraxial Mesodermal Domain in the Epiblast

The cell population that is destined to form the embryonic mesoderm has been mapped in the epiblast of mouse and chick embryos shortly before gastrulation commences (mouse: Lawson *et al.,* 1991, pre-streak stage of Downs and Davies, 1993; chick: Hatada and Stern, 1994, stage XIV of Eyal-Giladi and Kochav, 1976). In the mouse, prospective mesoderm is localized to the lateral region of the epiblast, extending as a strip from the proximal to the posterior–distal region of the epiblast (Fig. 2A). It is not possible at this early stage to delineate the respective precursors of paraxial mesoderm (which includes the cranial mesoderm and the somites) and lateral plate mesoderm, whereas a distinction may be made in the chick embryo. In the chick, the precursor cells of paraxial mesoderm are localized to the central region of the epiblast and overlap with those of the heart and the lateral plate mesoderm (Fig. 2B).

After the initiation of gastrulation (mouse: early-streak stage, chick: stage 2–3a), the embryonic mesodermal precursors become more localized and the specific populations destined for the paraxial mesoderm can now be readily distinguished. In the mouse, paraxial mesodermal precursor cells are found in the distal area of the mesodermal domain in the lateral epiblast (Lawson and Pedersen, 1992; Parameswaran and Tam, 1995) (Fig. 2C). In the chick, the prospective paraxial mesoderm occupies the epiblast lateral to the rostral half of the primitive streak

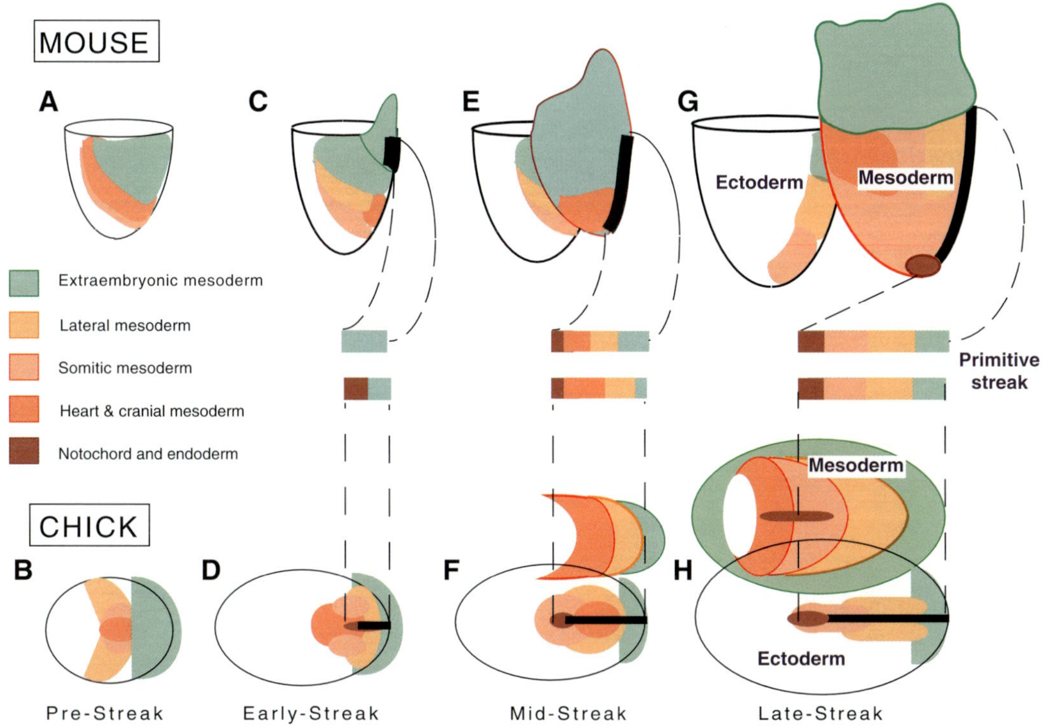

MOUSE
A
C
E
G
Ectoderm
Mesoderm
Extraembryonic mesoderm
Lateral mesoderm
Somitic mesoderm
Heart & cranial mesoderm
Notochord and endoderm
Primitive streak
CHICK
B
D
F
H
Mesoderm
Ectoderm
Pre-Streak
Early-Streak
Mid-Streak
Late-Streak

(Fig. 2D). Despite the different shape and topography of the chick and mouse embryo, it seems that the prospective paraxial mesoderm is localized to similar sites in the epiblast of the early gastrulas of the two organisms. An implication of this finding is that whatever the mechanism that leads to the specification and regionalization of the prospective paraxial mesoderm may be, it is likely to be common to both type of embryos. This finding also implies that the tissue movement involved in the recruitment of the prospective mesoderm to the primitive streak is also similar in both species.

C. Recruitment and Ingression through the Primitive Streak

In the mouse, the primitive streak is formed on the posterior side of the cup-shaped embryo at the junction of the epiblast and the extraembryonic ectoderm (Fig. 2C). During gastrulation, the primitive streak elongates toward the distal tip of the embryo (Lawson and Pedersen, 1992). Such distal extension is, in effect, equivalent to an anterior extension along the anterior–posterior embryonic axis (Tam and Behringer, 1997). In the discoidal chick embryo, the primitive streak first forms near the posterior margin of the blastoderm, and the primitive streak extends anteriorly during gastrulation (Selerio *et al.,* 1996; Shah *et al.,* 1997). The distal end of the primitive streak of the mouse gastrula is, therefore, analogous to the anterior (rostral) end of the streak in the chick embryo (Fig. 2D), and it shall be referred to as such hereafter for consistency and simplicity of nomenclature.

The pattern of cell movement in the epiblast during gastrulation can be reconstructed from the spatial distribution of the clonal descendants of labeled or grafted cells. In the mouse, there is, in general, a posterior displacement of the

Figure 2 Fate maps showing the localization of prospective somitic mesoderm in gastrulating mouse and chick embryos. The stages of development shown are pre-streak [mouse, 6.0-day (A); chick, stage XIV (B)], early-streak [mouse, 6.5-day (C); chick, stages 2–3a (D)], mid-streak [mouse, 7.0-day (E); chick, stages 3b and 3c (F)] and late-streak [mouse, 7.5-day (G); chick, stages 3d, 4 (H)]. In the mouse embryo, the mesoderm is formed superficially to the epiblast and spreads anteriorly (to the left of the diagram) and proximally (toward the top of the diagram). The mesoderm is shown in its normal position for the early-streak embryo, but is shown pulled away in the mid- and late-streak embryos to reveal the epiblast/ectoderm. The endoderm is not shown. The primitive streak, which is found at the posterior margin of the cylindrical embryo, is depicted in an anterior–posterior orientation to align with the primitive streak of the chick embryo. For the chick embryo, only the epiblast at the pre- and early-streak stages and the ectoderm and mesoderm at the mid- and late-streak stages are shown. For the purpose of illustration, the mesoderm which is normally found deep into the ectoderm is lifted above the ectoderm. The anterior–posterior embryonic axis is from left to right on the diagram. Fate maps of the mouse are based on Lawson *et al.* (1991), Lawson and Pedersen (1992), Smith *et al.* (1994), Tam and Quinlan (1996), Tam and Behringer (1997), and S. Kinder, G. A. Quinlan, and P. P. L. Tam (unpublished), and the chick on Hatada and Stern (1994), Smith and Schoenwolf (1997), and Lemaire and Kessel (1997).

prospective mesodermal cells toward the primitive streak (Lawson *et al.,* 1991; Tam and Behringer, 1997) as the cells closer to the primitive streak leave the epiblast by ingression through the streak (Poelmann, 1981; Tam *et al.,* 1993). However, the actual pattern of displacement of epiblast cells to the primitive streak has not yet been directly examined by tracking cell movement.

At early gastrulation, the primitive streak is formed adjacent to the region of the epiblast containing the prospective extraembryonic mesoderm. Cells in this region of the epiblast are likely to be the first to be recruited to the primitive streak. As the streak elongates anteriorly, it impinges on other mesodermal precursors in more distal locations in the epiblast, and these epiblast cells will now be recruited to the primitive streak (Fig. 2C–H). It would be reasonable to expect that the order of recruitment of epiblast cells to the primitive streak occurs in a posterior to anterior (i.e., proximal to distal) direction. The order of recruitment of cells to the primitive streak can be revealed by mapping the types of mesodermal precursors that are ingressing through the primitive streak (Lawson *et al.,* 1991), as well those that have been incorporated into the mesoderm (Section II,D) at different stages of gastrulation.

Fate mapping studies have been performed on the mouse gastrula in which the types of precursor cells that are present in different segments of the primitive streak during the first 24 hr of gastrulation have been examined (S. Kinder, G. A. Quinlan, and P. P. L. Tam, unpublished). At the early-streak stage, cells in transit through the streak are those destined for the extraembryonic mesoderm (Fig. 2C). At the mid-streak stage, the posterior segment of the streak contains the extraembryonic mesoderm, the midportion contains the lateral mesoderm (lateral plate and intermediate mesoderm), and the anterior portion contains the heart and cranial paraxial mesoderm and the gut endoderm (Fig. 2E). The primitive streak is at its full length by the late-streak stage (Lawson and Pedersen, 1992). The types of cells ingressing through the streak at this stage are, in posterior to anterior order, the precursors of extraembryonic mesoderm, lateral mesoderm, somitic mesoderm, axial mesendoderm, and gut endoderm (Fig. 2G) (Tam and Beddington, 1987, 1992; Lawson *et al.,* 1991; Smith *et al.,* 1994).

The sequential order of cellular ingression through the mouse primitive streak is remarkably similar to that of the chick gastrula (reviewed by Smith and Schoenwolf, 1997; Lemaire and Kessel, 1997) with one significant exception. In the chick, cells ingressing through the early primitive streak are those of the endoderm (and hypoblast, before formation of the streak), as well as those of the extraembryonic mesoderm, as in the mouse. Thereafter, the cellular composition is similar for the primitive streak of mouse and chick gastrulas at equivalent stages. At the early-to-mid-streak stage (stage 3), the streak contains in posterior to anterior sequence, precursors of the extraembryonic mesoderm, lateral plate mesoderm, intermediate mesoderm, heart mesoderm, head mesoderm, and endoderm (Fig. 2D,F). By the late-streak stage (stage 4), the posterior segment of the primitive streak contains essentially the same precursor cell populations of extraembryonic mesoderm, lateral plate, and intermediate mesoderm, but the anterior segment now contains the somitic mesoderm and the prechordal mesoderm (Fig. 2H). The notochord precursor cells ingress later at the most anterior sites, beginning just after stage 4. The similar order of recruitment of the various precursor cell populations in the mouse and the chick strongly suggests that a similar morphogenetic program may operate during germ layer formation in the two classes of vertebrates. It is not known in either organism, however, whether the order in which mesodermal cell types emerge from the primitive streak may be an indication of the order in which mesodermal lineages are specified.

It is clear from the fate mapping results that the prospective somitic mesoderm ingresses after the cranial paraxial mesoderm. Clonal analysis of the epiblast cells, however, reveals that cells that are allocated to the cranial mesoderm are distributed throughout the prospective paraxial mesoderm and are not always localized closer to the primitive streak (Lawson and Pedersen, 1992). Thus, there is no evidence for a craniocaudal regionalization of these mesodermal precursors in the epiblast. Fate mapping studies also suggest that by the late-streak stage, the bulk of prospective cranial mesoderm has already exited the streak, and the cells in the anterior segment of the primitive streak at this stage are principally destined for the first 6–10 pairs of somites and the presomitic mesoderm of the early-somite stage embryo (Tam and Beddington, 1986, 1987). This finding has now been confirmed by examining the effect of the ablation of the anterior one-third of the primitive streak and the adjacent germ layer tissues of the late-streak stage mouse embryo (Fig. 3). Microsurgical extirpation of this segment of the primitive streak results in the truncation of the paraxial mesoderm on either one side or both sides of the embryo. The embryo contains the normal amount of cranial mesoderm and the first four to five somites in the truncated paraxial mesoderm (Fig. 3). The embryonic axis continues to develop, but the missing somites are not replaced. Results of the ablation experiment show that by the late-streak stage (1) the prospective paraxial mesoderm that has previously been recruited during early gastrulation is allocated to the cranial mesoderm and the first four to five somites, and (2) the precursor cells of the rest of the paraxial mesoderm are localized in the anterior segment of the primitive streak and the neighboring germ layer tissues. These findings are particularly relevant to the understanding of the somite phenotype produced by mutating genes that are expressed in the prospective somitic mesoderm within the primitive streak. Somitogenesis after the formation of the first five to six somites is disrupted in embryos mutant for *Brachyury, Wnt3a,* or *Tbx6* (Takada *et al.,* 1994; Wilson *et al.,* 1995; Chapman and Papaioannou, 1997; Yoshikawa *et al.,* 1997). Thus, in each of the three mutants, it seems that the cranial mesoderm and the first few somites that are already formed are spared from the effects of the mutation, suggesting that activity of these genes is required specifically by the more posterior somitic mesoderm. In the case of the *Wnt-3a*

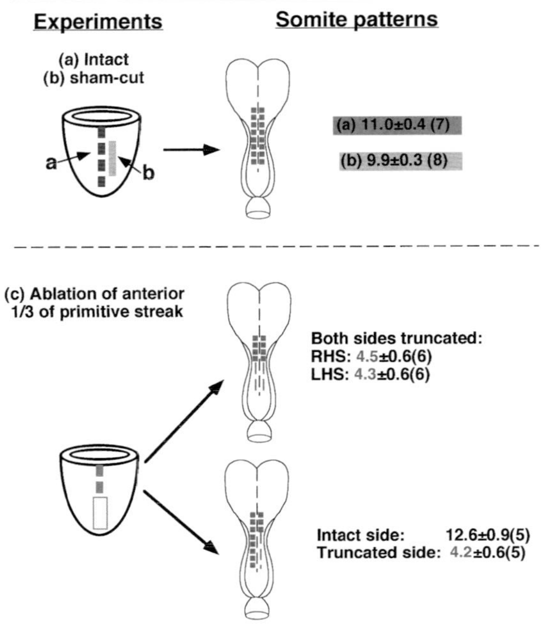

Figure 3 Experiments showing the effect on somite formation of ablation of the anterior one-third segment of the primitive streak in the late-streak mouse embryo (shown as viewed from the posterior side). The somite patterns were examined after 24 hr of *in vitro* development. Control experiments: (a) primitive streak left intact and (b) longitudinal cut made lateral to the primitive streak. (c) Ablation of the primitive streak results in the deletion of somites. The series of somites may be truncated on one side (n = 5) or both sides (n = 6). The cranial mesoderm is always intact, and about four to five somites are left in the truncated series. In another five cases not shown here, all somites on one side are lost. The variation in the extent of truncation is likely to be due to the amount of primitive streak tissue removed in different embryos.

and *Tbx6* mutations, this phenotpye makes sense given the expression patterns of these two genes. Specifically, the timing of their expression may preclude them from playing a role in specifying the most anterior paraxial mesoderm. *Wnt-3a* is expressed throughout the primitive streak (Takada *et al.,* 1994) and *Tbx6* is expressed in the anterior end of the primitive streak and adjacent epiblast at gastrulation (Chapman *et al.*, 1996); neither of these genes is expressed until the mid-to-late-streak stages. On the other hand, *Brachyury* (Wilkinson *et al.,* 1990; Herrmann, 1991) is expressed throughout the primitive streak, beginning at early gastrulation when the anterior paraxial mesoderm ingresses through it. Overall, these mutant phenotypes point to the presence of differential genetic requirments for the formation of the paraxial mesoderm at different segmental positions along the body axis.

The truncation of the somitic pattern following the ablation of the precursor population in the streak and the neighboring germ layers implies that this is the sole source of somitic mesoderm for the embryo. That this population is not reconstituted from other cells suggests that cells outside of the somitic precursor population are restricted in their differentiative potency from being respecified for somitic fate. It has been shown that in chick embryos, ablation of the node elicits the reconstitution of the organizer and the notochord from the surrounding non-node tissues (Yuan *et al.,* 1995a,b; Psychoyos and Stern, 1996a; Yuan and Schoenwolf, 1998). In the chick, defined populations of blastoderm cells are known to interact as a responder and an inducer, leading to the reconstitution of the body axis. Because of this, it is possible that the lack of regulation after the loss of the somitic precursor population is due to the ablation of not only the precursor population but also of other cell populations essential for the reconstitution of the somitic precursors. It must also be noted that, in the experiment described above, the node of the mouse gastrula is left intact. In the chick embryo, the presence of the node will suppress the induction of the extra organizer cells from the epiblast (Yuan *et al.,* 1995a,b; Yuan and Schoenwolf, 1998). Full reconstitution of somites occurs in the chick when both the node and the anterior segment of the primitive streak have been removed (Yuan and Schoenwolf, 1998). The presence of the node in the mouse embryo with anterior-streak ablations may, therefore, suppress the respecification of somites from the remaining germ layer tissues, which possess the potential for somitic differentiation as revealed by heterotopic transplantation (Garcia-Martinez and Schoenwolf, 1992). It is also possible that the lack of regulation for the loss of the somitic precursor cells is also due to the ablation of much if not all of the tissues that are critical for the *de novo* induction of extra somitic cells. A proper test of these hypotheses would be to perform selective extirpation of the node, the anterior streak, or adjacent germ layers to identify the cell populations essential for the reconstitution of the somitic precursors.

Circumstantial evidence suggests that the somitic precursor population in the primitive streak may act as a self-renewing pool of stem cells for somitogenesis. When the precursor population is labeled *in situ,* its descendants are found in the somites formed after labeling as well as in the primitive streak (Stern *et al.,* 1988; Selleck and Stern, 1992a; Psychoyos and Stern, 1996b). Similarly, in the mouse, somitic precursor cells transplanted to the primitive streak are found later not only in the somites but also in the primitive streak (Tam and Beddington, 1987; Tam and Tan, 1992). The presence of a resident population in the primitive streak despite the continuous cellular contribution to new somites is consistent with the presence of a self-renewing cell population. If the somite precursor cells behave like a stem cell population, then descendants of individual cells will make a periodic contribution to the somites as the parental cell goes through its proliferative cycles. A periodicity of cell allocation might be revealed if a small number of cells of the pool are labeled and followed over time. It has indeed been observed in both chick and mouse embryos that the descendants of tracked cells do not colonize a continuous series of somites, but instead colonize a group of three to four consecutive somites, sometimes with a discernible periodicity of five to six segments (Packard, 1986; Selleck and Stern, 1991, 1992b; Tam and Tan, 1992; Lawson and Pedersen, 1992; Tam and Beddington, 1986, 1987). A periodicity of cellular distribution to the somites is also found during the analysis of marked myotomal clones, generated by random activation of a *lacZ* transgene in a small number of somitic precursor cells (Nicolas *et al.,* 1996, and Chap. 17 by Eloy-Trinquet *et al.,* this volume).

D. Sequential Allocation of Somitic Precursor Cells to the Embryonic Axis

The mesoderm of the early-streak stage mouse embryo consists entirely of cells destined for the extraembryonic mesoderm. As gastrulation proceeds, cells destined for the cranial and heart mesoderm appear in the nascent mesoderm adjacent to the anterior segment of the elongating primitive streak (Fig. 2B,C). Cells destined for the first few somites are found in the mesoderm of the late-streak embryo following the ingression of the somitic mesoderm from the mid-streak stage onward (Parameswaran and Tam, 1995) (Fig. 2D). There is, therefore, an anterior to posterior sequence of allocation of cells to the paraxial mesoderm during gastrulation. As the somites are formed, cells derived from the precursor population are recruited to successively more posterior somites (Tam and Beddington, 1986, 1987). There is, however, no overt restriction in the segmental address for the somitic precursor population. Cells that will normally colonize more anterior somites can contribute to more posterior somites (and vice versa) after they are transplanted among the somite precursor population of embryos at different developmental stages (Tam and Tan, 1992). It is not known, however, whether the cells that are found in somites differ from those they will normally colonize may alter their characteristics in accordance with their new segmental address. It will be interesting to test whether they may change their expression profile of *Hox* genes (Kessel, 1991, 1992) or contribute to skeletal or myogenic derivatives of a totally different segmental level (Wachlter *et al.,* 1982).

E. Somites Are Assembled from Cells of Different Sources during Axis Development

In the chick embryo, somitic precursor cells are localized in the caudolateral part of the Hensen's node and the anterior 2/3 segment of the primitive streak. Cells derived from the node tend to occupy a more medial position in the paraxial mesoderm and, later, a more medial part of the somites. Cells from the more posterior primitive streak region acquire a more lateral position in the paraxial mesoderm and tend to congregate to the more lateral part of the somite (Selleck and Stern, 1991). However, cells that colonize the medial portion of the somites may be derived also from the primitive streak, resulting in considerable overlapping of the distribution of the node and streak-derived cells to the medial and lateral parts of the somite (Schoenwolf *et al.,* 1992; Psychoyos and Stern, 1996b). The initial positions of the medial and lateral streams of somitic cells may result from the difference in the order of movement of cells from the node versus those from the streak to the paraxial mesoderm. At a specific segmental level in the body axis, cells from the primitive streak ingress earlier and initially occupy both the lateral and the medial positions in the transverse plane. As the node regresses to the same segmental level, cells leaving the node now occupy the more medial position in the paraxial mesoderm, laterally displacing the cell population already in the paraxial mesoderm. This mode of cellular recruitment from the precursor population that spans the anterior–posterior length of the primitive streak will result in the apparent mediolateral segregation of cells in the paraxial mesoderm (Selleck and Stern, 1992b). Because a mixing of cells occurs during the development of the presomitic mesoderm (Stern *et al.,* 1988; Selleck and Stern, 1991, 1992b), a persistent segregation of medial and lateral somitic cells in the segmental plate is unlikely and may not be critical for the specification of the medial and lateral somitic cell populations.

The anterior primitive streak and the adjacent epiblast of the late-streak mouse embryo contains the precursor cells for at least the next 18 somites (Lawson and Pedersen, 1992). Concomitant with the elongation of the neural tube, the primitive streak becomes situated beneath the posterior neuropore (Tam, 1981; Tam *et al.,* 1982; Wilson and Beddington, 1996). Lineage analysis by cell labeling and cell transplantation shows that the primitive streak of the early-somite to the forelimb bud stage embryo continues to act as a major source of somitic mesoderm (Tam and Beddington, 1987; Tam and Tan, 1992). A detailed map of the prospective fate of cells in the primitive streak within the posterior region of the mouse embryo (at the 4–7 somite stage) shows that the somitic precursor population is localized in the medial region of the epiblast immediately posterior to the prospective neuroectoderm and the axial mesoderm (Wilson and Beddington, 1996) (Fig. 4A). A similar localization of the prospective somites in the caudal region is also found in the chick by analyzing chick/quail chimeras (Catala *et al.,* 1996) (Fig. 4B). Cells grafted to the mouse primitive streak at the 4–7 somite stage are not distributed evenly to the whole somite, but there is no consistent trend of mediolateral distribution (Tam and Beddington, 1987; Tam and Tan, 1992). However, in a finer mapping of the primitive streak, it has been possible to show that cells originating from the anterior to posterior parts of the primitive streak tend to distribute respectively

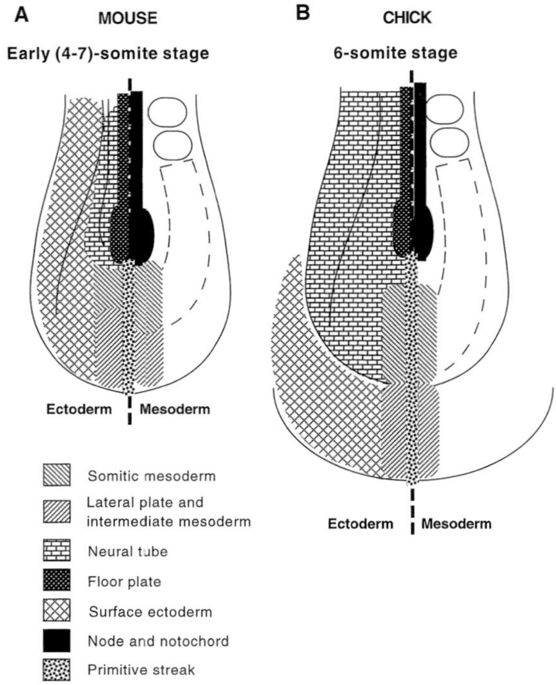

Figure 4 The localization of tissue precursors in the ectoderm and mesoderm of the posterior neuropore region of the mouse and chick embryo. The midline tissue consists of the primitive streak, displaced caudally from its position in the early gastrula. The outline of the neuropore is shown on the left-hand side of each diagram and that of the paraxial mesoderm on the right. The caudal neural tissue is more extensive in the chick than in the mouse, and in the chick but not in the mouse, prospective lateral neural plate lies in the midline, caudal to the prospective floor plate. As the notochord and overlying floor plate extend caudally, the prospective lateral neural plate lying in the midline is cleaved into right and left lateral sides. Fate maps are based on Wilson and Beddington (1996) and Catala *et al.* (1996).

to the medial and lateral parts of the somite (Wilson and Beddington, 1996). It seems, therefore, that the pattern of cell movement from the primitive streak to the paraxial mesoderm is similar between avian and mouse embryos.

A transition in cellular recruitment to the paraxial mesoderm from the primitive streak to the tail bud mesenchyme occurs at about the stage of the closure of the posterior neuropore. In the chick embryo, orthotopic transplantation of a labeled tail bud at stages just after closure of the posterior neuropore provides direct evidence of the somitogenic potential of the tail bud (Schoenwolf, 1977; Catala *et al.,* 1995). A direct test of the somitogenic potential of cells in the tail bud mesenchyme of the mouse embryo is technically not feasible because of our inability to achieve normal growth of the late-organogenesis stage embryo *in vitro*. Instead, tail bud cells are tested by transplanting them to the primitive streak of the early-somite stage embryo (Tam and Tan, 1992). Despite the difference in the developmental age of the donor cells and the host environment, tail bud cells can colonize somites in the host. Even the cells taken from the tail bud of a 13.5-day embryo, which will soon cease to form somites, can participate in somite formation for a period extended well beyond their normal somitogenic life span. This finding shows that the cessation of somitogenesis is not due to the loss of potency by the tail bud somitic precursor cells, but rather is due to the constraint of the local tissue environment, perhaps by the withdrawal of essential growth factors for cell proliferation and tissue modeling.

III. Developmental Plasticity and Commitment to Somitic Fate

An important question about the somitic precursor cells is their state of commitment. Although the prospective paraxial mesoderm is regionalized to a specific region of the epiblast and the primitive streak at gastrulation, this does not mean that cells are irreversibly allocated to the somitic lineage. Clonal analysis reveals that epiblast cells have descendants in both ectoderm and mesoderm and have not committed to derivatives of only one germ layer. Less than 30% of the epiblast population contributes exclusively to one embryonic structure identifiable at the early-somite stage (Lawson *et al.*, 1991; Lawson and Pedersen, 1992). Heterotopic transplantation of prospective paraxial mesoderm from the posterior-lateral epiblast to the anterior epiblast (prospective ectoderm) results in the acquisition of a nonmesodermal fate, indicating the developmental plasticity of this tissue (Parameswaran and Tam, 1995). In the chick, prospective paraxial mesoderm from the epiblast can differentiate to neural tissues that express *En-2* and *Hoxb1* (Garcia-Martinez *et al.*, 1997). Even in the primitive streak, the somitic precursor cells are not committed. Anterior primitive streak cells transplanted to the posterior streak differentiate into extraembryonic or lateral plate mesoderm (Beddington, 1982; Garcia-Martinez and Schoenwolf, 1992). Prospective mesodermal cells from the anterior segment of the primitive streak can also contribute to cells in the neural plate when they are transplanted to the prospective neuroectoderm in the epiblast (Garcia-Martinez *et al.*, 1997); additionally, they can form notochordal (*Not1*-expressing) cells when grafted into Hensen's node of the mid-streak stage embryo (Selleck and Stern, 1992a).

Cells that are already incorporated into the presomitic mesoderm are still not fully committed to a somitic fate. Lineage analysis of *in situ* labeled presomitic cells and of presomitic cells transplanted heterotopically shows that they can also contribute to lateral plate mesoderm (reviewed by Tam and Trainor, 1994). Recently, it has been shown that the switch from somitic to lateral mesodermal fate may be influenced by the activity of growth factors of the Transforming Growth Factor β superfamily. Implanted COS cells producing Bone Morphogenetic Protein-4 may influence the differentiation of the somite in a dose-dependent manner. High doses of BMP-4 convert the somites to lateral plate mesoderm and low doses permit somite differentiation. There is a reduction in the *Pax3* expressing sclerotome, however, and some somitic cells display inappropriate differentiation, as revealed by the expression of *cSim1,* which is characteristic of the lateral somite (Tonegawa *et al.,* 1997). Commitment to a somitic fate appears to be irreversible only after the epithelialization of the mesenchymal cells into somites (Tam and Trainor, 1994).

IV. Does a Prepattern of Segmentation Exist in the Presomitic Mesoderm?

A. Experimental Evidence

The concept of a meristic (i.e., presegmental) organization of cells in the mouse presomitic mesoderm was first suggested in a study of somite formation in the *amputated* mutant embryo (Flint *et al.,* 1978). The posterior body axis of the mutant is shorter than the wild-type sibling and the axis is truncated at the lower sacral level. Although the first 30–35 somites form, they are smaller than their normal counterparts. The presomitic mesoderm is also shorter in the mutant, suggesting that the smaller somites might be the result of a deficiency in somitic precursor cells. The mutant embryo, nevertheless, seems to regulate for the tissue deficiency by adjusting the number of cells allocated to each segment so that a normal number of somites is produced before truncation of the axis occurs. A computer simulation was used to generate a model to account for this regulative mode of somite formation (Flint *et al.,* 1978). Two characteristic features of this model are that a consistent number of oscillating "cellular states" is present in the presomitic mesoderm, and the number of oscillations is approximately the same in the mutant and wild-type embryo. There is some experimental evidence that the cell cycle may potentially define the cellular state of the model. Based on the finding of the synchrony of mitotic activity and the predictable localization of disrupted segmentation in response to heat shock, it has been proposed that mitosis is the periodic cellular state that coordinates the establishment of a segmental unit in the chick segmental plate (Primmett *et al.,* 1988, 1989; Stern *et al.,* 1992). If the number of oscillation of cell state can be interpreted as the periodicity of a meristic pattern, the simulation is suggestive of the existence of presegmental units in the presomitic mesoderm.

The pattern of allocation of cells from the primitive streak to the paraxial mesoderm reveals that prospective somites are already established in the presomitic mesoderm. In the mouse, colonization of the paraxial mesoderm by streak-derived cells usually begins with the fifth or sixth somite that is formed after either labeling the streak or grafting cells to the streak (Tam and Beddington, 1986, 1987; Tam and Tan, 1992). In the chick embryo, quail cells grafted to the tail bud begin to colonize the twelfth somite formed after grafting (Catala *et al.*, 1995). When the mouse or avian presomitic mesoderm is explanted and cultured *in vitro,* about 5–6 or 10–12 somites, respectively, are formed from the isolated tissue (Tam *et al.*, 1982; Packard and Meier, 1983). Moreover, the same number of somites are formed from isolates of presomitic mesoderm at several stages even though the length of the presomitic mesoderm varies during development (Tam, 1986). These findings are consistent with the notion that a fixed number of prospective somites is already established in the presomitic mesoderm and that new presegmental units are added to the posterior end of this tissue at the same rate as cells are presegmented off into somites at the anterior end (Tam, 1981; Tam and Trainor, 1994).

B. A Meristic Pattern of Somitomeres

A somitomeric organization of mesenchymal cells has been described in the presomitic mesoderm of several vertebrates (reviewed by Jacobson and Meier, 1986; Jacobson, 1988; Tam and Trainor, 1994). The somitomere is identified (best by stereo scanning electron microscopy) by the concentric alignment of the somas and filopodia of the mesenchymal cells facing the neural primordium (Meier, 1979; Tam *et al.*, 1982). The packing density of the presomitic mesenchyme increases from the posterior to the anterior region of the presomitic mesoderm, but there are no abrupt changes that demarcate the domain of each somitomere (Tam and Beddington, 1986; Freund *et al.*, 1996). The number of somitomeres that are present in the presomitic mesoderm matches that of the prospective somites. In an explant of presomitic mesoderm, the formation of somites is accompanied by the loss of an equal number of somitomeres (Packard and Meier, 1983; Tam, 1986). A somitomere, however, does not seem to behave as a closed tissue compartment, and some mixing of cells between adjacent somitomeres does occur (Tam, 1988).

The experimental evidence to date has been strongly in favor of the hypothesis that a meristic organization of cells is present in the mesenchyme of the presomitic mesoderm. It is tempting to infer that the somitomeres are the morphological manifestation of the presegmental pattern. However, the morphological criteria for delineating the somitomere are subjective and controversial. The generation of a regular number of somites in the tissue explants does not require the pre-existence of a somitomeric organization. Whatever cellular organization may exist in the presomitic mesoderm, it has been shown to be plastic. Because the quail embryo is smaller than the chick embryo, the quail segmental plate is found to contain more somitomeres than the chick counterpart of the same size. When a portion of the chick segmental plate is replaced by the quail segmental plate of an equivalent size, the number of somites that is formed by the quail transplant is the same as the number of chick somitomeres that have been removed and not the same as the number of quail somitomeres that have been transplanted (Packard *et al.,* 1993). The meristic pattern of the segmental plate is, therefore, not fully determined and is subject to regulation until overt segmentation occurs.

C. Revelations from Gene Expression and Mutational Studies

There is a rapidly expanding list of genes that are expressed in a regionalized manner in the presomitic mesoderm of the four common vertebrate experimental models. Broadly, two patterns of gene expression are recognized. First, the gene is expressed in localized segments of the presomitic mesoderm, as if its activity defines specific segmental units (Fig. 5A). Second, the gene is expressed in bands of tissue in the presomitic mesoderm and later in either the anterior or posterior half of the somite, giving the impression that the gene activity is related to the formation of either the prospective compartments or the future boundary of the somites (Fig. 5B). In addition, a recently discovered homologue of the *hairy* gene is found to be expressed in a dynamic manner that parallels the temporal sequence of somite segmentation (Palmeirim *et al.,* 1997; see also Chap. 4 by Pourquié, this volume). It is pertinent to note that the activity of the genes in the presomitic mesoderm has been assayed by the detection of the steady-state level of transcripts. The *in situ* hybridization pattern does not necessarily reflect the synthesis and the localization of the functional protein (e.g., *twist,* Gitelman, 1997). This discrepancy may account for the apparent disparity between the expression pattern in the presomitic mesoderm and the tissues that are affected by the loss-of-function mutation of the gene (e.g., *Mesp2,* Saga *et al.,* 1997; *Fgfr1,* Yamaguchi *et al.,* 1994).

Some clues to the organization of cells in the presomitic mesoderm may be sought by studying patterns of gene expression. If gene activity has any bearing on the meristic organization of prospective somites, it is expected that it will reveal (1) the physical dimensions, (2) the position of the boundary, and/or (3) the prospective rostrocaudal compartments of the segmental units. The activity of some genes may also reflect the sequential and temporal order of somite segmentation. For simplicity of terminology in the following discussion, the prospective somite in the presomitic mesoderm is referred to as somitomere, irrespective of whether such an anatomical unit has been described for the species or not. Somitomeres are referred to by their positions in the presomitic mesoderm: the most anterior one is somitomere I and the last one is somitomere VII for *Xenopus,* XII for the bird, and VI for the mouse (Tam *et al.,* 1982; Tam, 1986; Tam and Bedding-

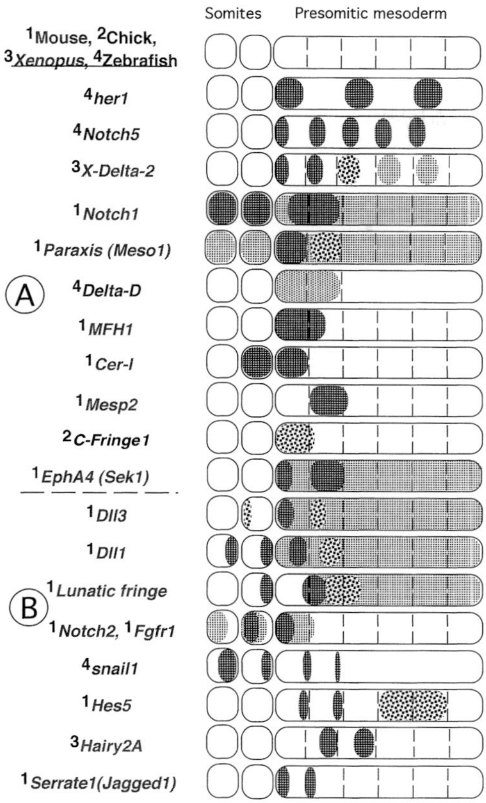

Figure 5 Patterns of gene expression in the presomitic mesoderm and the two most recently segmented somites in mouse, chick, *Xenopus,* and zebrafish embryos. Gene expression patterns in (A) indicate the meristic patterns in the presomitic mesoderm. Those in (B) demarcate the boundaries and the rostrocaudal compartments of the prospective somites. See text for details and the references of the expression pattern of specific genes. The genes are coded for the species (1, mouse; 2, chick; 3, *Xenopus;* 4, zebrafish) and the expression domain in the paraxial mesoderm is blue. The level of expression as revealed by the signal intensity of *in situ* hybridization is indicated as low (light stipple), moderate (coarse stipple), and strong (dense stipple). For alignment of the expression pattern to the meristic prepattern, the presomitic mesoderm is subdivided into 6 somitomeres. A different number of somitomeres are found in the presomitic mesoderm of different vertebrate embryos: 5–6 in the mouse, 10–12 in the bird, and 5–7 in the *Xenopus.* Somitomeres have not been described in the zebrafish so the matching of gene expression domains to somitomeres is based on results of the fate mapping study by Müller *et al.* (1996).

ton, 1986; reviewed by Tam and Trainor, 1994). The number of somitomeres in the zebrafish is not known, but 10 somitomeres have been found in one fish species, *Oryzias latipes* (Martindale *et al.,* 1987).

1. Regionalized Gene Expression Implies a Meristic Prepattern

If the activity of the gene is reflecting the organization of the meristic unit in the presomitic mesoderm, its expression pattern is expected to correlate closely with the position and the number of somitomeres present in the presomitic mesoderm. In zebrafish, expression of the hairy homologue (*her1*) in the presomitic mesoderm is displayed as two or three discrete transverse bands separated by nonexpressing bands of similar size (Müller *et al.,* 1996). The size and position of the *her1* expressing population suggests that these cells may correspond to somitomeres I, III, and V. The segmental position and the number of stripes are maintained during the development of the embryo from the 3- to 21-somite stage (Müller *et al.,* 1996), suggesting that *her1* activity undergoes an alternating phase of up- and downregulation as groups of cells in the presomitic mesoderm mature and segment into somites. Fate mapping studies, conducted by labeling the cells in the anterior two *her1*-expressing domains at the three-somite stage, reveal that these cells are fated for the second and fourth somites formed after labeling (somite five and seven, respectively) (Müller *et al.,* 1996). Therefore, *her1* expression may represent a molecular manifestation of the meristic prepattern in the zebrafish. A similar stripped pattern of expression is found for the *Notch5* gene in the zebrafish (Westin and Lardelli, 1997). *Notch5* expression is generally weak and up to five stripes can be found in the presomitic mesoderm. The position of the first two stripes may correspond to the first two somitomeres. The precise number of somitomeres in the zebrafish, if they exist in the presomitic mesoderm, is not known. Until we have better knowledge of the somitomeric organization of the zebrafish presomitic mesoderm, it is certainly premature to suggest that the activity of these genes in the zebrafish has any relation to the somitomeres.

In *Xenopus,* the *X-Delta-2* gene is expressed in a series of stripes (each of which seems to fill a somitomere) that spread through the presomitic mesoderm as the axis develops (Jen *et al.,* 1997). *X-Delta-2* is likely to map to the first five somitomeres in the *Xenopus* presomitic mesoderm (Jen *et al.,* 1997), which has five to seven somitomeres (Jacobson and Meier, 1986). Besides the genes in the zebrafish and *Xenopus,* no genes have been identified to date that are expressed in a meristic pattern that matches the number of somitomeres in the chick and the mouse presomitic mesoderm (12 and 6 somitomeres, respectively). A possible reason for this may be related to the different mode of somite formation between the *Xenopus* and the chick and mouse. In *Xenopus,* cells in the presomitic mesoderm are initially aligned dorsoventrally. During somite formation, a group of cells in the anterior part of the presomitic mesoderm are rotated *en bloc* to become a stack of rostrocaudally aligned cells that constitute one somite (Hamilton, 1969). Cell mixing, therefore, would be limited in the *Xenopus* presomitic mesoderm and it would be more likely that a meristic pattern can be maintained more effectively as revealed by the regionalized pattern of gene expression. In chick and mouse embryos, the mesenchymal architecture of the presomitic mesoderm may render it less likely that cells will maintain a stable position within the tissues (but see observation of C. Stern cited by Tajbakhsh and Spörle, 1998). Instead of a meristic pattern of gene expression highlighting the somitomeres, many genes that appear to mark the anteriormost somitomeres have been identified and, in some cases, their function has been revealed by targeted disruption in the mouse. A detailed discussion of these genes will follow in the next section.

The discovery that hairylike segmentation genes are expressed during somitogenesis (e.g., *her1* in fish, *Hairy2A* in *Xenopus, c-hairy1* in chick, *Hes5,* and *Uncx4.1* in mouse) indicates that the gap and pair-rule gene system may participate in the segmentation of vertebrate paraxial mesoderm (Müller *et al.,* 1996; Jen *et al.,* 1997; Neidhardt *et al.,* 1997; Palmeirim *et al.,* 1997; R. A. Conlon, unpublished). Expression of *her1* in the presomitic mesoderm is highly reminiscent of the pair-rule expression pattern of hairy in *Drosophila* embryos. In *Xenopus, Hairy2A* expression is upregulated in the presumptive posterior portion of some anterior somitomeres (Jen *et al.,* 1997) (Fig. 5B). A more dynamic pattern of hairy gene expression is found in the chick segmental plate (see Chap. 4 by Pourquié, this volume). The *c-hairy1* gene is expressed in a rhythmic manner, and the activity spreads from the posterior toward the anterior part of the segmental plate, culminating in the convergence of the expression to the posterior border of the somitomere that is destined to become the next somite. Therefore, *c-hairy1* activity defines the posterior border of a prospective somite. It is of particular interest that the periodicity of the pulse of gene activity in cells at a particular position in the segmental plate corresponds to the time taken to generate a somite during segmentation (Palmeirim *et al.,* 1997). This rhythmic activity has been taken to be coupled to a developmental timing mechanism that regulates the rate of somitogenesis. It is not clear at this stage if this clock works by counting the number of rounds of *c-hairy1* transcription the cells have gone through or by titrating the amount of protein product that has been accumulated in the cell after each round of gene transcription.

2. Gene Activity and the Establishment of Prospective Somites

A number of genes are expressed in anterior regions of the presomitic mesoderm in patterns that seem to match the expected domains occupied by the first two somitomeres (Fig. 5A). In the zebrafish, expression of the *Delta-D* gene is restricted to two bands in the anterior region of the presomitic mesoderm (Dornseifer *et al.,* 1997). The tissue that expresses *Delta-D* is roughly enough to fill somitomeres I and II based on an estimate of size relative to that of the newly formed somite. The mouse *cerberus-like* (*Cer-l*) and the chick *C-Fringe1* genes are expressed mainly in somitomere I (Belo *et al.,* 1997; Sakamoto *et al.,* 1997). The mouse Eph receptor gene (*EphA4*) is expressed at a low level in the presomitic mesoderm. It is more strongly expressed in two discrete bands that seem to match the anterior half of somitomere I and the whole of somitomere II (Nieto *et al.*, 1992). Likewise, the mouse *MFH1* gene is expressed in tissues that may be equivalent to the most anterior 1.5 somitomeres (Winnier *et al.*, 1997). However, null mutation of the *MFH1* gene does not affect somitogenesis, suggesting that the expression of this gene in the presomitic mesoderm may not have a critical role for somite formation.

Mutational analysis in the mouse of some genes expressed in such a somitomeric pattern has partially uncovered their function in somitogenesis. The *Notch1* gene is expressed throughout the presomitic mesoderm, but it is most strongly expressed in somitomeres I and II (Reaume *et al.*, 1992; Williams *et al.*, 1995). Its disruption leads to epithelialization defects and uncoordinated segmentation; however, other aspects of somite differentiation are unaffected (Conlon *et al.*, 1995). The *Mesp2* gene is expressed specifically in the tissue corresponding to somitomere II (Saga *et al.*, 1997), whereas *paraxis* expression is upregulated in the anterior part of the presomitic mesoderm and is strongly expressed in somitomeres I and II (Burgess *et al.*, 1996a; Blanar *et al.*, 1995; Barnes *et al.*, 1997). The activity of these two basic helix-loop-helix transcription factors is clearly required for somite formation in the mouse embryo. Mutation of the *paraxis* gene results in the failure of epithelialization of the somites, though myogenic and chondrogenic differentiation of the somitic cells are not affected (Burgess *et al.*, 1996b), and *Mesp2* mutation leads to absence of segmentation (Saga *et al.*, 1997). The *paraxis−/−* and *Mesp−/−* phenotypes suggest that epithelialization or segmentation is not a prerequisite for the differentiation of somitic lineages. The expression of both genes in the anterior presomitic mesoderm overlaps with that of N-cadherin and fibronectin (Duband *et al.*, 1987; Takeichi, 1988). The loss of N-cadherin results in abnormal somite shape, but like the *paraxis* mutation, it has no effect on somite segmentation (Radice *et al.*, 1997). In contrast, loss of fibronectin results in failure to form somites (George *et al.*, 1993; Georges-Labouesse *et al.*, 1996). Although these mutant phenotypes relate to defects in different aspects of somite formation, they do not, however, clarify the functional role of these genes in somite segmentation.

3. Defining the Rostrocaudal Compartments

During development of the paraxial mesoderm, some genes are initially expressed uniformly in the presomitic mesoderm but later the expression is restricted to either the rostral or the caudal half of the somites (e.g., *cVg1*, Shah *et al.*, 1997). Cells in these two parts of the somite display different fates (Keynes and Stern, 1988) and different properties of contact-dependent cell interactions (Krull *et al.*, 1997; Orioli and Klein, 1997). A restriction of gene activity to the anterior or posterior half of the somites may occur prior to segmentation. Thus, the expression of some genes (e.g., *Dll3, Dll1, Lunatic fringe, Notch2, Fgfr1,* and *snail1;* Thisse *et al.,* 1993; Yamaguchi *et al.,* 1994; Lardelli, 1995; Williams *et al.,* 1995; Cohen *et al.,* 1997; Dunwoodie *et al.,* 1997; Forsberg *et al.,* 1998; Fig. 5B) is progressively upregulated such that strong expression is found in one or two stripes in the anterior presomitic mesoderm and is subsequently localized either to the rostral or caudal half of the somite. By correlating the size and the position of their expression domains, it seems that these genes are expressed in the rostral or caudal part of the somitomere. Finally, some genes are expressed in a regionalized pattern in the presomitic mesoderm but are absent from the somites (e.g., *X-Delta-2, Hes5, Hairy2A,* and *Serrate1/Jagged1;* R. A. Conlon, unpublished; Shawber *et al.,* 1996; Cohen *et al.,* 1997; Jen *et al.,* 1997; Mitsiadis *et al.,* 1997) (Fig. 5). The spacing between strips of expressing tissue is very close to the rostrocaudal length of the somite. In *Xenopus,* as the presomitic mesoderm matures, *X-Delta-2* expression is downregulated in the posterior part of the somitomeres, where a concomitant upregulation of the *Xenopus* hairy homologue (*Hairy2A*) takes place. This, therefore, results in a pattern of meristic expression domains comprising an anterior *X-Delta-2* and a posterior *Hairy2A* zone. This pattern is reminiscent of the alternating regions of *Dll3* and *Dll1-lunatic fringe* expression in the somitomeres of the mouse presomitic mesoderm (Dunwoodie *et al.,* 1997; Johnston *et al.,* 1997; Cohen *et al.,* 1997; Forsberg *et al.,* 1998) (Fig. 5). By extrapolating from the somitomere to the somite, the expression pattern, therefore, seems to mark the position of the boundary between the somitomeres. This raises an intriguing possibility that the specification of both the rostrocaudal subdivision and the boundary of the segmental unit occurs prior to segmentation (reviewed by Keynes and Stern, 1988; Tam and Trainor, 1994).

Of the genes that are known to be expressed in the presomitic mesoderm, many encode molecules of the Notch signaling pathway (reviewed by Artavanis-Tsakonas *et al.,* 1995; Artavanis-Tsakonas, 1997). The *Delta* genes (e.g., *Delta-D, X-Delta-2, Dll2, Dll3*) and the *Serrate/Jagged* genes encode the ligands that activate the receptors encoded by the *Notch* genes (e.g., *Notch1* and *Notch2*). The selective regulation of gene transcription through Notch signal-transduction is instrumental in the choice of cell fate during lineage differentiation (Kopan and Turner, 1996; Seugnet *et al.,* 1997). In *Drosophila,* and presumably in vertebrates (Robey, 1997), Notch signaling represses the activity of some *E(spl)* class basic-helix-loop-helix transcription factors (Kopan and Cagan, 1997; Seugent *et al.,* 1997) and activates expression of genes that are the targets of the *Su(H)* DNA binding protein (Artavanis-Tsakonas *et al.,* 1995; de Celis *et al.,* 1996; Nye, 1997). Apparently, the two ligands have differential expression and signaling activity in *Drosophila* and their juxtaposition results in the formation of a tissue boundary (Doherty *et al.,* 1996). It has recently been found that the interaction of the Notch receptor with its ligands may be potentially modulated by the glycosylation activity of a class of secreted proteins encoded by the *Fringe* genes (Wu *et al.,* 1996; Panin *et al.,* 1997). Specifically, the activity of the *Fringe* product may block the activation of the Notch receptor by the Serrate ligand and enhance activation by the Delta ligand (Fleming *et al.,* 1997; Panin *et al.,* 1997), providing an explanation for their differing signaling activites.

It is therefore interesting to note that two *Delta* genes (*Dll1* and *Dll3*) are differentially expressed in the mouse presomitic mesoderm and somites (Dunwoodie *et al.,* 1997), whereas the *Notch1* gene is widely expressed in the presomitic mesoderm (Reaume *et al.,* 1992; Williams *et al.,* 1995). It is of further interest that the *Fringe*-related gene, *lunatic fringe,* is expressed broadly in the posterior presomitic mesoderm of the mouse. As somitogenesis proceeds, this band of expression is narrowed first to the equivalent of a whole somitomere and then to occupy the posterior portion of the somitomere. The posterior border of the expression domain appears to coincide with the posterior margin of the somitomere, and it is later found in the posterior border of the somite (Cohen *et al.,* 1997; Johnston *et al.,* 1997; Forsberg *et al.,* 1998). The expression of *lunatic fringe* partly overlaps that of the *Dll1* gene (Bettenhausen *et al.,* 1995). Both genes are absent from the anterior region of the segment that expresses *Dll3* (Dunwoodie *et al.,* 1997). Thus, in the mouse presomitic mesoderm, the specificity of Notch signaling is likely to be derived from the varieties of ligands and receptors that are expressed in a tissue- or temporal-specific manner.

The critical requirement for Notch signaling in meristic organization and polarization of the somite has been tested by analyzing the impact of mutation of the ligand and the receptor genes. As previously mentioned, mutation of the *Notch1* gene in the mouse does not affect segmentation but results in uncoordinated somite formation (Conlon *et al.,* 1995; Ciruna *et al.,* 1997). Likewise, mutation of the *Su(H)* homologue, *RBP-Jϰ* gene results in a similar phenotype (Oka *et al.,* 1995); however, it is unclear from the data whether somites in homozygous mutants can establish normal rostrocaudal polarity. The overlapping expression of *Notch1* and *Notch2* in the somites suggests that these receptors are at least partially functionally redundant and that there are additional functions of Notch signaling during somitogenesis that remain to be uncovered. A loss of *Dll1* activity results in the absence of anterior polarity as well as the lack of epithelialization of the somites (Hrabe de Angelis *et al.,* 1997). The loss of segmental borders in the *Dll1*−/− embryo suggests that juxtaposition of cells with anterior and posterior identities is necessary for the maintenance of the somite boundary. Similarly, loss of function of *presenilin 1* (*PS1*), a molecule that helps to transduce the activated the Notch signal, can lead to diminished expression of *Dll1* and *Notch1* in the presomitic mesoderm. In these mutants, somites with irregular shape and disorganized epithelialization are formed (Wong *et al.,* 1997). Whether this reflects a defect in tissue patterning before segmentation or in maintaining segmentation is unclear. Finally, mouse embryos lacking the activity of *Mesp2* gene do not form segmented somites and the somites that do form lack rostrocaudal patterning. Furthermore, these defects occur concomitantly with changes in the expression of *Dll1* and a loss of *Notch1* and *Notch2* expression (Saga *et al.,* 1997).

To summarize, in general, it appears that any disruption in the activity of the

Notch signaling pathway will result in abnormal somitogenesis. However, it is not possible to separate the effects of the mutations on epithelialization and specification of rostrocaudal polarity of the somite. The mutant phenotypes seem to suggest that the primary function of Notch signaling is to refine rather than to specify a presegmental pattern in the presomitic mesoderm.

V. Unanswered Questions

Through the fate mapping of mouse and avian embryos, the localization of the somite precursors in the germ layers, the primitive streak, and the tail bud has been extensively elucidated. A stage-by-stage study of the localization of the precursor population and the distribution of the clonal descendants of labeled or transplanted cells has enabled the reconstruction of the morphogenetic movement of the prospective somitic cells during gastrulation and early organogenesis. The tracking of the somitic precursors at late organogenesis has been less successful in the mouse than in the chick because of the technical difficulties of maintaining normal development *in vitro.* However, studies on these two higher vertebrates have revealed substantial homology in the allocation of cell lineages, the morphogenesis of the paraxial mesoderm, and the specification of the somites, such that some generalization may be warranted for both species. Despite this, several outstanding issues are still not resolved. First, very little is known about the developmental process and the timing of induction of the paraxial mesoderm. Currently, the focus has been on the primitive streak, where differential gene activity has been demonstrated as different population of mesodermal tissues emerge during gastrulation (Sasaki and Hogan, 1993; Lemaire and Kessel, 1997). Mutational analysis so far has identified some genes, such as *Fgfr1, Tbx6,* and *Wnt3a,* whose activity may be critical for somite formation (reviewed by Tam and Trainor, 1994; Tam and Behringer, 1997), but it is not clear whether they are involved in the induction of the somitic mesoderm. Further analysis of the inductive mechanism may begin with the streak-ablation approach, which could reveal the tissue components that are involved in somite induction, in a manner similar to what has been obtained from the node-ablation studies (Psychoyos and Stern, 1996a; Yuan and Schoenwolf, 1998; Yuan *et al.,* 1995a,b).

Results of fate mapping experiments have shown that the primitive streak and the tail bud are continuous sources of new cells for somitogenesis. It is presently not known how this self-renewing population is maintained and how the prospective somitic cells are incorporated into the paraxial mesoderm. Of significant interest is how the somitogenic potential of this self-renewing population is regulated, especially toward the cessation of somitogenesis. As the first step of investigation, detailed maps of the cell fates and the molecular/genetic activity of the primitive streak and tail bud should be constructed for formulating testable hypotheses about the regulatory mechanisms of somitogenesis.

Gene expression in the presomitic mesoderm has been intensively studied for evidence of presegmental tissue organization and for hints of delineation of segmental boundaries and lineage compartments. It must be emphasized that in none of these studies has the gene expression pattern been matched directly with the somitomeric pattern in the same embryo. It, therefore, remains conjectural as to whether the regionalized gene activity is related to the organization of a prepattern for segmental units. Further evidence is needed to show that gene activity is defining the membership of cells to a specific somitomere and that the boundary of the expression domain corresponds to the boundary of the somitomere that subsequently becomes that of the somite. Mutational studies of genes that are expressed in the presomitic mesoderm have produced confounding results. The lack of a somite phenotype in some cases and the wider spectrum of somite defects that is unrelated to gene expression patterns strongly suggest that somite patterning is subject to multifactorial regulation, as exemplified by the complex phenotypic outcome of mutations of the *Notch* signaling pathways.

Acknowledgments

We thank Ron Conlon, Gabriel Quinlan, and Shipeng Yuan for allowing us to cite their unpublished results, and Bruce Davidson and Peter Rowe for reading the manuscript. Our work is supported by the National Health and Medical Research Council (NHMRC) of Australia, the National Institutes of Health, USA, and Mr. James Fairfax. D. G. is supported by a *Development* Traveling Fellowship from the Company of Biologists Ltd. P. T. is a Principal Research Fellow of the NHMRC.

References

Arkell, R., and Beddington, R. S. P. (1997). BMP-7 influences pattern and growth of the developing hindbrain of mouse embryos. *Development* **124,** 1–12.

Artavanis-Tsakonas, S. (1997). Alagille syndrome—a notch up for the Notch receptor. *Nat. Genet.* **16,** 212–218.

Artavanis-Tsakonas, S., Matsuno, K., and Fortini, M. E. (1995). Notch signaling. *Science* **268,** 225–232.

Barnes, G. L., Alexander, P. G., Hsu, C. W., Mariani, B. D., and Tuan, R. S. (1997). Cloning and characterization of chicken paraxis: A regulator of paraxial mesoderm development and somite formation. *Dev. Biol.* **189,** 95–111.

Beddington, R. S. P. (1982). An autoradiographic analysis of tissue potency in different regions of the embryonic egg cylinder. *J. Embryol. Exp. Morphol.* **69**(2), 65–285.

Belo, J. A., Bouwmeester, T., Leyns, L., Kertesz, N., Gallo, M., Follettie, M., and De Robertis, E. M. (1997). *Cerberus-like* is a secreted factor with neuralizing activity expressed in the anterior primitive endoderm of the mouse gastrula. *Mech. Dev.* **68,** 45–57.

Bettenhausen, B., Harbe de Angelis, M., Simon, D., Guénet, J.-J., and Gossler, A. (1995). Transient and restricted expression during mouse embryogenesis of Dll1, a murine gene closely related to *Drosophila* Delta. *Development* **121,** 2407–2418.

Blanar, M. A., Crossley, P. H., Peters, K. G., Steingrímsson, E., Copeland, N. G., Jenkins, N. A., Martin, G. R., and Rutter, W. J. (1995). Meso1, a basic helix-loop-helix protein involved in mammalian presomitic mesoderm development. *Proc. Natl. Acad. Sci. U.S.A.* **92,** 5870–5874.

Burgess, R., Cserjesi, P., Ligon, K. L., and Olson, E. N. (1996a). Paraxis: A basic helix-loop-helix protein expressed in paraxial mesoderm and developing somites. *Dev. Biol.* **168,** 296–306.

Burgess, R., Rawles, A., Brown, D., Bradley, A., and Olson, E. N. (1996b). Requirement of paraxis gene for somite formation and musculo-skeletal patterning. *Nature* **384,** 570–573.

Catala, M., Teillet, M. A., and Le Douarin, N. M. (1995). Organization and development of the tail bud analysed with chick-quail chimaera system. *Mech. Dev.* **51,** 51–66.

Catala, M., Teillet, M. A., De Robertis, E. M., and Le Douarin, N. M. (1996). A spinal cord fate map in the avian embryo: While regressing, Hensen's node lays down the notochord and floor plate thus joining the spinal cord lateral walls. *Development* **122,** 2599–2610.

Chapman, D. L., and Papaioannou, V. E. (1998). Three neural tubes in mouse embryos with mutations in the T-box gene *Tbx6. Nature* **391,** 695–697.

Chapman, D. L., Agulnik, I., Hancock, S., Silver, L. M., and Papaioannou, V. E. (1996). *Tbx6,* a mouse *T*-Box gene implicated in paraxial mesoderm formation at gastrulation. *Dev. Biol.* **180,** 534–542.

Christ, B., and Ordahl, C. (1995). Early stages of chick somite development. *Anat. Embryol.* **191,** 381–396.

Ciruna, B. G., Schwartz, L., Harpal, K., Yamaguchi, T. P., and Rossant, J. (1997). Chimaeric analysis of *fibroblast growth factor receptor-1* (*Fgfr1*) function: A role for FGFR1 in morphogenetic movement through the primitive streak. *Development* **124,** 2829–2841.

Cohen, B., Bashirullah, A., Dagnino, L., Campbell, C., Fisher, W. W., Leow, C. C., Whiting, E., Ryan, D., Zinyk, D., Boulianne, G., Hui, C. C., Gallie, B., Phillips, R. A., Lipshitz, H. D., and Egan, S. E. (1997). Fringe boundaries coincide with Notch-dependent patterning centres in mammals and later Notch-dependent development in *Drosophila. Nat. Genet.* **16,** 283–288.

Conlon, R. A., Reaume, A. G., and Rossant, J. (1995). Notch1 is required for coordinate segmentation of somites. *Development* **121,** 1533–1545.

de Celis, J. F., de Celis, J., Ligoxygakis, P., Preiss, A., Delidakis, C., and Bray, S. (1996). Functional relationships between Notch, Su(H) and the βHLH genes of the E(spl) complex: The E(spl) genes mediate only a subset of Notch activities during imaginal development. *Development* **122,** 2719–2728.

Darnell, D. K., and Schoenwolf, G. C. (1997). Vertical induction of Engrailed-2 and other region-specific markers in the early chick embryo. *Dev. Dyn.* **209,** 45–58.

Doherty, D., Feger, G., Younger-Shepherd, S., Yeh Jan, L., and Nung Jan, Y. (1996). Delta is a ventral to dorsal signal complementary to Serrate, another Notch ligand, in *Drosophila* wing formation. *Genes Dev.* **10,** 421–434.

Dornseifer, P., Takke, C., and Campos-Ortega, J. A. (1997). Overexpression of a Zebrafish homologue of the *Drosophila* neurogenic gene Delta perturbs differentiation of primary neurons and somite formation. *Mech. Dev.* **63,** 159–171.

Downs, K. M., and Davies, T. (1993). Staging of gastrulating mouse embryos by morphological landmarks in the dissecting microscope. *Development* **118,** 1255–1266.

Duband, J.-L., Dugour, S., Hatta, K., Takeichi, M., Edelman, G. M., and Thiery, J. P. (1987). Adhesion molecules during somitogenesis in the avian embryo. *J. Cell Biol.* **104,** 1361–1374.

Dunwoodie, S. L., Henrique, D., Harrison, S. M., and Beddington, R. S. P. (1997). Mouse *Dll3:* A novel divergent *Delta* gene which may complement the function of other *Delta* homologues during early pattern formation in the mouse. *Development* **124,** 3065–3076.

Eyal-Giladi, H., and Kochav, S. (1976). From cleavage to primitive streak formation: A complementary normal table and a new look at the first stages of the development of the chick. I. General morphology. *Dev. Biol.* **49,** 321–337.

Fleming, R. J., Gu, Y., and Hukriede, N. A. (1997). Serrate-mediated activation of *Notch* is specifically blocked by the product of the gene fringe in the dorsal compartment of the *Drosophila* wing imaginal disc. *Development* **124,** 2973–2981.

Flint, O. P., Ede, D. A., Wilby, O. K., and Proctor, J. (1978). Control of somite number in normal and *amputated* mutant mouse embryos: An experimental and a theoretical analysis. *J. Embryol. Exp. Morphol.* **45,** 189–202.

Forsberg, H., Crozet, F., and Brown, N. A. (1998). Waves of mouse lunatic fringe expression, in four-hour cycles at two-hour intervals, precedes somite boundary formation. *Curr. Biol.* **8,** 1027–1030.

Freund, R., Dörfler, D., Popp, W., and Wachtler, F. (1996). The metameric pattern of the head mesoderm—does it exist? *Anat. Embryol.* **193,** 73–80.

Garcia-Martinez, V., and Schoenwolf, G. C. (1992). Positional control of mesoderm movement and fate during avian gastrulation and neurulation. *Dev. Dyn.* **193,** 249–256.

Garcia-Martinez, V., Darnell, D. K., Lopez-Sanchez, C., Sosic, D., Olson, E. N., and Schoenwolf, G. C. (1997). State of commitment of prospective neural plate and prospective mesoderm in late gastrula/early neurula stages of avian embryos. *Dev. Biol.* **181,** 102–115.

George, E. L., Georges-Labouesse, E. N., Patel-King, R. S., Rayburn, H., and Hynes, R. O. (1993). Defects in mesoderm, neural tube and vascular development in mouse embryos lacking fibronectin. *Development* **119,** 1079–1091.

Georges-Labouesse, E. N., George, E. L., Rayburn, H., and Hynes, R. O. (1996). Mesodermal development in mouse embryos mutant for fibronectin. *Dev. Dyn.* **207,** 145–156.

Gitelman, I. (1997). Twist protein in mouse embryogenesis. *Dev. Biol.* **189,** 205–214.

Hamilton, L. (1969). The formation of somites in *Xenopus. J. Embryol. Exp. Morphol.* **22,** 253–264.

Hatada, Y., and Stern, C. D. (1994). A fate map of the epiblast of the early chick embryo. *Development* **120,** 2879–2890.

Herrmann, B. G. (1991). Expression pattern of the *Brachyury* gene in whole mount T^{Wis}/T^{Wis} mutant embryos. *Development* **113,** 913–917.

Hrabe de Angelis, M., McIntyre II, J., and Gossler, A. (1997). Maintenance of somite borders in mice requires the delta homolog Dll1. *Nature* **386,** 717–721.

Jacobson, A. G. (1988). Somitomeres: Mesodermal segments of vertebrate embryos. *Development* **104**(Suppl.), 209–220.

Jacobson, E. R. G., and Meier, S. (1986). Somitomeres: The primordial body segments. *In* "Somites in Developing Embryos" (R. Bellairs, D. A. Ede, and J. W. Lash, Eds.), NATO ASI Series, Vol. 118, pp. 1–16. Plenum, New York.

Jen, W.-C., Wettstein, D., Turner, D., Chitnis, A., and Kintner, C. (1997). The Notch ligand, *X-Delta-2,* mediates segmentation of the paraxial mesoderm in *Xenopus* embryos. *Development* **124,** 1169–1178.

Johnston, S. H., Rauskolb, C., Wilson, R., Prabhakaran, B., Irvine, K. D., and Vogt, T. F. (1997). A family of mammalian *Fringe* genes implicated in boundary determination and the *Notch* pathways. *Development* **124,** 2245–2254.

Kessel, M. (1991). Molecular coding of axial positions by *Hox* genes. *Semin. Dev. Biol.* **2,** 367–373.

Kessel, M. (1992). Respecification of vertebral identities by retinoic acid. *Development* **115,** 487–501.

Keynes, R. J., and Stern, C. D. (1988). Mechanisms of vertebrate segmentation. *Development* **103,** 413–430.

Kopan, R., and Cagan, R. (1997). Notch on the cutting edge. *Trends Genet.* **13,** 465–467.

Kopan, R., and Turner, D. L. (1996). The Notch pathway: Democracy and aristocracy in the selection of cell fate. *Curr. Biol.* **6,** 594–601.

Krull, C. E., Lansford, R., Gale, N. W., Collazo, A., Marcelle, C., Yancopoulos, G. D., Fraser, S. E., and Bronner-Fraser, M. (1997). Interactions of Eph-related receptors and ligands confer rostrocaudal pattern to trunk neural crest migration. *Curr. Biol.* **7,** 571–580.

Lardelli, M. (1995). Complementary and combinatorial patterns of *Notch* gene family expression during early mouse development. *Mech. Dev.* **53,** 357–368.

Lawson, K. A., and Hage, W. J. (1994). Clonal analysis of the origin of primordial germ cells in the mouse. Ciba Found. Symp., **182,** 68–91.

Lawson, K., and Pedersen, R. A. (1992). Early mesoderm formation in the mouse embryo. *In* "Formation and Differentiation of Early Embryonic Mesoderm" (R. Bellairs, E. J. Sanders, and J. W. Lash, Eds.), pp. 33–46. Plenum, New York.

Lawson, K. A., Meneses, J. J., and Pedersen, R. A. (1991). Clonal analysis of epiblast fate during germ layer formation in the mouse embryo. *Development* **113,** 891–911.

Lemaire, L., and Kessel, M. (1997). Gastrulation and homeobox genes in chick embryos. *Mech. Dev.* **67,** 3–16.

Martindale, M. Q., Meier, S., and Jacobson, A. G. (1987). Mesodermal metamerism in the teleost, *Oryzias latipes* (the Medaka). *J. Morphol.* **193,** 241–252.

Meier, S. (1979). Development of the chick mesoblast. Formation of embryonic axis and establishment of the metameric pattern. *Dev. Biol.* **73,** 25–45.

Mitsiadis, T. A., Henrique, D., Thesleff, I., and Lendahl, U. (1997). Mouse *Serrate-1* (*Jagged-1*): Expression in the developing tooth is regulated by epithelial–mesenchymal interactions and fibroblast growth factor-4. *Development* **124,** 1473–1493.

Müller, M., Weizsäcker, E. V., and Campos-Ortega, J. A. (1996). Expression domains of a Zebrafish homologue of the *Drosophila* pair-rule gene hairy correspond to primordia of alternating somites. *Development* **122,** 2071–2078.

Neidhardt, L. M., Kispert, A., and Herrmann, B. G. (1997). A mouse gene of the paired-related homeobox class expressed in the caudal somite compartment and in the developing vertebral column, kidney and nervous system. *Dev. Genes Evol.* **207,** 330–339.

Nicolas, J. F., Mathis, J., and Bonnerot, C. (1996). Evidence in the mouse for self-renewing stem cells in the formation of a segmented longitudinal structure. *Development* **122,** 2933–2946.

Nieto, M. A., Gilardi-Hebenstreit, P., Charnay, P., and Wilkinson, D. G. (1992). A receptor protein tyrosine kinase implicated in the segmental patterning of the hindbrain and mesoderm. *Development* **116,** 1137–1150.

Nye, J. S. (1997). Developmental signaling: Notch signals Kuz it's cleaved. *Curr. Biol.* **7,** R716–R720.

Oka, C., Nakano, T., Wakeham, A., de la Pompa, J. L., Mori, C., Sakai, T., Okazaki, S., Kawaichi, M., Shoita, K., Mak, T. W., and Honjo, T. (1995). Disruption of the mouse RBP-Jk gene results in early embryonic death. *Development* **121,** 3291–3301.

Orioli, D., and Klein, R. (1997). The Eph receptor family: Axonal guidance by contact repulsion. *Trends Genet.* **13,** 354–359.

Packard, D. S., Jr. (1978). Chick somite determination: The role of factors in young somites and the segmental plate. *J. Exp. Zool.* **203,** 295–306.

Packard, D. S., Jr. (1980). Somitogenesis in cultured embryos of the Japanese quail, *Coturnix coturnix japonica. Am. J. Anat.* **158,** 83–91.

Packard, D. S., Jr. (1986). The epiblast origin of the avian somite cells. *In* "Somites in Developing Embryos" (R. Bellairs, D. A. Ede, and J. W. Lash, Eds.), pp. 37–45. Plenum, New York.

Packard, D. S., Jr., and Meier, S. (1983). An experimental study of the somitomeric organisation of the avian segmental plate. *Dev. Biol.* **97,** 191–202.

Packard, D. S., Jr., Zhang, R.-Z., and Turner, D. C. (1993). Somite pattern regulation in the avian segmental plate mesoderm. *Development* **117,** 779–791.

Palmeirim, I., Henrique, D., Ish-Horowicz, D., and Pourqie, O. (1997). Avian *hairy* gene expression identifies a molecular clock linked to vertebrate segmentation and somitogenesis. *Cell* **91,** 639–648.

Panin, V. M., Papayannopoulos, V., Wilson, R., and Irvine, K. D. (1997). Fringe modulates Notch–ligand interactions. *Nature* **387,** 908–912.

Parameswaran, M., and Tam, P. P. L. (1995). Regionalisation of cell fate and morphogenetic movement of the mesoderm during mouse gastrulation. *Dev. Genet.* **17,** 16–28.

Poelmann, R. E. (1981). The formation of the embryonic mesoderm in the early postimplantation mouse embryo. *Anat. Embryol.* **162,** 29–40.

Primmett, D. R. N., Stern, C. D., and Keynes, R. J. (1988). Heat shock causes repeated segmental anomalies in the chick embryo. *Development* **104,** 331–338.

Primmett, D. R. N., Norris, W. E., Carlson, G., Keynes, R. J., and Stern, C. D. (1989). Periodic segmental anomalies induced by heat shock in the chick embryo are associated with the cell cycle. *Development* **105,** 119–130.

Psychoyos, D., and Stern, C. D. (1996a). Restoration of the organizer after radical ablation of Hensen's node and the anterior primitive streak in the chick embryo. *Development* **122,** 3263–3273.

Psychoyos, D., and Stern, C. D. (1996b). Fates and migratory routes of primitive streak cells in the chick embryo. *Development* **122,** 1523–1534.

Radice, G. L., Rayburn, H., Matsunami, H., Knudsen, K. A., Takeichi, M., and Hynes, R. O. (1997). Developmental defects in mouse embryos lacking N-cadherin. *Dev. Biol.* **181,** 64–78.

Reaume, A. G., Conlon, R. A., Zirngib, R., Yamaguchi, T. P., and Rossant, J. (1992). Expression analysis of a *Notch* homologue in the mouse embryo. *Dev. Biol.* **154,** 377–387.

Robey, E. (1997). Notch in vertebrates. *Curr. Opin. Genet. Dev.* **7,** 551–557.

Saga, Y., Hata, N., Koseki, H., and Taketo, M. M. (1997). Mesp2: A novel mouse gene expressed in the presegmented mesoderm and essential for segmentation initiation. *Genes Dev.* **11,** 1827–1839.

Sakamoto, K., Yan, L., Imai, H., Takagi, M., Nabeshima, Y., Takeda, S., and Katsube, K. (1997). Identification of a chick homologue of fringe and C-fringe 1: Involvement in the neurogenesis and the somitogenesis. *Biochem. Biophys. Res. Commun.* **234,** 754–759.

Sasaki, H., and Hogan, B. L. M. (1993). Differential expression of multiple fork head related genes during gastrulation and axial pattern formation in the mouse embryo. *Development* **118,** 47–59.

Schoenwolf, G. C. (1977). Tail (end) bud contributions to the posterior region of the chick embryo. *J. Exp. Zool.* **201,** 227–246.

Schoenwolf, G. C., Garcia-Martinez, V., and Dias, M. S. (1992). Mesoderm movement and fate during avian gastrulation and neurulation. *Dev. Dyn.* **193,** 235–248.

Selerio, E. A. P., Connolly, D. J., and Cooke, J. (1996). Early developmental expression and experimental axis determination by the chicken *Vg1* gene. *Curr. Biol.* **6,** 1476–1486.

Selleck, M. A. J., and Stern, C. D. (1991). Fate mapping and cell lineage analysis of Hensen's node in the chick embryo. *Development* **112,** 615–626.

Selleck, M. A. J., and Stern, C. D. (1992a). Commitment of mesoderm cells in Hensen's node to the chick embryo to notochord and somite. *Development* **114,** 403–415.

Selleck, M. A. J., and Stern, C. D. (1992b). Evidence for stem cells in the mesoderm of Hensen's node and their role in embryonic pattern formation. *In* "Formation and Differentiation of Early Embryonic Mesoderm" (R. Bellairs, E. J. Sanders, and J. W. Lash, Eds.), pp. 23–29. Plenum, New York.

Seugnet, L., Simpson, P., and Haenlin, M. (1997). Transcriptional regulation of *Notch* and *Delta:* Requirement for neuroblast segregation in *Drosophila. Development* **124,** 2015–2025.

Shah, S. B., Skromne, I., Hume, C. R., Kessler, D. S., Lee, K. J., Stern, C. D., and Dodd, J. (1997). Misexpression of chick Vg1 in the marginal zone induces primitive streak formation. *Development* **124,** 5127–5138.

Shawber, C., Boulter, J., Lindsell, C. E., and Weinmaster, G. (1996). Jagged2: A Serrate-like gene expressed during rat embryogenesis. *Dev. Biol.* **180,** 370–376.

Smith, J. L., and Schoenwolf, G. C. (1997). Neurulation: Coming to closure. *Trends Neurosci.* **20,** 510–517.

Smith, J. L., Gesteland, K. M., and Schoenwolf, G. C. (1994). Prospective fate map of the mouse primitive streak at 7.5 days of gestation. *Dev. Dyn.* **201,** 279–289.

Stern, C. D., and Keynes, R. J. (1987). Interactions between somite cells: The formation and maintenance of segmental boundaries in the chick embryo. *Development* **99,** 262–272.

Stern, C. D., Fraser, S. E., Keynes, R. J., and Primmett, D. R. N. (1988). A cell lineage analysis of segmentation in the chick embryo. *Development* **104**(Suppl.), 231–244.

Stern, C. D., Hatada, Y., Selleck, M. A. J., and Storey, K. G. (1992). Relationships between mesoderm induction and the embryonic axes in chick and frog embryos. *Development Suppl.* 1992, 151–156.

Tajbakhsh, S., and Spörle, R. (1998). Somite development: Constructing the vertebrate body. Meeting review. *Cell* **92,** 9–16.

Takada, S., Stark, K. L., Shea, M. J., Vassileva, G., McMahon, J. A., and McMahon, A. P. (1994). *Wnt-3a* regulates somites and tail-bud formation in the mouse embryo. *Genes Dev.* **8,** 174–189.

Takeichi, M. (1988). The cadherins: Cell–cell adhesion molecules controlling animal morphogenesis. *Development* **102,** 639–655.

Tam, P. P. L. (1981). The control of somitogenesis in mouse embryos. *J. Embryol. Exp. Morphol.* **65**(Suppl.), 103–128.

Tam, P. P. L. (1986). A study on the pattern of prospective somites in the presomitic mesoderm of mouse embryos. *J. Embryol. Exp. Morphol.* **92,** 269–285.

Tam, P. P. L. (1988). The allocation of cells in the presomitic mesoderm during somite segmentation in the mouse embryo. *Development* **103,** 379–390.

Tam, P. P. L., and Beddington, R. S. P. (1986). The metameric organisation of the presomitic mesoderm and somite specification in the mouse embryo. *In* "Somites in Developing Embryos" (R. Bellairs, D. A. Ede, and J. W. Lash, Eds.), NATO ASI Series, Vol. 118, pp. 17–36. Plenum, New York.

Tam, P. P. L., and Beddington, R. S. P. (1987). The formation of mesodermal tissues in the mouse embryo during gastrulation and early organogenesis. *Development* **99,** 109–126.

Tam, P. P. L., and Beddington, R. S. P. (1992). Establishment and organization of germ layers in the gastrulating mouse embryo. *In* "Postimplantation Development in the Mouse" Ciba Foundation Symp., Vol. 165, pp. 27–49. Wiley, Chichester.

Tam, P. P. L., and Behringer, R. R. (1997). Mouse gastrulation: The formation of a mammalian body plan. *Mech. Dev.* **68,** 3–25.

Tam, P. P. L., and Quinlan, G. A. (1996). Mapping vertebrate embryos. *Curr. Biol.* **6,** 106–108.

Tam, P. P. L., and Tan, S.-S. (1992). The somitogenetic potential of cells in the primitive streak and the tail bud of the organogenesis-stage mouse embryo. *Development* **115,** 703–715.

Tam, P. P. L., and Trainor, P. A. (1994). Specification and segmentation of the paraxial mesoderm. *Anat. Embryol.* **189,** 275–305.

Tam, P. P. L., and Zhou, S. X. (1996). The allocation of epiblast cells to ectodermal and germ-line lineages is influenced by the position of cells in the gastrulating mouse embryo. *Dev. Biol.* **178,** 124–132.

Tam, P. P. L., Meier, S., and Jacobson, A. G. (1982). Differentiation of the metameric pattern in the embryonic axis of the mouse II. Somitomeric organisation of the presomitic mesoderm. *Differentiation* **21,** 109–122.

Tam, P. P. L., Williams, E. A., and Chan, W. Y. (1993). Gastrulation in the mouse embryo: Ultrastructural and molecular aspects of germ layer morphogenesis. *Microsc. Res. Tech.* **26,** 301–328.

Tam, P. P. L., Parameswaran, M., Kinder, S. J., and Weinberger, R. P. (1997). The allocation of epiblast cells to the embryonic heart and other mesodermal lineages: The role of ingression and tissue movement during gastrulation. *Development* **124,** 1631–1642.

Thisse, C., Schilling, T. F., and Postlethwait, J. H. (1993). Structure of the Zebrafish snail1 gene and its expression in wild type, *spadetail* and *no tail* mutant embryos. *Development* **119,** 1203–1215.

Tonegawa, A., Funayama, N., Ueno, N., and Takahashi, Y. (1997). Mesodermal subdivision along the mediolateral axis in chicken controlled by different concentration of BMP-4. *Development* **124,** 1975–1984.

Wachlter, F., Christ, B., and Jacob, H. J. (1982). Grafting experiments on determination and migratory behaviour of presomitic, somitic and somatopleural cells in avian embryos. *Anat. Embryol.* **164,** 369–378.

Westin, J., and Lardelli, M. (1997). Three novel Notch genes in Zebrafish. Significance for vertebrate evolution. *Dev. Genes Evol.* **207,** 51–63.

Wilkinson, D. G., Bhatt, S., and Herrmann, B. (1990). Expression pattern of the mouse *T* gene and its role in mesoderm formation. *Nature* **343,** 657–659.

Williams, R., Lendahl, U., and Lardelli, M. (1995). Complementary and combinatorial patterns of *Notch* gene family expression during early mouse development. *Mech. Dev.* **53,** 357–368.

Wilson, V., Manson, L., Skarnes, W. C., and Beddington, R. S. P. (1995). The *T* gene is necessary for normal mesodermal morphogenetic cell movements during gastrulation. *Development* **121,** 877–886.

Wilson, V., and Beddington, R. S. P. (1996). Cell fate and morphogenetic movement in the late mouse primitive streak. *Mech. Dev.* **55,** 79–90.

Winnier, G. E., Hargett, L., and Hogan, B. L. M. (1997). The winged helix transcription factor MFH1 is required for proliferation and patterning of paraxial mesoderm in the mouse embryo. *Genes Dev.* **11,** 926–940.

Wong, P. C., Zheng, H., Chen, H., Becher, M. W., Sirinathsinghji, D. J. S., Trumbauer, M. E., Chen, H. Y., Price, D. L., Van der Ploeg, L. H. T., and Sisodia, S. S. (1997). Presenilin 1 is required for Notch 1 and DII1 expression in the paraxial mesoderm. *Nature* **387,** 288–292.

Wu, J. Y., Wen, L., Zhang, W.-J., and Rao, Y. (1996). The secreted product of *Xenopus* gene lunatic fringe, a vertebrate signaling molecule. *Science* **273,** 355–358.

Yamaguchi, T. P., Harpal, K., Henkemeyer, M., and Rossant, J. (1994). *fgfr-1* is required for embryonic growth and mesodermal patterning during mouse gastrulation. *Genes Dev.* **8,** 3032–3044.

Yoshikawa, Y., Fujimoto, T., McMahon, A. P., and Takada, S. (1997). Evidence that absence of *Wnt-3a* signaling promotes neuralization instead of paraxial mesoderm development in the mouse. *Dev. Biol.* **183,** 234–242.

Yuan, S., and Schoenwolf, G. C. (1998). De novo induction of the organizer and formation of the primitive streak in an experimental model of notochord reconstitution in avian embryos. *Development* **125,** 201–213.

Yuan, S., Darnell, D. K., and Schoenwolf, G. C. (1995a). Mesodermal patterning during avian gastrulation and neurulation: Experimental induction of notochord from non-notochordal precursor cells. *Dev. Genet.* **17,** 38–54.

Yuan, S., Darnell, D. K., and Schoenwolf, G. C. (1995b). Identification of inducing, responding, and suppressing regions in an experimental model of notochord formation in avian embryos. *Dev. Biol.* **172,** 567–584.

2

Retrospective Tracing of the Developmental Lineage of the Mouse Myotome

Sophie Eloy-Trinquet, Luc Mathis, and Jean-François Nicolas
Unité de Biologie moléculaire du Développement
75724 PARIS Cédex 15, France

I. Introduction

The classic view of muscle development is that virtually all skeletal musculature, except certain craniofacial muscles, derives from the dermomyotome epithelium,

Current Topics in Developmental Biology, Vol. 47

0070-2153/00 $30.00

which represents a dorsolateral territory of the somites (Christ *et al.,* 1990; Wachtler and Christ, 1992; Christ and Ordahl, 1995). In the mouse, the 65 pairs of somites are produced during embryogenesis, between embryonic day 8 (E8) and E13.5 (Theiler, 1989; Kaufman, 1998), from the cranial end of the presomitic mesoderm, a region of the paraxial mesoderm also referred to as segmental plate or unsegmented paraxial mesoderm. The number of cells involved is enormous. Each somite comprises 300 to 1000 cells at the time of its segmentation from the presomitic mesoderm (Tam, 1981), and, by E11.5, the dermomyotome has already produced 55,000 myogenic cells (Nicolas *et al.,* 1996). These myogenic cells are arranged as 40 pairs of segments on either side of the developing neural tube, with each possessing a characteristic number of cells from 300 to 1500 (Fig. 1A, G). The production of myogenic cells is a highly dynamic process, whereby mononucleated myocytes (the early myotome) are rapidly joined by other myogenic cells (Christ and Ordahl, 1995). The dermomyotome also produces the dermis and the distal portion of the ribs (Kato and Aoyama, 1998). During development, the dermomyotome is flanked ventrally by the sclerotome, another derivative of the somite, dorsally by the ectoderm, medially by the neural tube and the notochord, and laterally by the intermediate and lateral mesoderm.

Our comprehension of muscle development is based on notions of territorial delimitation (parcelation) in part independent of extrinsic influences (Menkes and Sandor, 1977; Bellairs, 1979; Palmeirim *et al.,* 1997) and genetic specification. These events occur progressively in anteroposterior and dorsoventral directions through the interaction of precursors with adjacent structures and morphogens (Brand-Saberi *et al.,* 1993; Pourquié *et al.,* 1993; Münsterberg *et al.,* 1995; Stern *et al.,* 1995; Borycki *et al.,* 1998), followed by tissue determination and differentiation. These ideas have been supported by the following: fate mapping studies of the dermomyotome epithelium (Denetclaw *et al.,* 1997; Williams and Ordahl, 1997; Dietrich *et al.,* 1998; Kahane *et al.,* 1998a,b), grafting of tissues from these territories (Christ *et al.,* 1978; Ordahl and Le Douarin, 1992; Pourquié *et al.,* 1995; Kato and Aoyama, 1998), analysis of developmental gene expression patterns (Ott *et al.,* 1991; Lyons and Buckingham, 1992; Pownall and Emerson, 1992; Smith *et al.,* 1993; Goulding *et al.,* 1994; T. H. Smith *et al.,* 1994; Tajbakhsh and Buckingham, 1994; Pourquié *et al.,* 1996), and genetic manipulation in mice (Rudnicki *et al.,* 1993; Patapoutian *et al.,* 1995; Cossu *et al.,* 1996; Spörle *et al.,* 1996; Yoon *et al.,* 1997; Rawls *et al.,* 1998; McMahon *et al.,* 1998), as reviewed extensively in this book.

The developmental history of the precursor cells of muscle tissue is still not completely understood, and contrasting views have been presented. A complete description of the cellular basis of this system would help further our understanding of how the fundamental processes of development are coordinated, namely, cell proliferation, spatial distribution, and cell specification. Previous studies have addressed the late stages of development, particularly those of the somite and dermomyotome (Denetclaw *et al.,* 1997; Williams and Ordahl, 1997; Dietrich *et al.,* 1998; Kahane *et al.,* 1998b), whereas others have focused on the early stages of mesoderm formation, beginning at the pregastrulation epiblast (Tam and Beddington, 1987; Lawson *et al.,* 1991; Selleck and Stern, 1991; Lawson and Pedersen, 1992; Hatada and Stern, 1994; Psychoyos and Stern, 1996; J. L. Smith *et al.,* 1994; Tam and Trainor, 1994). Clearly, however, the link between these two distant stages is missing, and the most convincing of these experiments have been done in the chick, with little being known for mammals. Furthermore, there has been no attempt, so far, to obtain a complete description of the properties for all possible ancestral cells of the progenitors to the muscle system. This is because classic methods are not amenable to long-term studies (Gardner and Lawrence, 1985; Beddington and Lawson, 1990), nor do they permit one to follow groups of cells at all stages of development. These technical difficulties are even harder to overcome for a clonal description in mammals. Moreover, it is impossible to know whether the analysis has not favored certain territories and disregarded others.

Therefore, a search for novel methods that are completely independent of any subjectivity is justified. It is to this end that a genetic approach, using saturated mutagenesis screens in *Drosophila* (Nüsslein-Volhard and Wieschaus, 1980), has successfully yielded an integrated view of the molecular aspects of certain stages of development (Scott and Carroll, 1987; Scott, 1987; Lawrence, 1992). The existence of techniques for genome manipulation allows not only for the investigation of molecular pathways but also for the elaboration of new analytical methods that considerably simplify the approach to these problems. Among vertebrates, the mouse is particularly favorable for genetic modifications; therefore, this organism has been the first one chosen for the application of these improvements (Bonnerot and Nicolas, 1993a; Zinyk *et al.,* 1998). In the future, these methods will certainly still be improved and adapted to other vertebrates.

The method of clonal analysis recently developed in our laboratory, which is critically described in this chapter, has been used to study cellular events that lead to the formation of the muscle (Nicolas *et al.,* 1996) and the central nervous system (Mathis *et al.,* 1997). From a theoretical view, this method has similarities with genetic mosaics obtained from aggregation chimeras (Tarkowski, 1963; Mintz, 1965, 1967; Lewis *et al.,* 1972; McLaren, 1972; Russell, 1978; Gardner and Lawrence, 1985; Rossant, 1987; Vogel and Herrup, 1993) and from retroviral labeling (Sanes *et al.,* 1986; Nicolas and Bonnerot, 1988; Cepko *et al.,* 1993) but has the additional advantage of restricting the analysis to a class of cells or to a territory (Bonnerot and Nicolas, 1993b). More importantly, the random nature of the labeling event at the origin of the clones allows for a complete description of any ancestral cell of a structure (Mathis and Nicolas, 1997), since cells can be marked at virtually all stages of development.

II. The LaacZ Method

The method (Bonnerot and Nicolas, 1993b; Mathis and Nicolas, 1998), illustrated in Fig. 2, exploits a *nls LacZ* reporter gene (Sanes *et al.,* 1986; Bonnerot *et al.,*

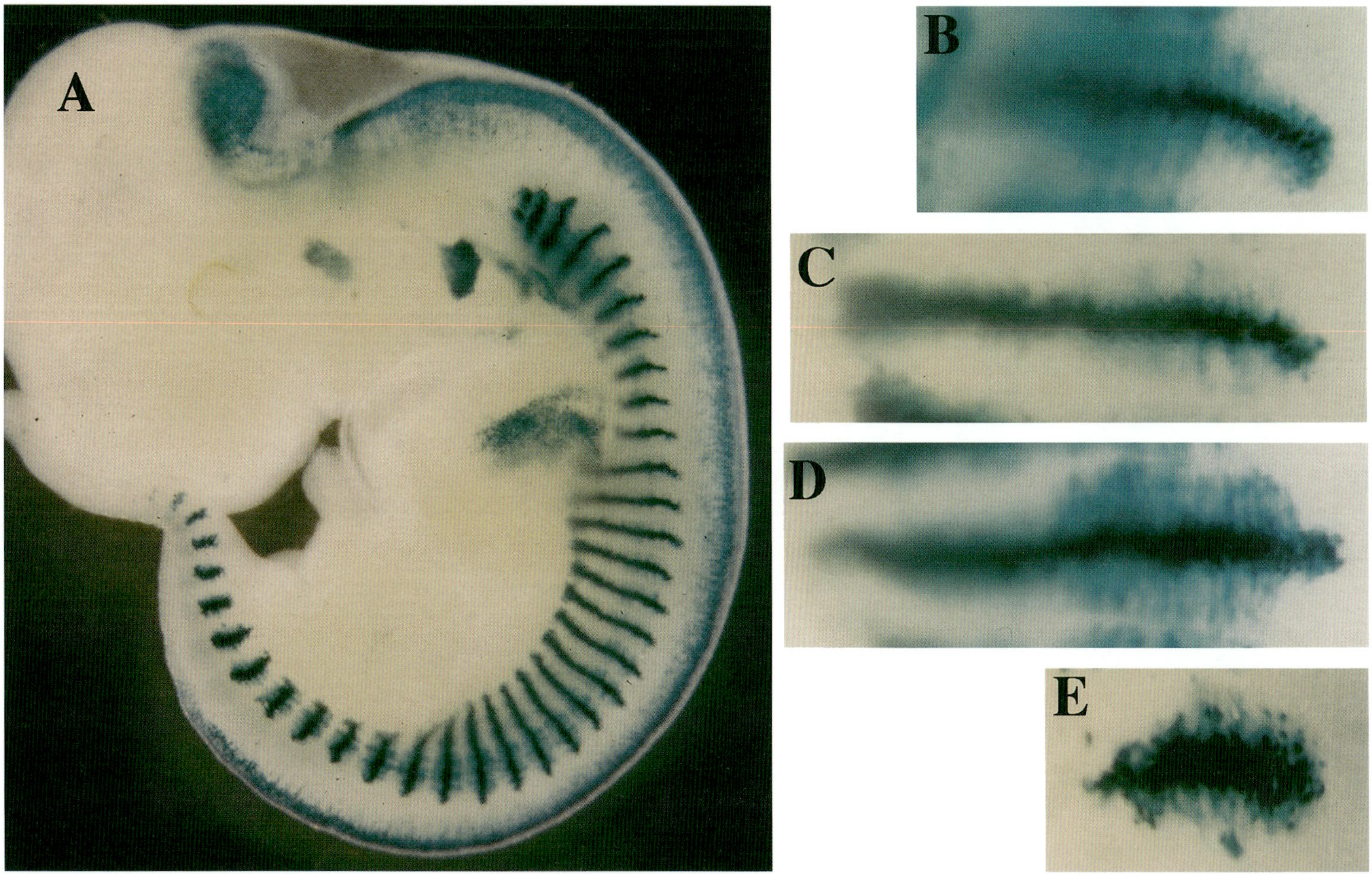
A
B
C
D
E

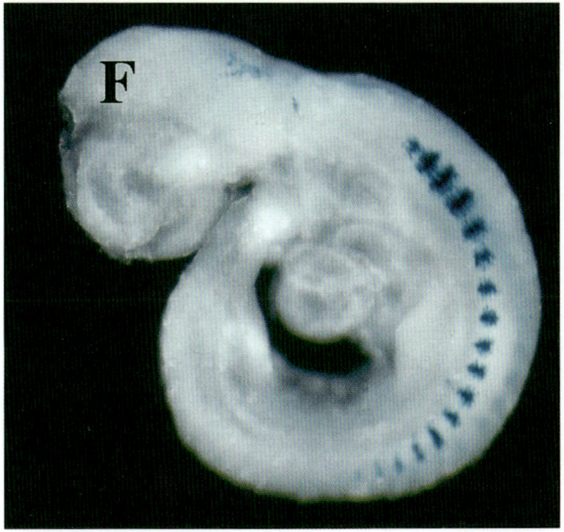

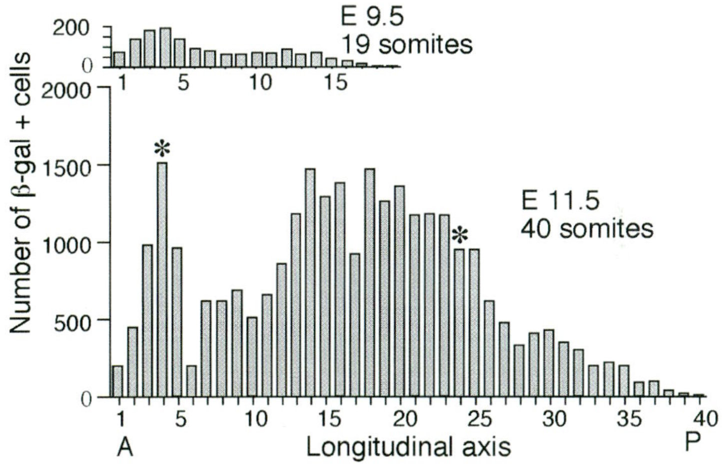

Figure 1 Segmental organization of myogenic cells at E9.5 and E11.5. *In toto* X-gal staining of RαNLZ2 embryos. (A) At E11.5, the 40 segments are distributed longitudinally, from a rostral position adjacent to rhombomere 7 (not visible) to the most caudal part of the embryo. The forelimbs and hind limbs showing migrating β-gal$^+$ myogenic cells. The *in toto* characteristics of the segments suggest a three group classification: group I from segments 1 to 10 (cervical), group II from segments 11 to 24 (thoracic), and group III from segments including and caudal to 25 (lumbar, sacral, and caudal). In RαNLZ2 mice, there is an ectopic labeling of cells in the neural tube. (B–E) High magnification of one segment from each group, with medial to the right: segment 4, group I (B); segments 14 (C) and 24 (D), group II; segment 31, group III (E). At E11.5, the number of β-gal$^+$ cells in these segments is 1501, 1467, 943, and 349, respectively. (F) At E9.5, the 19 segments are distributed longitudinally. The nine most caudal segments have still not differentiated muscle cells. The different groups of somites are not yet obvious. (G) Pattern of myogenic cell numbers for the 19 segments at E9.5 and for the 40 segments at E11.5. Note the progressive rostrocaudal maturation of the muscle system and the individual differences between segments. The asterisks indicate the two segments analyzed in this chapter.

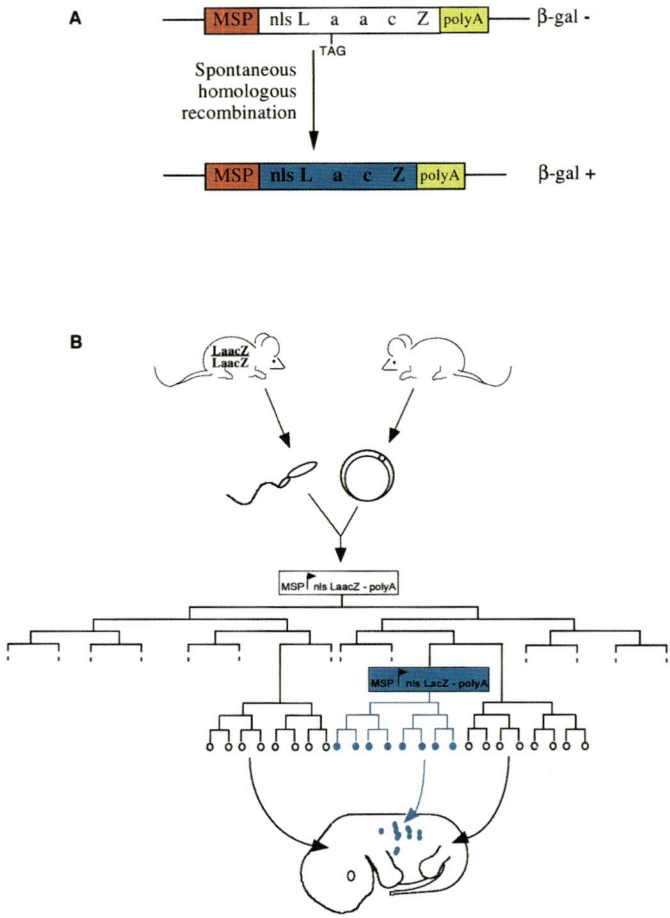

Figure 2 The LaacZ labeling system of clones. (A) The *nls LaacZ* reporter gene. A *nls LacZ* reporter gene, in which a duplication has rendered the enzyme inactive, is linked to a muscle specific promoter, MSP. A spontaneous homologous recombination between the duplicated "aa" sequences re-establishes the β-galactosidase coding frame (in blue). (B) Production of clones in the progeny of a cross between a *LaacZ* transgenic male and a wild-type female. During development, a spontaneous homologous recombination occurs between the duplicated sequences of the *LaacZ* gene, thus creating a functional *LacZ* gene in the cell. This event initiates the clonal labeling (colored in blue) in the embryo. In the α-2 transgenic line, the frequency of myogenic β-gal$^+$ clones at E11.5 is 5×10^{-2} per embryo. Thus far, 315 independently marked embryos have been analyzed from the 6254 embryos generated.

1987; Bonnerot and Nicolas, 1993a) that has been inactivated by an internal duplication (*nls LaacZ*) of 289 bp. The clonal labeling is initiated by a spontaneous homologous recombination between the duplicated sequences. This can occur at any stage during development (Fig. 2B). The recombination event re-establishes the coding frame for the reporter gene, and the clone that arises can then be visualized by histoenzymatic or immunochemical methods. The recombination event in the *LaacZ* reporter gene generates an intrinsic long-term neutral marker (β-galactosidase) that allows for visualization of the descendants of the labeled ancestral cell, even in a mature differentiated structure (Mathis *et al.,* 1997). In order to analyze the lineage of cells in a particular cell type or domain, the expression of *LaacZ* is driven by specific promoters.

A transgenic line is established and clones are produced by crossing transgenic males with normal female mice (Fig. 2B). Clones are found in the progenies from these crosses (Fig. 2B). So far the *LacZ* gene remains the most convenient reporter for this purpose mainly because its detection is possible in whole mount embryos and, therefore, the description of the geometrical parameters of the clones is very precise and easy. The absence of background enzymatic activity and the fact that the labeling is strictly cell autonomous (when the histochemistry is properly done) (Bonnerot and Nicolas, 1993a; Mathis and Nicolas, 1998) are additional important advantages. However, this method is adaptable to other reporter genes, such as alkaline phosphatase (Yoon *et al.,* 1988), and clonal development could potentially be followed *in vivo* using "live" reporter systems, such as the green fluorescent protein (GFP) (Hadjantonakis *et al.,* 1998; Zernicka-Goetz, 1999). This method should permit labeling of any cell in the intact embryo and, thus, the detection of all cells whose descendants will ultimately participate in the formation of a particular structure that has been targeted by the reporter gene. In this way, a detailed description is obtained for, what we call in this chapter, the "ancestral cells" of a structure.

III. The Basic Logic of Clonal Analysis

Before summarizing the results obtained for the development of the muscle system, we will discuss a theoretical example of what would be the genealogical history of the ancestral cells of a structure, beginning from the time at which there is only one ancestral cell in the embryo (Fig. 3A). This model is not intended to describe any particular structure but simply to demonstrate several basic properties of cell systems that constitute the lineages involved in the formation of any structure in the organism (Mathis and Nicolas, 1997). It will also serve to clarify the terminology used in this chapter (McLaren, 1972; West, 1978; Vogel and Herrup, 1993; Mathis and Nicolas, 1997).

A. Ancestral and Founder Cells

In the example presented in Fig. 3A, at $t = 14$ of development, the structure, M, comprises 32 cells (8 cells are drawn in Fig. 3A) belonging to three different cell

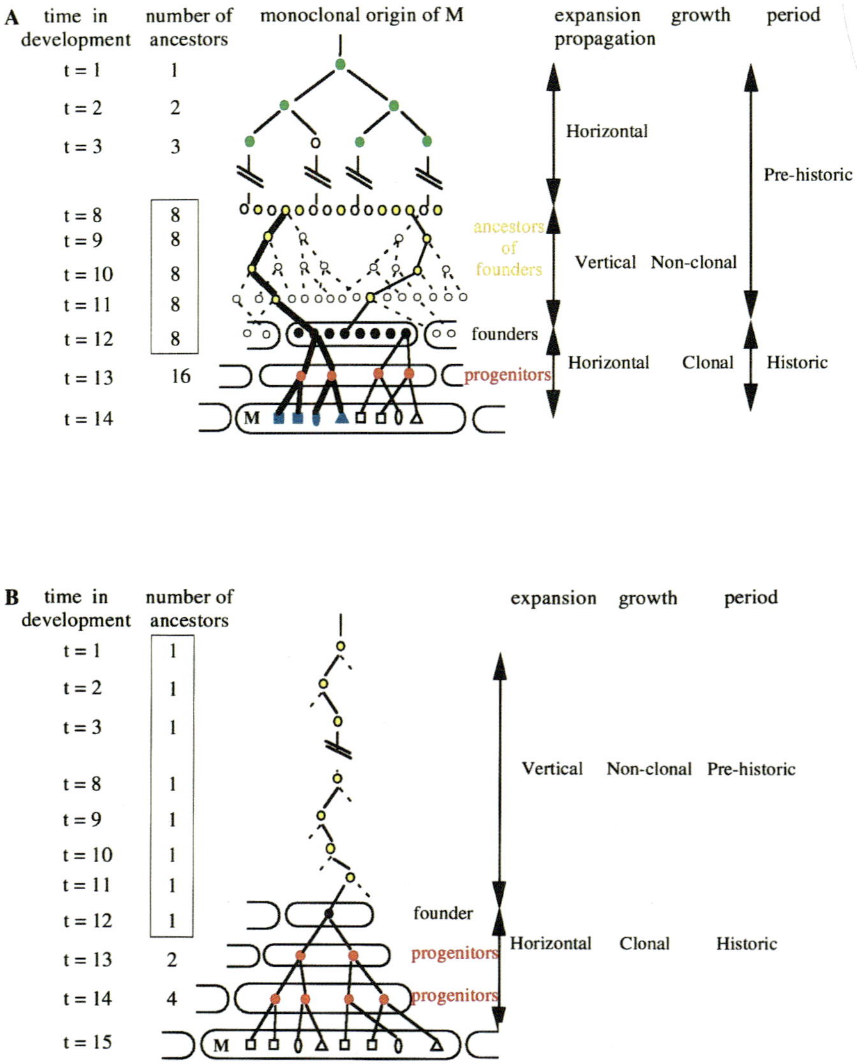

Figure 3 Genealogic trees of the ancestral cells for simple model structures. (A) Polyclonal origin from a single pool of founder cells. Schematically represented are the following: in bold, the complete lineage of a genealogical cohort in M (the four blue cells on the left at t = 14; the clonal relationships of a cell cohort trace back to a single cell, called a founder cell of M, and during this period the growth is clonal and represents the period of production of the progenitor (in red) and differentiated cells of M); in black, the eight founder cells; in yellow, at t = 11 to t = 8, the ancestors of two founder cells and, at t = 8, the eight ancestors of the eight founder cells (during the period of vertical propagation, between t = 8 and t = 12, the number of ancestral cells remains constant at each generation, and the growth is nonclonal); in white, the lineage of two cells whose descendants will never participate in M (at each cell generation, their number doubles); and in green, the period of horizontal expansion of ancestral cells, during which the ancestral cells have clonal relationships that trace back to a unique cell, which,

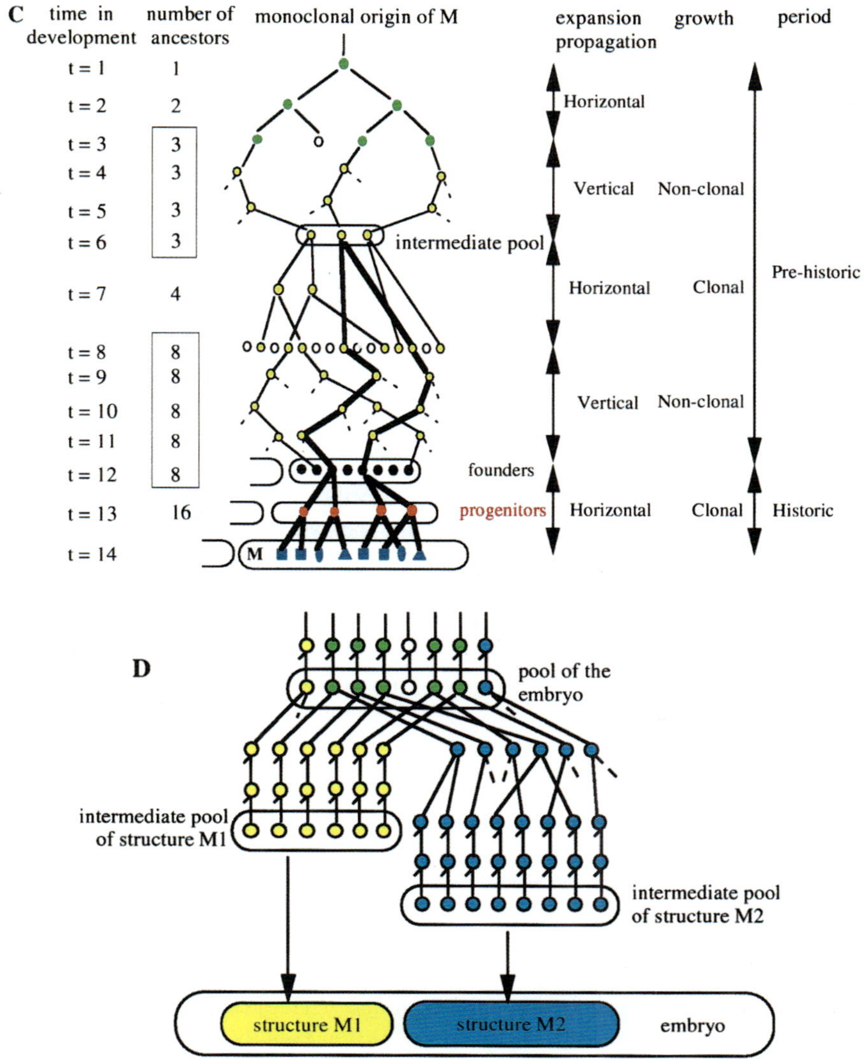

itself, is the monoclonal origin of M. (B) Monoclonal origin from a single founder cell. The period of horizontal propagation involves only one cell. M is produced by the horizontal expansion of the single founder cell. (C) Polyclonal origin of a pool of founder cells derived from an intermediate pool. This situation occurs when a period of clonal growth intercalates between the two periods of horizontal expansion of the ancestral and founder cells. An intermediary pool (at $t = 6$) of three founder cells of M is thus defined. For the cells of this intermediary pool, there is a corresponding cohort in M with a 2-fold greater number of cells (follow the bold lineage line from $t = 6$) compared to the cohorts from the founder cells at $t = 12$. (D) The founder cells of a structure are not necessarily derived from all the founder cells of the embryo proper: the eight founder cells of the embryo are represented on the top of the figure. The five cells in green contribute to both the intermediate pools of M1 and M2. The cell in yellow contributes to only M1, the blue cell to only M2, and the white cell to neither M1 nor M2.

types. At any given stage in development (except the earliest), there are cells whose descendants will never participate in M (drawn in white on Fig. 3A) and cells of which at least some descendants will participate in M (drawn in different colors in Fig. 3A–C). These latter cells are the ancestral cells of M. At each cell division, these ancestral cells can be characterized by the fate of their two daughter cells at the next generation. If the two daughter cells contribute to M, growth is called clonal. If only one daughter cell contributes to M, growth is nonclonal. Thus, in this example, from $t = 12$, the ancestral cells display clonal growth in relation to M. We call these cells the founder cells of M (black cells in Fig. 3). Therefore, $t = 12$ divides the history of the ancestral cells of M into two fundamentally different periods.

B. Genealogical Cohorts, Clonal Growth

First, there is the period during which the descendants of the founder cells form genealogical cohorts, defined as the progenitors of M. Their growth is clonal, and each genealogical cohort groups all cells in M whose clonal relationship goes back to a founder cell (one has been colored in blue in Fig. 3A). If the number of founder cells is f, M is formed of f genealogical cohorts. In the idealized example, the cohorts are all equal and each contains four cells. We call this period the historical period of M, because the history of the cells corresponds to the history of the structure.

C. Nonclonal Growth

Second, the period of the ancestors to the founder cells corresponds to the period during which, at each cell generation, only one daughter cell of the ancestral cell will have descendants contributing to M. Therefore, the growth is nonclonal and, at each generation during this period, the number of ancestral cells remains equal to f. For instance, f ancestral cells at $t = 8$ produce f ancestral cells at $t = 9$, which still produce only f ancestral cells at $t = 10$, so that even though the total number of cells in the embryo may have increased severalfold (follow the lineages of the two white cells in Fig. 3A), the number of ancestors, f, remains constant. We call this period the prehistoric period of the structure, because the history of the cells has—in principle—little consequence on the properties of the clones analyzed in M (that is to say, has little consequence on the founder cells of M).

During the prehistoric period of the ancestral cells, the pool of cells contributing to M remains the same size at each generation; therefore, we characterize this period as the period of vertical propagation of the ancestors to the founder cells. During the historic period, the size of the pool contributing to M (that is, the pool of the precursor cells of M) varies at each cell generation in relation to the mode

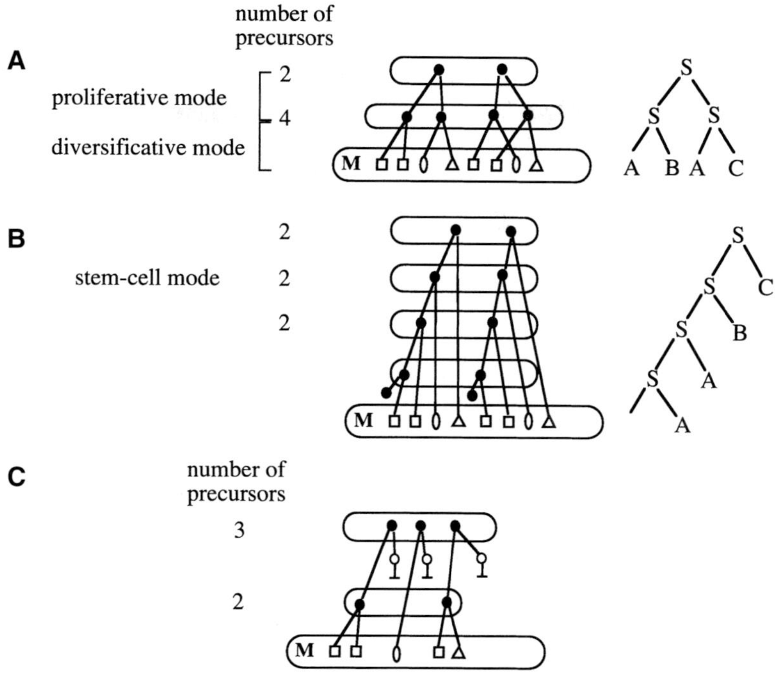

Figure 4 Modes of cell division during the period of horizontal expansion of the descendants of the founder cells to M. (A) The proliferative/diversificative mode. During the period of horizontal expansion of the descendants of founder cells, the pool of progenitor cells (in black) increases at each generation. (B) The stem cell mode. During the period of horizontal expansion of the descendants of the founder cells, the pool of progenitors (in black) remains constant at each generation. (C) A special hypothetical case: during the period of horizontal expansion, the pool of progenitors decreases. The progenitors whose lineage is interrupted do not participate in M.

of production of these precursors and of the differentiated cells in M (Fig. 4). For example, if this mode is proliferative then its size increases exponentially (Fig. 4A). In contrast if this mode corresponds to a stem cell mode, the size of the precursor pool at each generation remains constant (Fig. 4B). It is possible to imagine situations in which the pool size could also decrease at a given cell generation because of a lack of participation by some cells of the lineage to the final structure, due to processes such as apoptosis, for example. This scenario is depicted as an interrupted line in Fig. 4C.

D. The Primary Expansion Period

The last period to consider is the expansion period of the pool of the ancestral cells (Fig. 3A,B in green). In contrast to the period of vertical propagation, during which the ancestor pool propagates (at each cell generation) without changing its size, this period is characterized by a horizontal expansion, because, at each cell generation, the size of the pool increases. This period begins at a time when only one ancestral cell exists and ends at the time when f ancestral cells exist. Stated another way, the horizontal expansion occurs from the monoclonal origin of the structure ($t = 1$) to the establishment of its definitive polyclonal size, which marks the beginning of the period of vertical propagation of the pool of ancestral cells.

This model does not state the existence of invariant lineages between embryos and is still valid when these lineages are variable. Nor does the model link this historical description to the ancestral cells of any particular stage or developmental event, such as cell determination or specification. It concerns only the main patterns of cell growth. For instance, the period during which the number of ancestral cells equals one (monoclonal period) is not specified; it could be restricted to the fertilized egg or could last longer. The most extreme situation would be a monoclonal origin at the time of the constitution of the pool of founder cells ($f = 1$). In this situation, the genealogical tree of the ancestral cells of M is simplified (Fig. 3B), because the period of horizontal expansion does not exist and the period of vertical propagation involves only one cell that produces a single genealogical cohort in M.

E. Intermediary Pool of Founders

More realistic variations of the example presented in Fig. 3A are easy to imagine. One additional example is illustrated in Fig. 3C. In this scenario, the ancestors of the founder cells go through an initial, intermediary period of clonal growth, that is, a period during which f founder cells will be produced from a smaller pool of cells. It should be noted that the size of this intermediate pool cannot be larger than the size of the final pool of founder cells, and this is because the initial period of clonal growth increases the number of founder cells. During this intermediary period of clonal growth, the ancestral cells are genealogically related; descendants of two daughter cells of the intermediary pool will contribute to M. The genealogical tree of the ancestral cells is thus modified as indicated in Fig. 3C. This intermediate phase of clonal growth corresponds to an intermediate pool of founder cells (whose size is inferior to f) and later to a period of horizontal expansion that increases this size to f. More generally, if several intermediate periods exist, each phase of clonal growth will correspond to a period of horizontal expansion. These periods of clonal expansion interrupt the period of vertical propagation.

This more complex variation of the model has been described because it may correspond to what occurs for any structure in the mammalian embryo and probably in the embryos of many other amniotes. Indeed, the embryo of this class of vertebrates does have a polyclonal origin, as was first demonstrated by the chimerism of tetra- and hexaparental mice (Tarkowski, 1963; Mintz, 1965; Rossant, 1987) and next by retrospective analysis after retroviral infection at preimplantation stages (Soriano and Jaenisch, 1986). It is from the descendants of this first pool of embryonic cells that pools for the different structures are derived. More generally, it is possible that between the pool of founder cells of a structure and the first period of horizontal expansion, several periods of clonal growth interrupt the period of vertical propagation and each period produces an intermediate pool of founder cells.

Finally, it is clear that the founder cells of a given structure do not necessarily derive from all founder cells of the embryo (Fig. 3D), and that the period of clonal growth may involve groups of cells larger than those corresponding to the founder cells of the structure studied.

In summary, from a clonal point of view, the genealogical history of the ancestral cells of a structure goes through a series of horizontal expansions followed by vertical propagation (clonal versus nonclonal growth). These periods end when the structure is formed from its progenitors, which are, themselves, the descendants of the founder cells of the structure (Fig. 4). The periods of horizontal expansion correspond to phases of clonal growth for at least some of the founder cells. However, clonal growth is distinct from the modes of spatial distribution of cells. Spatial distribution could be coherent, when daughter cells remain geometrically contiguous, or dispersive when extensive cell mingling randomizes their distribution. If the period of vertical propagation necessarily corresponds to a dispersive mode of distribution, the period of horizontal expansion (that is, clonal growth) may correspond to any mode of spatial distribution. This last point is also relevant to the period when the structure forms, beginning with the final pool of founder cells that serve as a source for the progenitors of the structure.

In the next part of this chapter, we will try to relate the real data from our clonal analyses (Nicolas *et al.,* 1996; Eloy-Trinquet and Nicolas, 1999) to these models and attempt to demonstrate how these studies modify and enrich our current understanding of muscle development and of the genealogical history of the ancestral cells of a structure in the mouse.

IV. The Myotome in the E11.5 Mouse Embryo

A. Longitudinal Organization

In the mouse, at E11.5, the myotome is formed of 36 to 42 discrete well separated segments (Fig. 1A). Segments 7 to 11 are positioned in front of the forelimbs and 25 to 29 are found in front of the hind limbs. The segments of the tail begin at 34–35. The four most caudal somites, which are the last to be produced from the presomitic mesoderm, are still not differentiating myogenic cells at this stage. At E10.5, the myotome is formed of 31–35 segments; at E9.5, there are 17–25 segments (Fig. 1F), and, at E8.5, 0–4 segments. The production of segments thus occurs at a rate of one every 2 hours. Myogenic cells first appear in the ninth (E9.5) or sixth (E11.5) somite from the most rostral end of the presomitic mesoderm, and there is a slight delay in the segmentation of the somites positioned between the left (in general retarded) and right sides of embryo (S. Eloy-Trinquet, C. Bonnerot, and J.-F. Nicolas, unpublished data).

B. Mediolateral Organization

Each myotome segment has its own characteristics (Fig. 1). At E11.5, it is easy to distinguish at least three main groups (S. Eloy-Trinquet and J.-F. Nicolas, unpublished data). Group I consists of the first 10 segments. Their morphology has a specific brush aspect due to the preferential anteroposterior elongation of the myocytes (Fig. 1B). The number of muscle cells in each of these segments varies from 200 (segments 1 and 6) to 1500 (segment 4), (Fig. 1G). They probably correspond to the cranial somites. Group II consists of segments 11 to 24. In each of these segments, in addition to the dorsal group of cells with a specific brush aspect, there is a more lateral group whose cells have not elongated anteroposteriorly (Fig. 1C,D). The number of muscle cells in segments 11 to 14 increases from 700 to 1500, and there are about 1000–1500 cells in segments 15 to 24 (Fig. 1G). Group II segments probably correspond to the 14 thoracic somites. Group III consists of segments 25 to 40. They probably correspond to the sacral and coccygeal somites. The first five are positioned in front of the hind limb. This group consists of cells with only a brush aspect (Fig. 1E), which also characterizes the dorsal cells of the two other groups. At E11.5, these segments are observed to be relatively immature.

V. The Questions

This description, although very superficial, still raises certain questions concerning the formation of this structure. First, there is the question of forming a longitudinal segmented structure and, in particular, the unique or multiple origin of the cells involved (Fig. 9D), (Nicolas *et al.,* 1996). Specifically, is there an early regionalization of the longitudinal structure that would, therefore, correspond to independent pools of intermediate founder cells for the muscle system (Fig. 9D, compare b and c to a and d)? Second, what is the basis for the complex composition of the segments of group II and, by contrast, the simpler structure for the segments of group III? Third, what is the cellular basis for the mediolateral organization of the segment?

A first explanation is that there are founder cells of independent origin for the three groups of segments, that is, a longitudinal organization in separate clonal domains leading to different mediolateral organizations of the somites (Fig. 9D b or c). Another possibility is that there is a common longitudinal organization for the dorsal part of all segments and a separate one for the ventral part of the segments. Finally, a third conception is that there is no fundamental cell organization until late in development (for instance, at the somite stage). These hypotheses predict highly contrasting properties for the clones that participate in the development of the muscle system.

Molecular genetic studies have already provided partial molecular views to some of these questions. For example, the longitudinal patterning of the axis is controlled by developmental genes, such as those of the Hox family (Kessel and Gruss, 1990; Krumlauf, 1992; Burke *et al.,* 1995), and the dorsoventral organization of the somites (sclerotome versus dermomyotome) is dependent on several developmental genes expressed in the notochord, neural tube, ectoderm, and lateral mesoderm (Brand-Saberi *et al.,* 1993; Pourquié *et al.,* 1993; Williams and Ordahl, 1994; Münsterberg *et al.,* 1995; Stern *et al.,* 1995; Cossu *et al.,* 1996; Borycki *et al.,* 1998). Therefore, in addition to the questions already mentioned, this raises the question of how the genetic and cellular organization are coordinated. A prerequisite for a discussion of this coordination is a better understanding of the cellular organization, namely, a detailed description of cell proliferation and positioning, and the mode of cell diversification and differentiation during myotome development.

VI. The Clones

If classified by their date of obtainment in the screen, the clones, whose characteristics are very different (Fig. 5), resemble a chaotic collection of elements (Fig. 6). But immediately evident is that the pattern does not contain only clones contributing to one segment. The majority of clones contribute to several segments (very frequently adjacent), and this, therefore, suggests a presomitic clonal organization for each segment of the dermomyotome. Independent clonal organization for each segment would generate clones contributing to only one segment or to several nonadjacent segments. In addition, labeled cell counts within clones indicate that none of them contribute to the labeling of a complete segment, even when the whole axis contains labeled myogenic cells. Therefore, each segment must have a polyclonal origin. Analysis of a very large collection of clones (Fig. 7) fails to reveal labeling of an ancestral cell during the period of vertical propagation of the monoclonal precursor to the dermomyotome of one segment (Fig. 3A). This pattern similarly shows the existence of very long clones, none of which, however, contributes to the 55,000 myogenic cells of the E11.5 embryo; so, again, the period of vertical propagation of a unique ancestral cell at the origin of the longitudinal structure is not revealed. Next, it is easy to calculate the size of the polyclone at the origin of one segment, as well as the size of the polyclone at the origin of the whole myotome structure, by using a modification of the minimum clone size estimation method (Rossant, 1984). They correspond to 152 and 137 ancestral

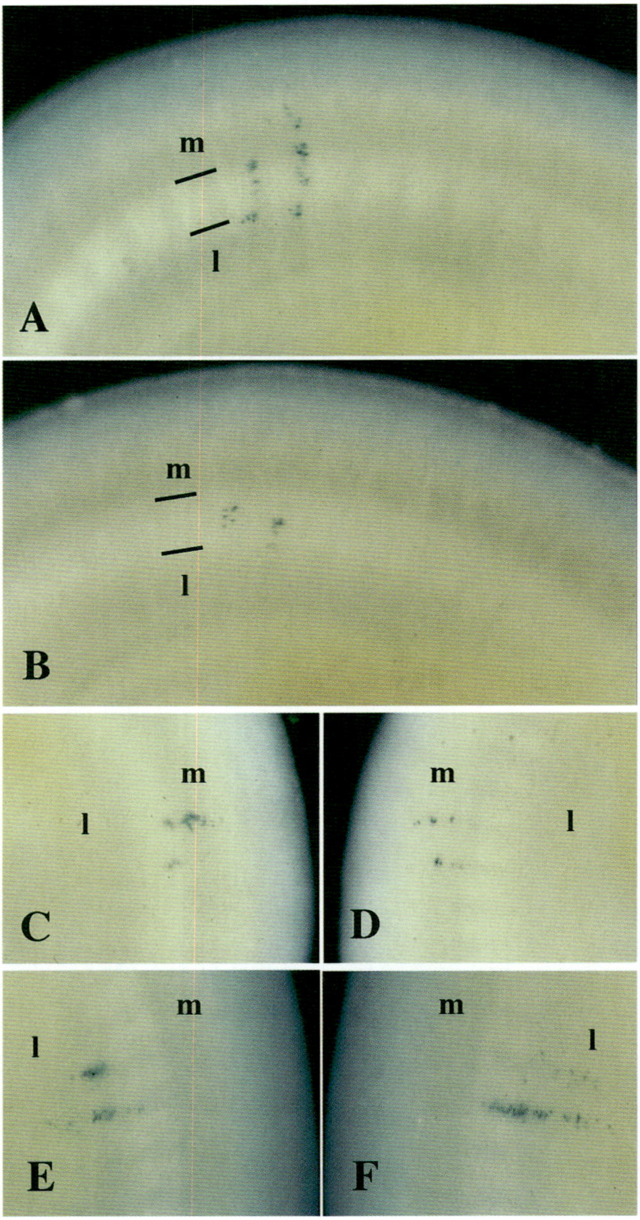

Figure 5 Examples of recombinant *LacZ* clones. (A, B) Two plurisegmented unilateral clones: VG 20 (A) and VG 27 (B). Anterior is to the right. (C–F) Two short bilateral clones: left (C, E) and right (D, F) sides of SC 186 (C, D) and VG 13 (E, F). Note the mirror symmetry between left and right.

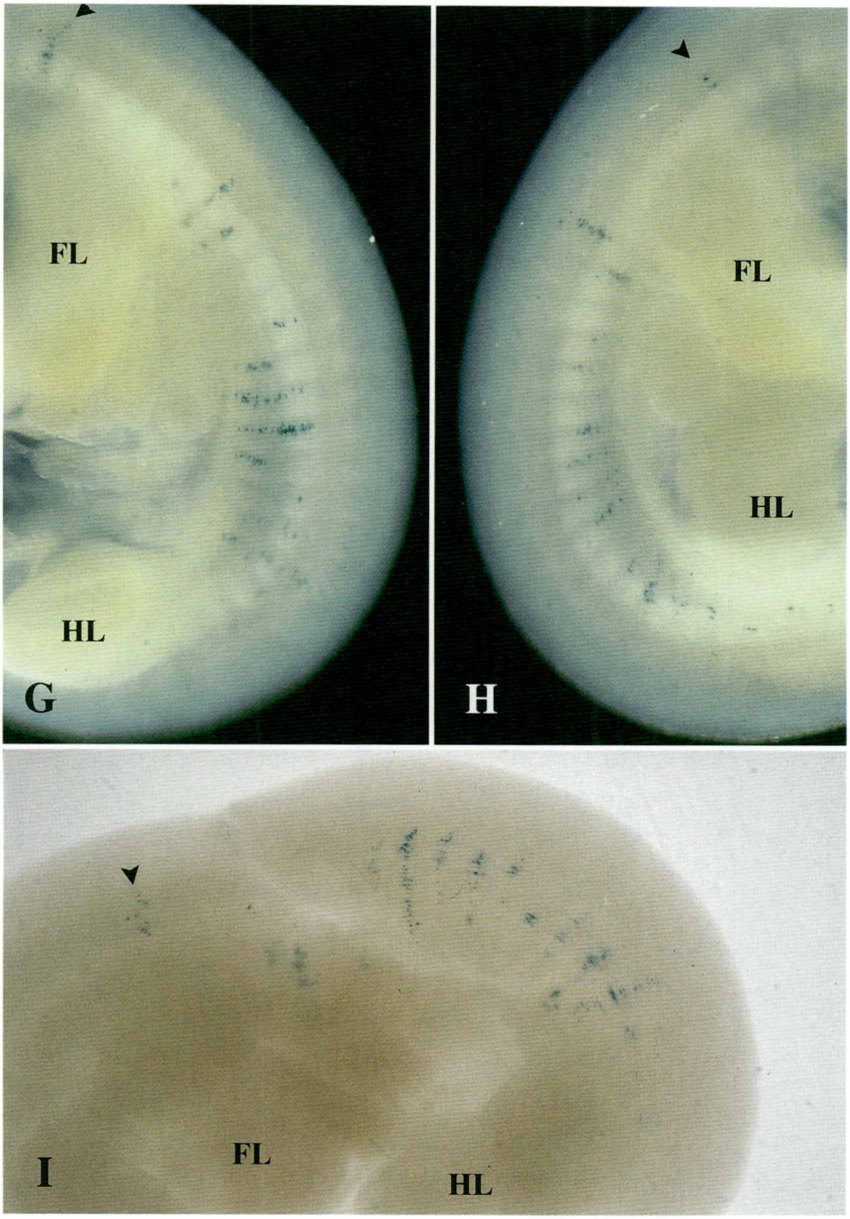

(G–I) Two long bilateral clones: left (G) and right (H) sides of VG 70. (I) Left side of SC 346, a clone participating essentially in the left side of the embryo. The arrowheads indicate the more anterior segment labeled. m, Median; l, lateral; FL, forelimb; HL, hind limb.

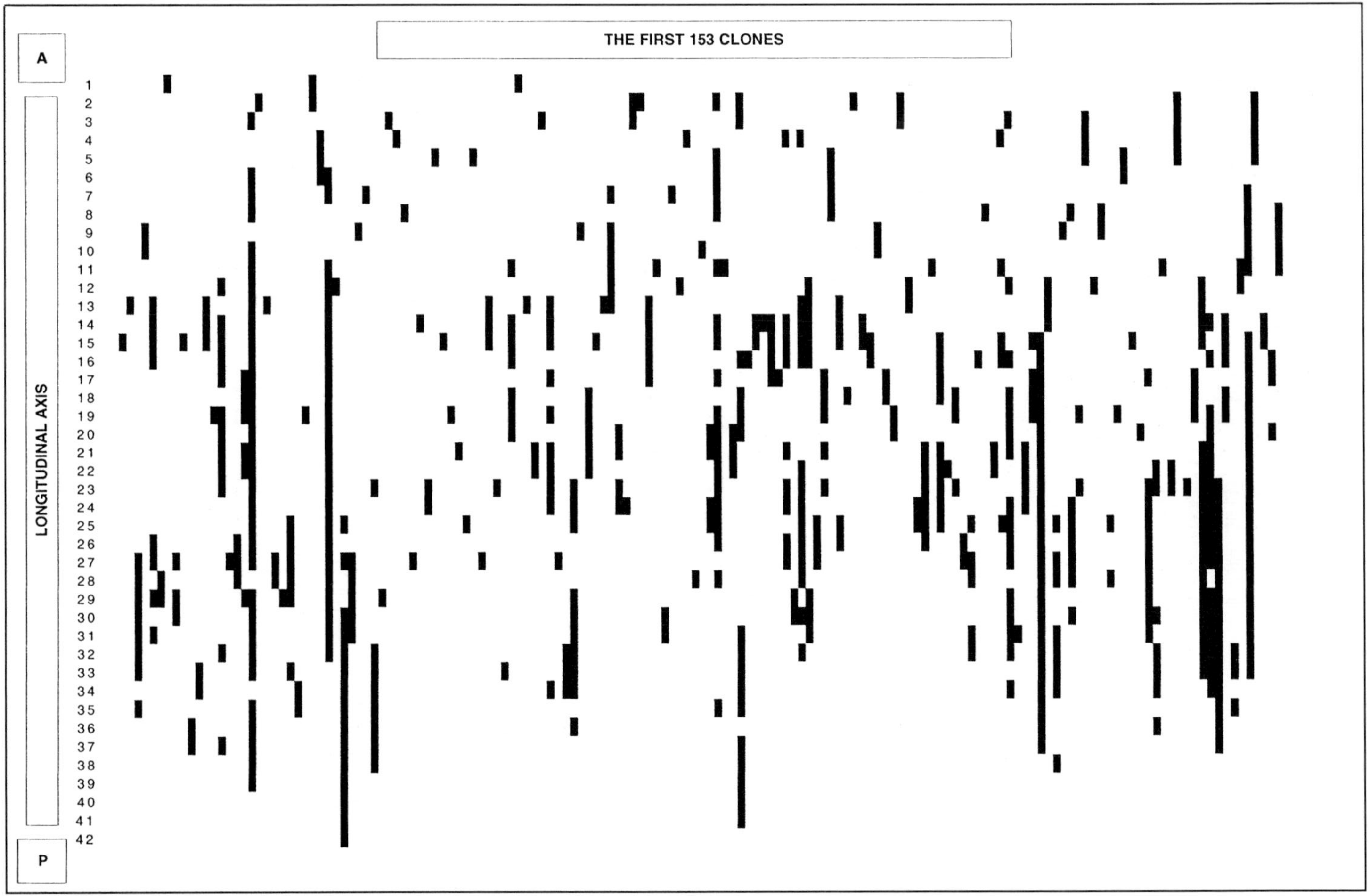
THE FIRST 153 CLONES
A
LONGITUDINAL AXIS
P
1 2 3 4 5 6 7 8 9 10 11 12 13 14 15 16 17 18 19 20 21 22 23 24 25 26 27 28 29 30 31 32 33 34 35 36 37 38 39 40 41 42

cells, respectively, and these values are not different (Nicolas *et al.,* 1996). Thus, an examination of one segment probably corresponds to the pattern of ancestral cell propagation illustrated in Fig. 3C, with production of an intermediate pool of founders and an early, yet undectectable, initial period of horizontal expansion. The next challenge is to understand the genealogical arrangement between the pool at the origin of the whole structure and the pools at the origin of each individual segment.

The criteria used to classify the clones is crucial to understanding the history of the structure studied. We have successfully used an approach based on the classification of clones by their most anterior border (Fig. 7), a parameter related to their date of birth (the time at which the spontaneous homologous recombination event occured in the embryo) (Nicolas *et al.,* 1996). This type of analysis has been instrumental to our understanding of the global pattern of all clones, the logic of longitudinal organization of the structure, and the way in which the segments are produced.

In this chapter, we introduce an approach based on the retrospective interpretation of all the clones participating in a single segment. The clonal history of two segments of different axial level, segments 4 and 24, will be described and then compared. This approach puts into perspective the developmental history of all possible ancestral cells of a given segment. By following specific clonal characteristics, we have arrived at similar conclusions regarding myotome development as those reported in Nicolas *et al.* (1996), and we have also discovered new properties of the system (Eloy-Trinquet and Nicolas, 1999).

VII. The Longitudinal Organization of the Segments

A. Several classes of Clones

In Fig. 8, the clones contributing to segments 4 (Fig. 8A) and 24 (Fig. 8B) are presented. Most of them (see the figure legend) have been arranged by their most anterior border, and this analysis reveals the following. First, in segment 4, group I, there are clones (4) containing more than one cell (up to 15) that participate only in this segment. Thus, there are precursor cells restricted to a single segment; therefore, this segment is a clonal entity, that is, there is **no systematic** clonal

Figure 6 Classification of the first 153 clones (identified among 3000 embryos) according to their order of production. Each vertical column corresponds to a clone in an E11.5 embryo. Its axial position is indicated by the number on the left, which corresponds to the number of segments from the most anterior (1) to the most posterior position (42). The diversity of the clones illustrates the random nature of the labeling event in a cell (that is, the spontaneous homologous recombination). For instance, some clones contribute to only one segment (the smallest rectangle) and others to a very large portion of the axis, either continuously (a single large rectangle) or discontinuously (several large rectangles). This schematic representation is the basis for other classifications according to different criteria (see Figs. 7 and 8). A, Anterior; P, posterior level of the embryos.

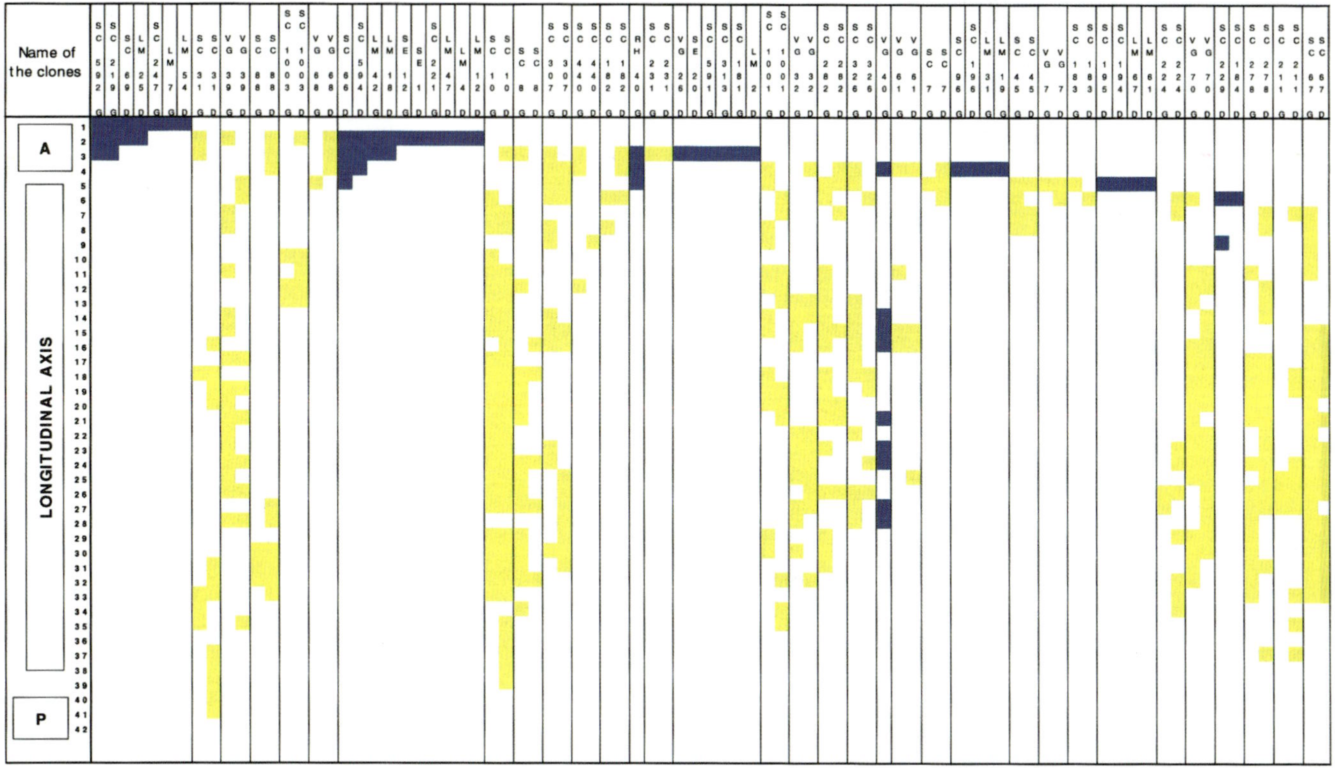

Figure 7 Classification of all clones according to the most anterior segment to which they contribute (longitudinal organization). Clones 1 to 315 (x-axis) were obtained from a total of 6524 E11.5 embryos analyzed. Each clone is represented in two adjacent columns, if bilateral (yellow), and in one column, if unilateral (blue). Their name is indicated above each column. Only the segments to which the clones contribute are colored. The pattern contains four main categories of clones: monosegmented unilateral clones, short unilateral clones contributing to two to four segments, short bilateral clones, and long bilateral clones. In addition, a few long monolateral clones are observed (in blue). The retrospective analysis of this pattern of clones provides much information concerning the way in which the structure forms during gastrulation and somitogenesis. A, Anterior; P, posterior level of the embryos.

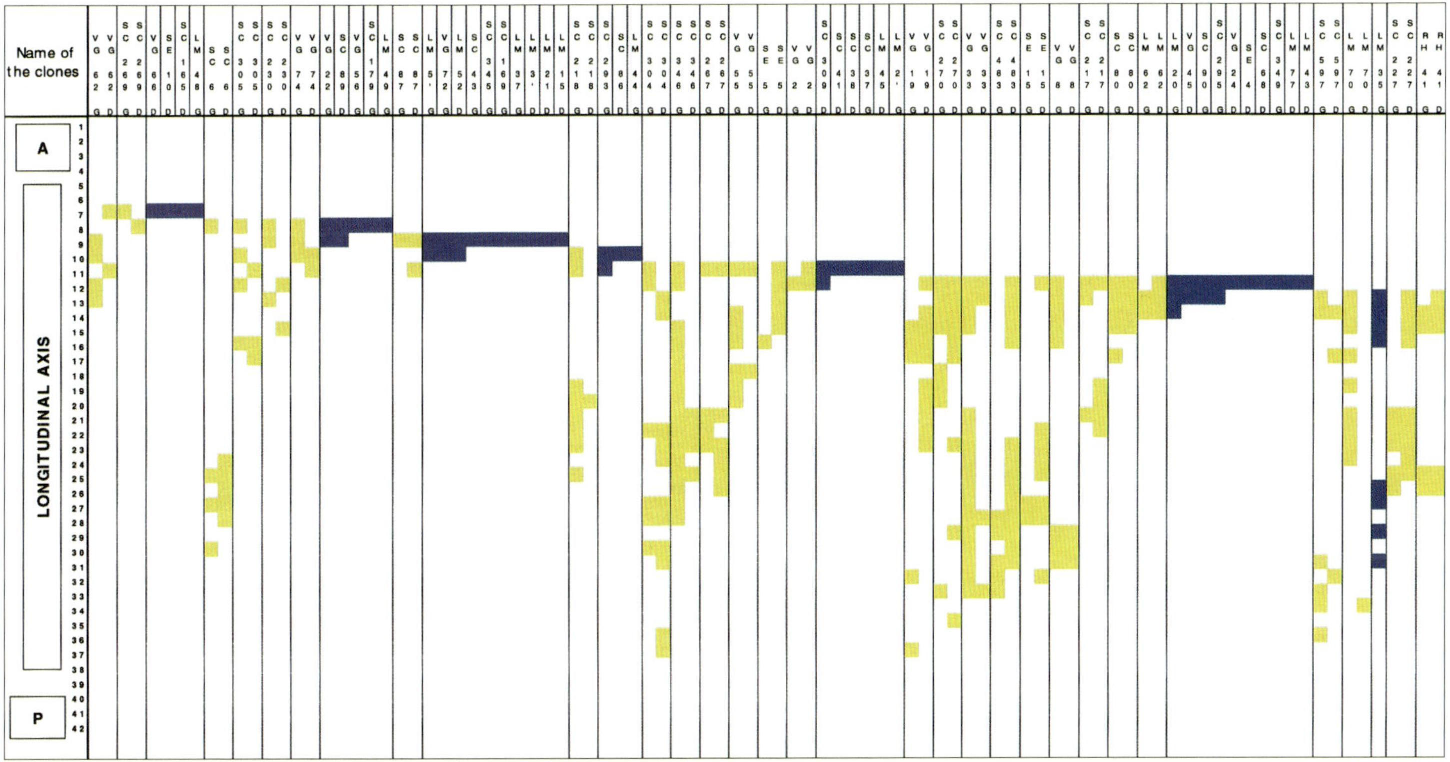

Figure 7 *Continued*

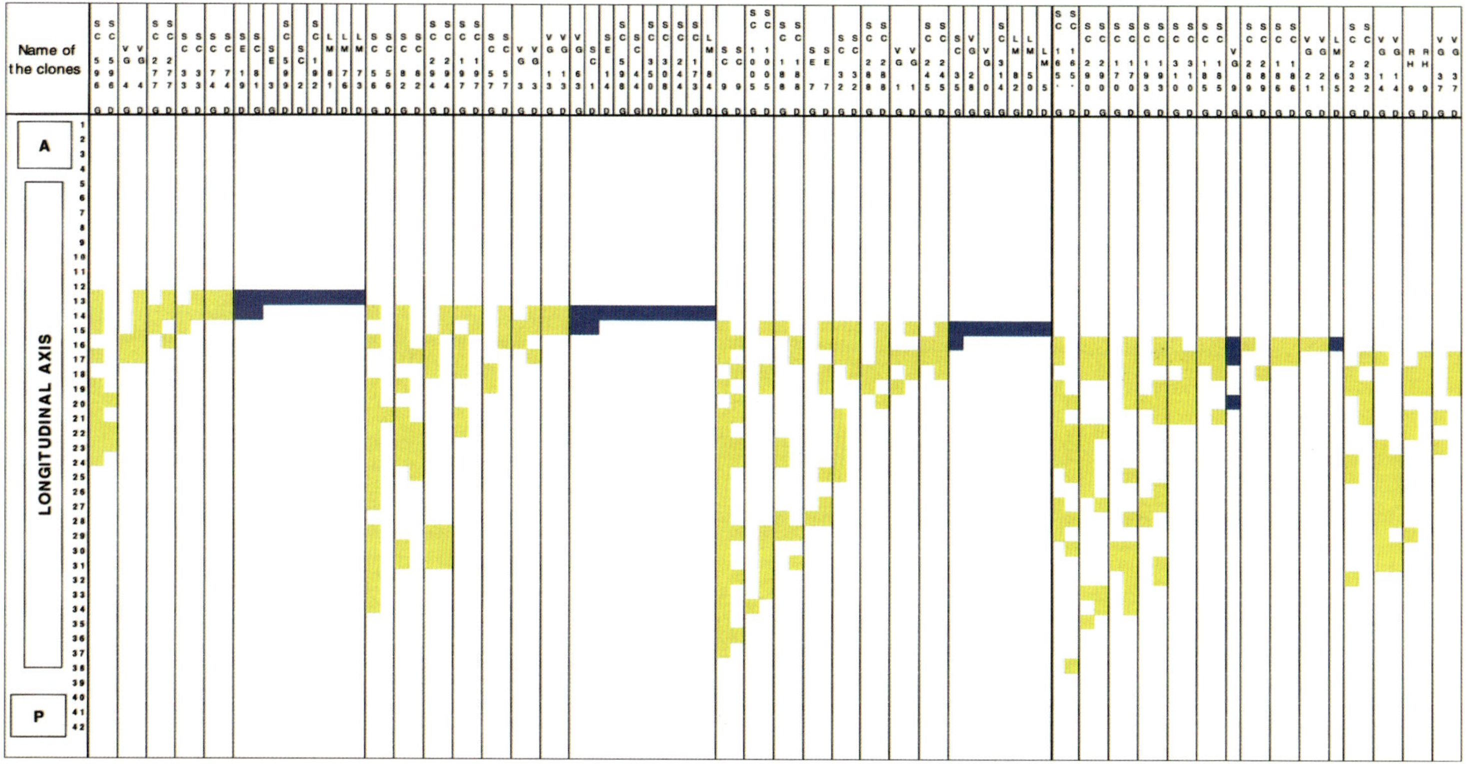

Figure 7 *Continued*

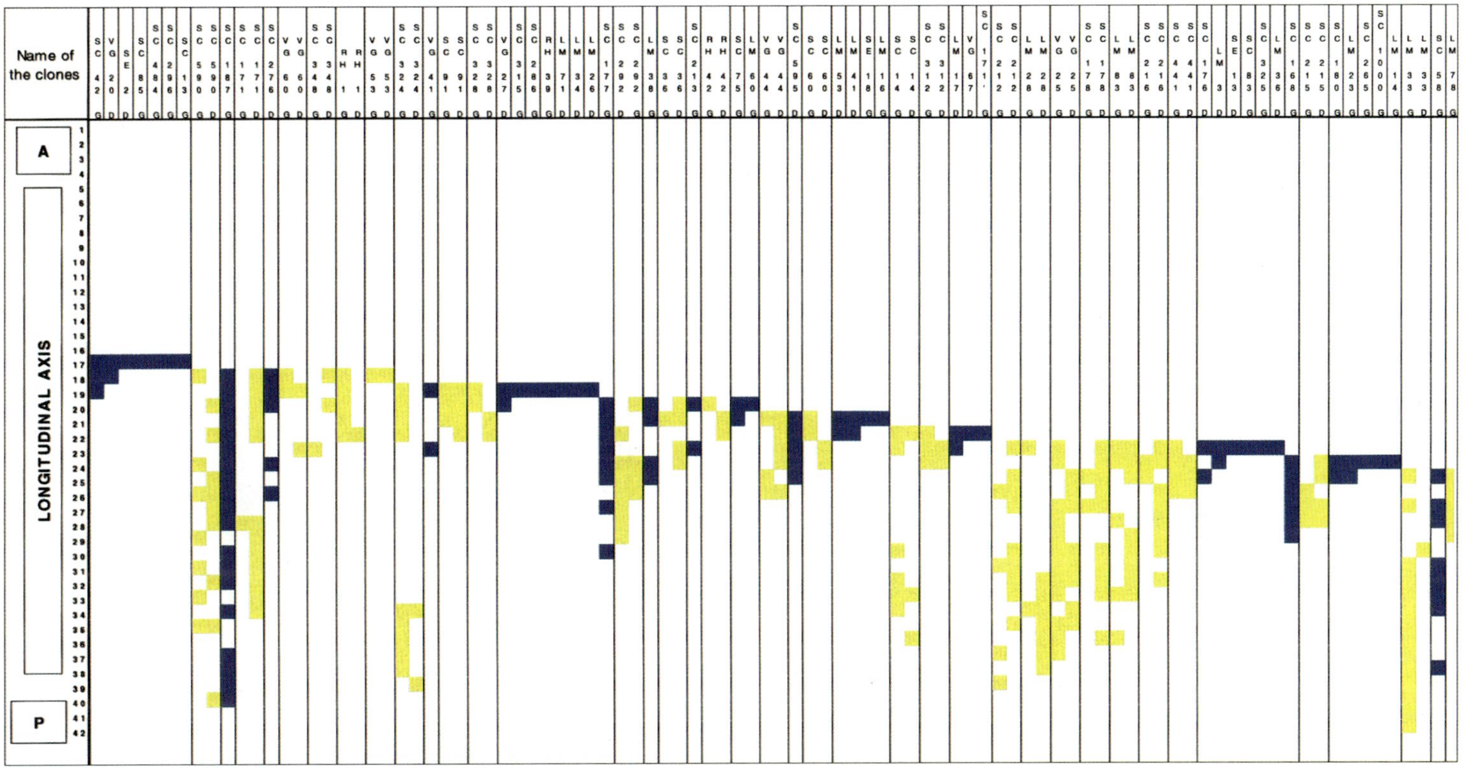

Figure 7 *Continued*

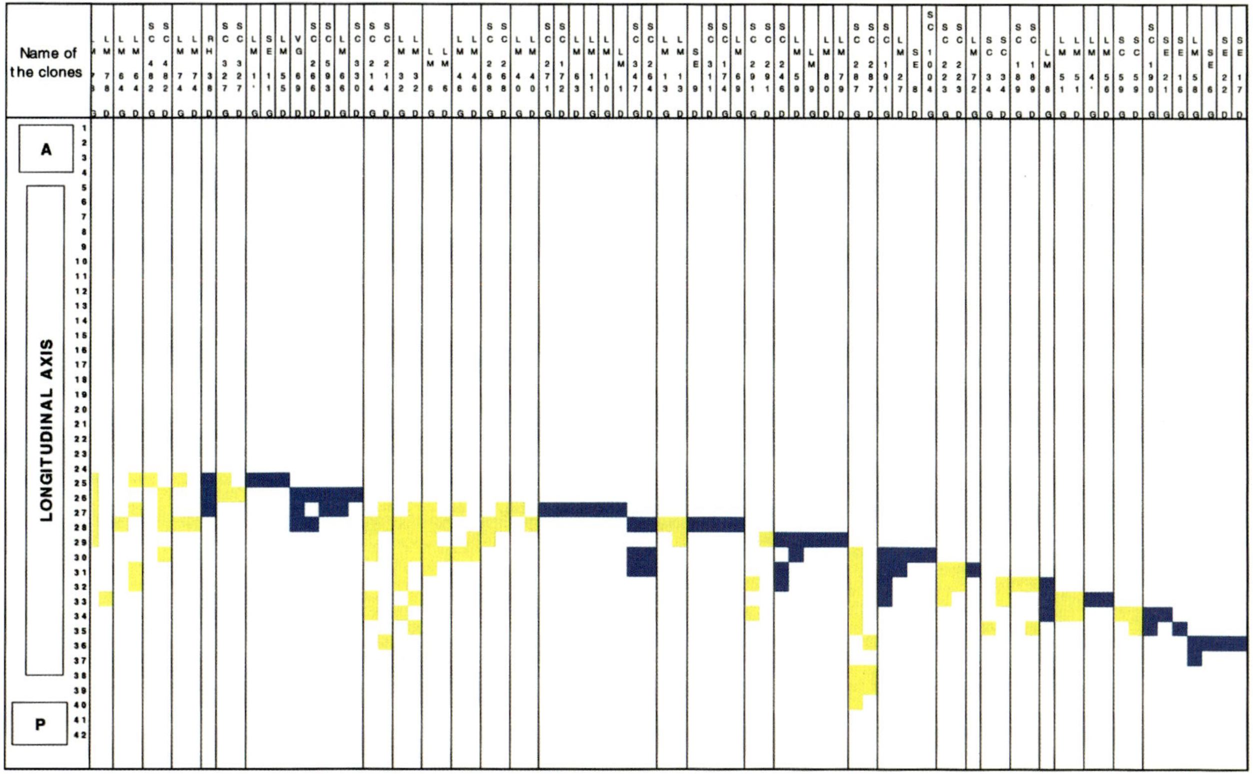

Figure 7 *Continued*

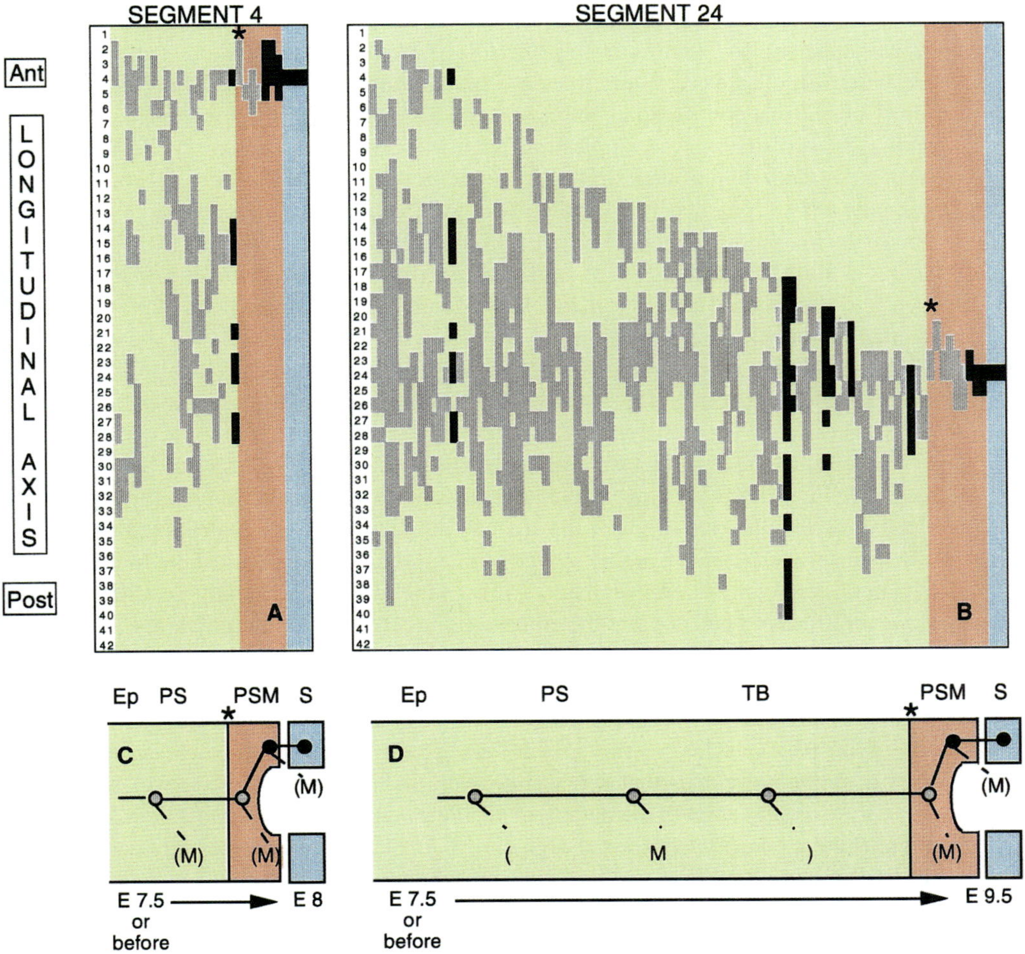

Figure 8 Schematic representation of the pattern of clones contributing to segments 4 (A), and 24 (B). Each bilateral clone corresponds to two adjacent gray vertical columns and the unilateral clones, to a single black vertical column. Clones are classified according to the most anterior segment to which they contribute, except for the shortest ones, for which the limit is indicated by the asterisk, where the unilateral clones have been grouped. The number of bilateral clones contributing to segments 4 and 24 is 11 and 42, respectively. (C, D) Under each pattern, the retrospective interpretation is schematically represented. The size of the structures represented is related to the number of observed clones, that is, to the size of the pool of cells involved. The size of the structures corresponding to the bilateral clones increases from the rostral segments (4) to the more caudal segments (24) because the total number of ancestral cells increases. The small white and black circles correspond to the genealogical tree of the ancestral cells of the segment, beginning from their initial localization in the primitive streak. M indicates that certain cells affiliated with these ancestral cells contribute to other segments. Ant, Anterior; Post, posterior level of the embryos. S, Somite (monosegmented unilateral clones); PSM, presomitic mesoderm (plurisegmented unilateral clones and short bilateral clones); PS, primitive streak; Ep, epiblast, and TB, tail bud (long bilateral clones).

arrangement in two segments. Second, there are also clones (3) that contribute to several adjacent ipsilateral segments (up to 4) (in black on Fig. 8). Therefore, there are ancestral cells that can participate in a small number of longitudinal segments. Third, clones that contribute to both ipsilateral and contralateral segments (in gray on Fig. 8) are also observed, suggesting that the left and right segments are formed from a common pool of ancestral cells. Some of these bilateral clones (2) spread into only a few segments (1 to 4), whereas others (9) contribute to a very large portion of the longitudinal structure (up to 30 segments). This is in contrast to the unilateral clones (in black on Fig. 8) among which only one clone spreads into more than four segments. Thus, four main classes of clones are detected: the monosegmented unilateral, the plurisegmented unilateral, the short bilateral, and the long bilateral. The first class corresponds mainly to progenitors and occasionally to ancestral cells of segment 4. The last three classes correspond to the labeling of myogenic ancestral cells of segment 4.

Very similar observations were obtained for the segments of group II. In segment 24 (Fig. 8B), which represents a caudal group II segment, there are some clones restricted to one segment (4), a small number of clones contributing to a few ipsilateral segments (2), and a large number of clones that contribute to both ipsilateral and contralateral segments (42), and which are frequently positioned along almost the whole axis of the embryo or, less frequently, in only a few segments. Again, unilateral clones almost never contribute to more than three segments and, when a clone does participate in more than four segments, it also contributes contralaterally to the muscular system. The clonal history of segment 24 is very like the clonal history of segment 4 except for one difference: there is a new class of unilateral clones that spread into more than 4 segments (up to 11 segments). An examination of the pattern of all clones (Fig. 7) indicates that this class of unilateral clones first appears at segment 18. Therefore, there are clearly strong similarities between the clonal histories of segment 4 and 24 but also significant differences. Apparently, the system of production of segments is modified at certain stages of development.

B. Localization in the Anatomical Structures of the Embryo

Labeled cells at the origin of the four main classes of clones can be localized in the embryo by comparing their geometric characteristics with those of the structures from which somites are derived (Fig. 8C–D). The monosegmented unilateral clones arise, for the most part, from the labeling of precursor cells after their segregation into the somite (in blue in Fig 8). The clones restricted to one side of the muscular system (left or right) correspond to the labeling of ancestral cells in the presomitic mesoderm (with the exception of the long unilateral clones, see below, in orange in Fig. 8). In confirmation of this interpretation, the mean segmental contribution of monosegmented unilateral clones is smaller (5.7 cells per segment) than the mean segmental contribution of plurisegmented unilateral clones (9.2

cells per segment). The short and long bilateral clones correspond to the labeling of ancestral cells localized in structures that contribute bilaterally to the muscular system (in yellow in Fig. 8). Among the anatomical structures which meet these criteria are the presomitic mesoderm (PSM, in orange)—during the brief period when it is still not bilateralized—the primitive streak (PS), and certain regions of the epiblast (Ep, in yellow in Fig. 8). The maintenance of bilateral clones that contribute to caudal segments and are produced after the regression of the primitive streak indicates an origin from the tail bud blastema (TB, also in yellow in Fig. 8) and shows that there is still a pool of nonbilateralized ancestral cells in this structure.

In summary, the clones shown in Fig. 8 demonstrate definitively that the lineage of at least some ancestral cells of the muscle system pass through these anatomical structures. These clones, therefore, provide information about the properties of cells at the origin of these structures and on their mode of propagation from one structure to another. Indeed, the genealogical tree of ancestral cells of the structure in Fig. 3A,C demonstrates how a comparison between two parameters from each class of clones (i.e., their mean contribution per segment and their frequency) reveals their mode of propagation.

C. A Long and Unique Period of Vertical Propagation

The mean contribution per segment of the plurisegmented unilateral clones and bilateral clones is nearly identical (Table I). The mode of propagation between these two pools (depicted as the filiation between the gray and black cells in Fig. 8C,D therefore corresponds to a direct vertical mode, meaning, a mode without any horizontal expansion. In other words, for each gray cell, there is only one corresponding black cell.

Similarly, for each cell in the pool of precursors to the long bilateral clones, there is only one corresponding cell in the pool of precursors to the short bilateral clones (Table I). Indeed, if this were not the case, the mean contribution per segment would be higher for the long bilateral clones than for the short bilateral clones. This finding establishes that for the precursor cells of **a** single segment, the

Table I Average Segmental Contribution By Unilateral and Bilateral Clones

Category of clones	Average segmental contribution[a]
Short unilateral plurisegmented clones	9.2 ± 1.0 (n = 118)
Short bilateral clones	12.6 ± 1.2 (n = 146)
Long bilateral clones	10.0 ± 0.4 (n = 1170)
All unilateral plurisegmented clones	8.1 ± 0.7 (n = 195)
All bilateral clones	10.3 ± 0.4 (n = 1316)

[a]Number of β-gal$^+$ cells divided by number of segments with β-gal$^+$ cells. n = Number of segments with β-gal$^+$ cells.

periods of horizontal expansion occur only at an early stage, prior to the formation of the first somites at E8, and at a later stage in the somite. In other words, the polyclonal size of the segment remains nearly unchanged from a very early stage.

The number of plurisegmented unilateral clones is smaller than the number of bilateral clones. This applies to segment 4 (3 and 11, respectively) as well as for segment 24 (8 and 42, respectively) and for all other segments (Fig. 7). This result indicates that the total number of precursors of the bilateral clones is higher than those of the plurisegmented unilateral clones. Because the pool of the precursors of these two classes of clones are necessarily physically connected, either partially or completely, this suggests that the pool of bilateral clones has a vertical organization. Indeed, it is not possible (Fig. 3C) to have an intermediary pool of early founders (those of the bilateral clones) whose size would be larger than the later pool of founders (those of unilateral clones).

The conclusions drawn from the latter two examples (a vertical mode of filiation between the pools of short unilateral clones and bilateral clones, and a long vertical organization of the pool of long bilateral clones) are schematically drawn in Fig. 8C,D. The excess of bilateral clones compared to unilateral clones indicates that the period of vertical propagation involves several generations of ancestral cells. In other words, during this period, the ancestral cells that contribute descendants to **a** single segment also contribute affiliated cells to structures other than the myogenic cells of this segment (indicated by the dotted lines from the ancestral cells in Fig. 8C,D). These other structures could be inside or outside the muscle system (depending on the degree of coherence of the muscle system, see below).

Finally, a comparison of the clonal history of the ancestral cells of segments 4 and 24 shows that the number of bilateral clones is higher in segment 24 (42) than in 4 (11). More generally, clonal complexity (that is, the number of times a segment is labeled by a clone) increases from anterior to posterior, even though the cellular complexity is nearly the same (Fig. 1G) (Nicolas *et al.,* 1996). This increase in complexity is not due to an increase in the size of the polyclone at the origin of the segments due to horizontal expansion, because there is no decrease in the mean number of clones per segment; therefore, this increase must be due to an increase in the number of generations of the ancestral cells during the period of vertical propagation, suggesting that the period of vertical propagation occurs during longitudinal axis formation (Fig. 8C,D, the length of the yellow areas).

D. Contribution to Other Segments by the Cells Affiliated with the Ancestral Cells of One Segment

The main feature of an ancestral cell during a period of vertical propagation is its nonclonal growth (Fig. 3A). In the case of segment 24, this means that, at each cell generation during this period, the descendants of only one daughter cell contribute to the myogenic cells in segment 24. Information on the fate of the descendants of the affiliated cells can be gathered by examining other contributions of the clone outside of segment 24. It is clear that at least some of the affiliated cells also contribute to myogenic cells in other segments (this is indicated by M after the dotted line from the ancestral cells in Fig. 8C,D). Other descendants may contribute to nonmyogenic structures, but they are not detected in this study.

Because cells affiliated with the ancestral cells of a segment contribute regularly to muscle cells in other segments, this suggests that the whole muscle system has some kind of clonal organization. This organization involves the more rostral segments, which derive from the ancestral cells of segment 24, for example, as well as more caudal segments that derive from descendants of segment 24 ancestral cells. Indeed, if the ancestral cells of a segment were constantly in a dispersive mode, cells affiliated with the ancestral cells of one segment would not contribute to myogenic cells of other segments; therefore, a majority of monosegmented clones would be expected, but this situation is not observed (see Figs. 7 and 8). Therefore, a question that remains is, what is at the origin of this clonal organization?

E. A Permanent Lineage in Nonbilateralized Structures of the Embryo

To answer this question, the localization of ancestors (rostral contribution) and descendants (caudal contribution) of ancestral cells of segment 24 that contribute to other segments of the muscle system must be examined. Since these contributions are ipsilateral and contralateral in both cases, this suggests that these "ancestors" and "descendants" are in nonbilateralized structures of the embryo. This conclusion is confirmed by the bilaterality of the short clones observed all along the axis of the embryo (Fig. 7). Indeed, they are probably derived from the direct labeling of these descendants and ancestors. Therefore, a labeled cell at the origin of a long bilateral clone produces not only cells that leave the bilateral structure, enter the presomitic mesoderm, and contribute to segment 24, but also cells that remain in these bilateral structures and later produce cells which will again leave the structure, enter the presomitic mesoderm, and contribute to segments caudal to segment 24. Obviously, this scenario occurs several times during axiogenesis, and it demonstrates the existence of a population of cells that are permanently maintained in the bilateralized structure of the embryo, and which contribute to the formation of the left and right presomitic mesoderm.

These findings apply to all segments (Fig. 7); in each case, the founder cells of each segment have distant ancestors that participate in more rostral segments and whose descendants will participate in caudal segments. Therefore, in all cases from segments 1 to 40, the population of cells that is maintained in nonbilateralized structures from E8 is detected in this analysis.

F. A Transient Lineage in the Presomitic Mesoderm

The demonstration of the existence of a pool of permanent cells in the nonbilateralized structure of the embryo is based on the observation of long bilateral clones (that is, clones contributing to more than four segments; Fig. 7). If the presomitic mesoderm were also housing a similar cell population, then a significant class of long unilateral clones would be expected. Because only 3 long unilateral clones (their interpretation is given later) are observed among 70 long bilateral clones (Fig. 7), the presomitic mesoderm can be formed only from cells that are constantly renewed. The descendants of ancestral cells of the muscle system in bilateralized structures of the embryo, thus, constitute a transient lineage in the presomitic mesoderm.

The fact that ancestral cells of a segment labeled in the presomitic mesoderm (short unilateral clones) contribute to several segments indicates that these ancestors divide at least once and then disperse. The growth along the longitudinal axis is not strictly coherent in the presomitic mesoderm. The alternative situation would predict only monosegmented unilateral clones but this is not observed.

Finally, the frequency of unilateral clones in more than one segment (52 cases) is almost 2-fold the frequency of short bilateral clones (38 cases; Fig. 7). This is expected if the period of vertical propagation of the precursors to short bilateral clones is very short. Alternatively, a greater number of short bilateral clones (corresponding to a higher number of precursors) would have been expected. Thus the propagation of the ancestral cells from one pool to the other is rapid.

VIII. A Model for the Longitudinal Organization of the Muscle System

Because the observations for segments 4 and 24 also apply to all other segments (Fig. 7), a simple model can be proposed for the developmental lineage of the myogenic ancestral cells for **one** segment. It links its history to the history of the other segments disposed all along the anterior–posterior axis of the embryo (Fig. 9).

Figure 9 A model of the longitudinal and mediolateral (dorsoventral) organization of myogenic cells in the first 40 segments of the embryo. (A) A unique lineage, schematized by the blue horizontal line, which genealogically links the S cells, remains permanently in the nonbilateralized structures of the embryo. It produces reiteratively the P cells that constitute a transient lineage in the presomitic mesoderm. S cells have the properties of stem cells involved in axiogenesis. The labeling of a cell in the somite generates a unilateral clone (in gray) restricted to one segment. The labeling of a P cell in the presomitic mesoderm generates a plurisegmented unilateral clone (in green). The labeling of a P cell in the nonbilateralized presomitic mesoderm generates a short bilateral clone (in yellow). The labeling of a S cell generates a long bilateral clone (in red). S and P cells are ancestors of the myogenic cells of a segment. Certain cells, which are clonally related to these ancestral cells, also contribute to other segments. (B) The S lineage is recruited in the paraxial regions of the young gastrula. These cells converge toward the primitive streak where they form a pool of stem cells that produce P cells during axiogenesis. Most of the recruited epiblast cells contribute to both ipsilateral and contralateral paraxial mesoderm. (C) The pool of S cells in the nonbilateralized structures remains relatively coherent during axiogenesis. The organization along the y-axis prefigures the mediolateral organization of the dermomyotome, remains unchanged during somitogenesis, and probably corresponds to the anterior–posterior axis of the primitive streak. The organization along the x-axis prefigures bilaterality and probably corresponds to the mediolateral axis of the primitive streak. The most lateral cells (in blue) contribute to only one side of the embryo. More axial cells (in red) contribute, in general unequally, to both

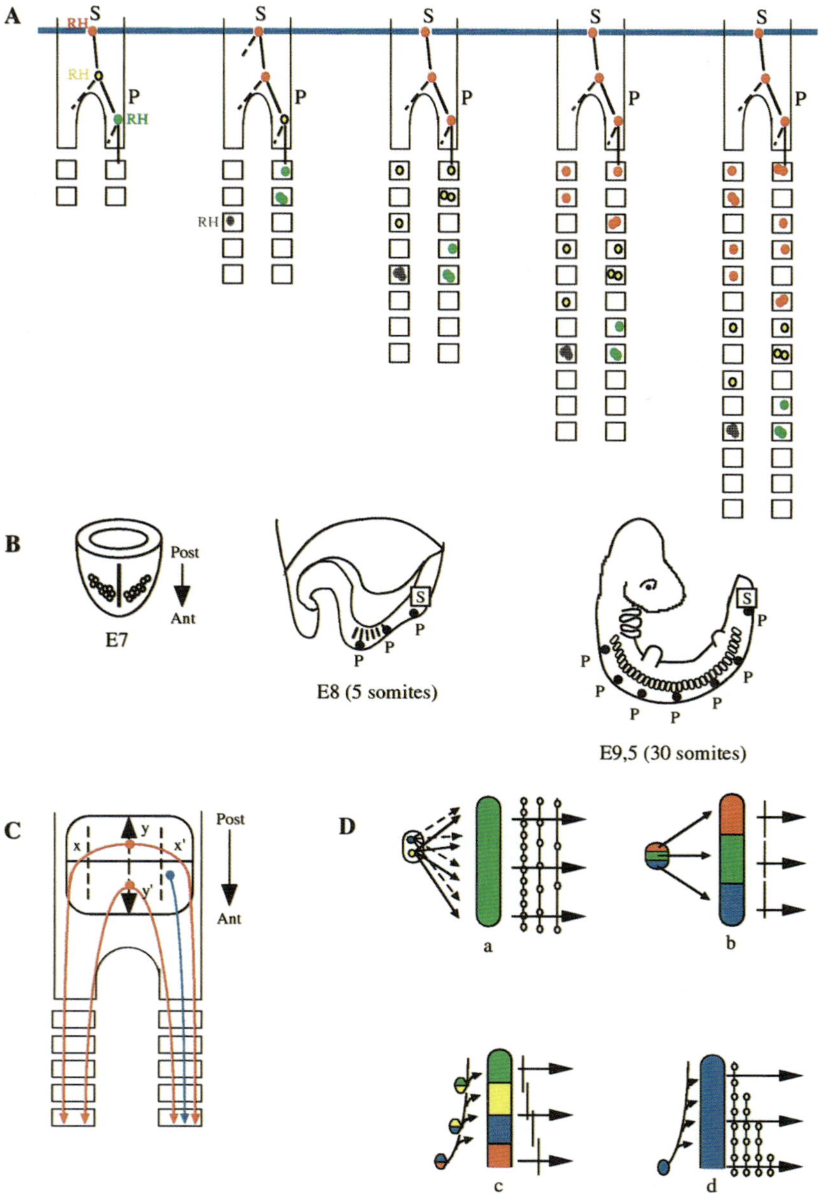

sides of the embryo. The unequal contribution remains the same during the period of axiogenesis. Note that axis orientation changes during this process. (D) Four principal representative models for the formation of a longitudinal structure. The models relate the longitudinal structure to its pool of precursor cells. In a and b, the pool is fixed early. In c and d, the pool is formed progressively. (a) Extension intercalation model, in which the descendants of precursor cells intermingle extensively. (b) Early regionalization. (c) Temporal production from a changing pool of cells. (d) Temporal production from a self-renewing pool of cells. On the left is a representation of the pool of precursors; in the middle, a representation of the longitudinal axis after its formation and, on the right, the longest clones expected from the labeling of a cell in the pool of precursors. The horizontal arrows represent the clonal complexity in relation to the axial level.

A pool of about 100–150 cells that correspond to distant ancestral cells for each segment contributes to all segments of the animal. This pool of cells (called S cells) is maintained in the nonbilateralized structures of the embryo involved in axiogenesis (Fig. 9A, the blue line). Therefore, these pools constitute a particular lineage of stem cells that reside first in the primitive streak (Fig. 9B,a) until the production of about the first 20 segments (from E8 to E9), then in the tail bud (Fig. 9B,c). There is clonal continuity within and between these two structures. Models based on a permanent, or perhaps even partial, recruitment of cells into the primitive streak (Fig. 9D,c), as well as models based on the spatial regionalization of precursors of the axial region (Fig. 9D,b), can be discarded by the demonstration of this clonal continuity in the muscle system (Nicolas *et al.,* 1996).

The S lineage is recruited in the regions of the epiblast that ultimately converge in the primitive streak between E6 and E7.5 (Fig. 9B,a), (Tam and Beddington, 1987; Lawson *et al.,* 1991). If the mesoderm produced by the primitive streak is initially nonbilateralized, it could house a pool of permanent stem cells; however, the properties of mesodermal cells, and, in particular, their mobility, makes this hypothesis unlikely. Therefore, the pool of S cells more likely resides in the primitive streak (Nicolas *et al.,* 1996). A recombination in the reporter gene in an S cell produces a long bilateral clone (in red in Fig. 9A). Its date of birth is related to the date of birth of the most anterior segment to which it contributes and precedes it by about 12 hr (i.e., the time required to pass through the presomitic mesoderm). The pool of S cells produces cells (called P cells) whose descendants will participate in a few adjacent segments. P cells are still true ancestral cells of the muscle cells of one segment. However, these cells rapidly pass into the nonbilateralized mesoderm (short bilateral clones) and then into the presomitic mesoderm (mainly bi- to quadrisegmented unilateral clones) but they do not stay there; the two structures are, instead, renewed by the constant recruitment of new descendants from the S pool. A recombination in the reporter gene of the P cell produces a unilateral clone (in green in Fig. 9A) or a short bilateral clone (in yellow in Fig. 9A). The descendants of the P cells end up in the somite from which the dermomyotome is formed.

The two main properties of S cells are self-renewal and reiterative production of another type of cells—the P cells—and these are similar to the properties of stem cells. Their localization suggests that they are, indeed, stem cells involved in the process of axiogenesis. This conclusion has the following implication with respect to clonal analysis: it is not possible to further divide regions of the fate maps (Lawson *et al.,* 1991; Tam and Behringer, 1997) that contain the precursors to these cells—at the young gastrula stage, for instance (Fig. 9B,a)—into subregions that would correspond to particular somites or to only group I or group II regions, and this even if one could analyze the system at the single cell level in labeling experiments. This is because each stem cell precursor will participate in all segments, and the regions demarcated in the fate map represent the entire structure. What will serve to define segments and subregions is a temporal process of production. The only alternative would be that the pool of stem cells at the origin of the long clones is itself temporally recruited (from E7 to E11) from epiblastic cells, and this is unlikely. Moreover, the apparent bilaterality of the presumptive territories in the fate maps of young gastrula (Fig. 9B,a) is unlikely to correspond to the future bilaterality of the presomitic mesoderm or the adult muscle system. These regions must be viewed as a unique territory, possessing almost no sign (but see below) of its future bilaterality. Bilaterality will only be definitively fixed when the S lineages produce the P lineages, and a single S cell frequently produces P cells that contribute to both sides of the embryo.

The model easily explains the four main classes of clones (monosegmented unilateral, short unilateral, short bilateral, and long bilateral) (Fig. 9, the gray, green, yellow, and red cells, respectively), and analysis of the clonal subclasses, mentioned earlier, suggests the following addition to the model: the long unilateral clones that appear in segment 18 (Fig. 7) point to a change in some of the characteristics of the S/P system during the transition between the primitive streak and tail bud formation (Eloy-Trinquet and Nicolas, 1999).

Clonal origins of muscles have been studied in allophenic mice, and it was found that each somite and each muscle must arise from at least two precursor cells and that neighboring somites may share some cellular ancestry as they were often correlated in GPI alloenzyme composition (Gearhart and Mintz, 1972). Prospective analysis using orthotopic grafts of [^{3}H]thymidine-labeled cells has suggested that the streak may act as a source of cells and not just as a route of relocation and that some cell mixing occurred within the presomitic mesoderm (Tam and Beddington, 1987). These observations are in agreement with our results. However, the notion that presomitic mesoderm comprises two populations, a resident population that is supplemented by an influx of cells from the streak (Bellairs, 1985; Tam, 1988) is not confirmed by our analysis.

IX. The Mediolateral Organization of the Segments

Up until now, in this chapter, only the longitudinal organization of the murine muscle system has been interpreted. But in the chick, a mediolateral and a dorsoventral organization of the somite have been described (Selleck and Stern, 1991; Ordahl and Le Douarin, 1992; Psychoyos and Stern, 1996; Kahane *et al.*, 1998a,b; Kato and Aoyama, 1998). Major questions concern the extension of this notion to the mouse, the clonal basis of these regionalizations, and, if such a clonal basis exists, the stage at which regionalization occurs. A very brief summary follows regarding the mediolateral organization of the group II segments, as determined by a retrospective clonal analysis of the randomly produced clones described below (Eloy-Trinquet and Nicolas, 1999).

A. Mediolateral Regionalization Precedes Bilateralization

The description of clones along the mediolateral axis of the segments (Fig. 10) reveals several remarkable properties. (1) The clones of cells labeled in the pre-

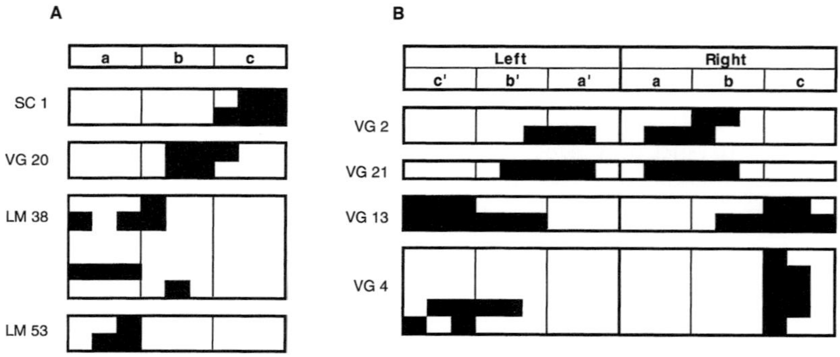

Figure 10 Mediolateral organization of the segments. Segments from group II (11 to 24) have been subdivided into medial (a), intermediate (b), and lateral (c) regions. (A) Schematic representation of plurisegmented unilateral clones. Each large rectangle corresponds to the segment to which the clone contributes. The name of the clone is on the left. The area occupied by the clone is in black. Clones contribute to only a portion of the mediolateral axis and keep the same mediolateral position in the different segments to which they contribute. (B) Short bilateral clones, same representation as in A. Clones are restricted to a portion of the mediolateral axis and have a left–right mirror symmetry.

somitic mesoderm (plurisegmented unilateral clones, labeling of a P cell) (Fig. 9) do not contribute to the whole mediolateral axis of the segment but only to a portion of the medial, lateral, or intermediate regions (Fig. 10A). Therefore, the pool of ancestral P cells is regionalized in a manner analogous to the model illustrated previously (Fig. 9D, b or c). (2) The descendants of a labeled P cell maintain the same position in each of the few segments to which they contribute (Fig. 10A). Therefore, mediolateral regionalization preceeds segmentation and it is presomitic. (3) These two characteristics are also found in short bilateral clones, which correspond to the labeling of P cells just before their entry into the unilateral pool (Fig. 10B) and also with long bilateral clones (labeling of an S cell, not shown). This means that the pool of precursor cells is mediolaterally regionalized before its entry in the presomitic mesoderm. This organization persists unchanged (for several cell generations) during somitogenesis, until the formation of the dermomyotome. Thus, the mediolateral regionalization clearly corresponds to a robust regionalization of its ancestral cells.

B. A Mirror Production of Left and Right Structures

Because ancestral cells are mediolaterally regionalized, the relationship between cells in the nonbilateralized structure and the organization of their descendants in the two presomitic structures is not evident. This problem has been analyzed by comparing the left and right contributions of bilateral clones. Remarkably, these clones exhibit a high degree of mirror symetry (Fig. 10B): clones positioned medially on the left are also positioned medially on the right. This mirror symmetry

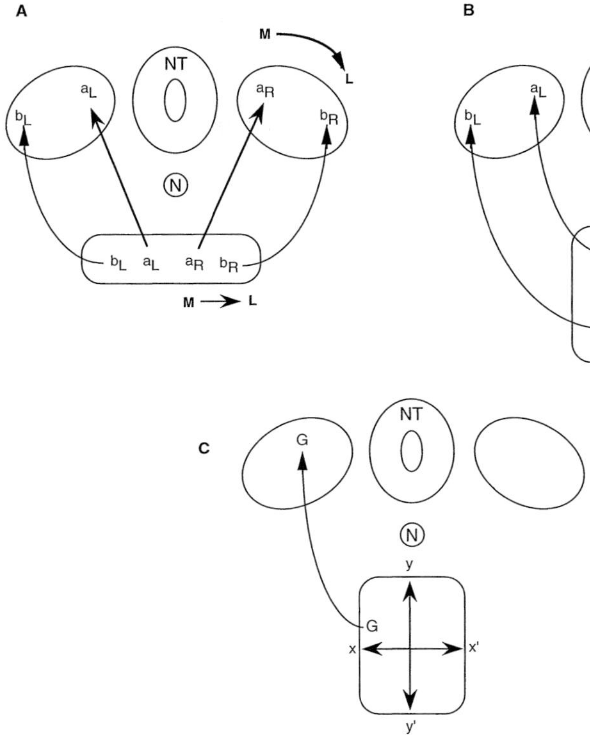

Figure 11 (A, B) Models for the derivation of the two pools of cells in the presomitic mesoderm for the nonbilateralized structure. (A) Direct relationship: both left and right sides are performed in the pool of cells before bilateralization. It is possible to explain the medial bilateral clones by the labeling of a_L or a_R, which would then contribute bilaterally to the presomitic mesoderm. However, it is not possible to explain the production and symmetry of the lateral bilateral clones because a labeling in b_L or b_R would not result in a contribution to both sides of the embryo. (B) Indirect relationship: both left and right have a unique representation in the nonbilateralized structure. a_L and a_R, as well as b_L and b_R of A, correspond to a unique a or b cell, respectively. Their labeling produces either medial (a) or lateral (b) clones contributing to both sides of the embryo. (C) Organization along the *x*-axis. The long unilateral clones indicate a coherent behavior of the cells in the pool before bilateralization occurs (see text). The G cell contributes to only the left side of the embryo, either medially or laterally. More medial cells along the *x*-axis would contribute a majority of cells to the left and a minority of cells to the right side of the embryo (not represented). N, Notochord; NT, neural tube; M, medial; L, lateral.

is also observed for clones restricted to the lateral part of the segments. These observations refute models postulating a direct and separate representation of the left and the right structures which would pre-exist in the nonbilateralized pool of precursors (Fig. 11A). Indeed, cells that will be arranged laterally in the presomitic mesoderm (as well as those that will be arranged medially) must occupy the same unique position in the pool of nonbilateralized precursors to allow the generation of a symmetric clone. Therefore, there is an indirect and unique representation of left and right structures in the pool of nonbilateralized precursors (Fig. 11B), and cells originating from a single pool will produce two pools by mirror duplication. Moreover, the existence of both lateral and medial symmetrical clones cannot be explained without postulating a change in orientation of one axis between the pool of nonbilateralized precursors and the two bilateralized pools.

Altogether, these observations show that the mediolateral organization of the founder cells for segments of the muscle system is relatively fixed in the nonbilateralized structure of the embryo, in other words, in the pool of S cells.

C. A Globally Coherent Organization in the Nonbilateralized Structures

Because the pools of P and S cells are relatively coherent along the axis that will eventually become the mediolateral dimension of the dermomyotome (y-axis, Fig. 11C), we next examined coherence along the other axis (the x-axis in Fig. 11C). The three long unilateral clones, whose anterior borders are found in the group I and II segments (see Fig. 7), suggest a coherent organization along the x-axis as well (Fig. 11C). The most lateral cells will produce the unilateral clones and the more axial cells will contribute to the bilateral clones. The high degree of coherence along the x-axis is demonstrated by the maintenance of the unilateral characteristic during axiogenesis. Thus, the three clones that remain strictly unilateral along 25, 29, and 23 segments, respectively, demonstrate maintenance of the corresponding S cells in the lateral region, even after their passage through the tail bud.

A more complete analysis of the parameters of all clones (not detailed here) (Eloy-Trinquet and Nicolas, 1999) strengthens these interpretations. In brief, there is a continuum in the characteristic of bilaterality, which is most easily explained by the position of the S cells along the x-axis; the more axial the S cell, the more equal is its contribution to both sides of the axis. We propose that the coherence of the S cell pool involves both the x- and y-axes.

X. A Model for the Mediolateral Organization of the Muscle System

The pool of S cells that are maintained in the nonbilateralized structure of the embryo still exhibit an organization along at least two of its axes. The organization along the y-axis prefigures the position of cells along the mediolateral axis in the dermomyotome, and the organization along the x-axis prefigures its bilaterality (Fig. 9C).

An analysis of global cell movements during gastrulation suggests that the most anterior region of the primitive streak contributes to the most medial region in the mesoderm and the more posterior region forms the more lateral mesoderm (Tam and Beddington, 1987; Lawson *et al.,* 1991; Lawson and Pedersen, 1992; J. L. Smith *et al.,* 1994). An interpretation of the clones in this context suggests that the *y*-axis is equivalent to the anteroposterior axis of the primitive streak oriented along the proximal–distal axis of the egg cylinder. The primitive streak would, therefore, exhibit an, as yet, unsuspected degree of coherence for the presumptive regions of the presomitic mesoderm. Perhaps even more surprising is the fact that this coherence apparently persists in all intermediary structures formed during somitogenesis (Christ *et al.,* 1998). Neither the passage through the presomitic mesoderm nor the formation of the somites (and later of the dermomyotome) completely obliterates the ancestral organization established in the primitive streak.

Furthermore, we propose that the organization along the *x*-axis also exhibits a degree of coherence: cells initially placed laterally remain there during the whole process of axiogenesis and maintain this position in all the involved structures, from the primitive streak to the tail bud. Labeling of these lateral cells produces the long unilateral clones; whereas, cells placed more medially produce bilateral clones. Unilateral contribution of labeled cells lateral to the primitive streak to paraxial mesoderm has been obtained in embryo at E8 (Tam and Beddington, 1987). Thus, in contrast to the situation in the young gastrula, where it is difficult if not impossible to recognize the bilaterality of the presomitic mesoderm in the context of the bilateral presumptive territory, it is possible to discern such bilaterality in the primitive streak (Fig. 9C).

The coherent organization along the two axes explains the mirror symmetry of the bilateral clones. The P cells, which can divide at least once and then disperse longitudinally into two to four prospective segments, maintain their mediolateral position in the presomitic mesoderm and the somites. These ancestors to the founder cells of the segments in the dermomyotome are therefore regionalized: they participate only to a median, lateral, or intermediary domain of the structure at E11.5.

The shared properties of regionalization and coherence attributed to the ancestral cells of a segment appear, at first glance, to be in contradiction to their necessary dispersive growth (i.e., their vertical propagation, whereby the pool of ancestors maintains a constant polyclonal value). However, the dispersive aspect of growth predicted by this model can be explained by a stem cell mode of production of P cells by S cells (Fig. 9A), in which one of the two cells leaves the structure and the other "stem cell" remains in the structure. Indeed, coherence and regionalization are attributes of stem cells, and this is maintained in the pool of P cells and in the somites. In other words, the S cells produce P cells reiteratively (temporal dispersion) but keep their position in the pool (spatial coherence). Therefore the genealogical tree of the ancestral cells of a segment in Fig. 3C must be modified, as indicated in Figs. 8C,D and 9A.

XI. Conclusion

The method of clonal analysis discussed in this chapter complements the information obtained from other techniques and, in particular, prospective analyses. Its specificity is that it can be used to retrospectively trace the lineage of all possible ancestral cells of a given structure, as shown in the example described. To our knowledge, it is the only method that is capable of tracing cells continuously backward from a differentiated structure to their oldest ancestors.

For the myotome, the oldest ancestral cells are likely to derive from the primitive streak (Fig. 12). Their number is about 100 to 150 cells. Other studies (mainly prospective) regarding cells of the epiblast have shown that growth is dispersive prior to gastrulation (Gardner and Lyon, 1971; Gearhart and Mintz, 1972; Soriano and Jaenisch, 1986), and this conclusion has been confirmed through prospective analyses (Lawson *et al.,* 1991; Gardner and Cockroft, 1998). After a phase of coherent growth of the embryo (Mintz, 1965; Garner and McLaren, 1974; Balakier and Pedersen, 1982), lasting until E6, there is a successive period of highly dispersive growth that precedes gastrulation (Gardner and Cockroft, 1998). Altogether, these data demonstrate that the cell history of the myotome begins at the primitive streak (but not before this stage), through the formation of a pool of founder cells (Fig. 13).

Next, if we consider a particular segment, then a major finding is that the mean segmental contribution of ancestral cells at a given cell generation is nearly the same in the primitive streak and presomitic mesoderm. Therefore, between these two stages of development, the ancestral cells must enter a period of vertical propagation that is not interrupted by a period of clonal growth (Fig. 13). The initial period of horizontal expansion of the ancestral cells of a segment, thus, occurs before and/or during gastrulation. Only two pools of founder cells are detected: an initial intermediary pool in the primitive streak and a second pool, which arises much later in the somite. All these conclusions apply to at least the first 40 segments.

During the period of vertical propagation, which occurs between the production of these two pools, cells affiliated with these ancestors also contribute regularly to other myotomal segments; therefore, we can conclude that the muscle system exhibits a presomitic cell "coherence" that, in fact, can be traced back to the intermediary pool in the primitive streak.

We propose that this "coherence" of the muscle system is due to two main factors: a system of stem cells and spatial coherence (Fig. 13). This spatial coherence persists during axiogenesis and lasts until somitogenesis. A stem cell system is instrumental in dispersing the cells along the axis of the embryo. As a consequence, it also obscures the strong spatiotemporal organization. More generally, it is important to realize that only a careful examination of the properties of a cell system will show whether it is organized or not. The mere examination of the spatial distribution of cells is only marginally informative.

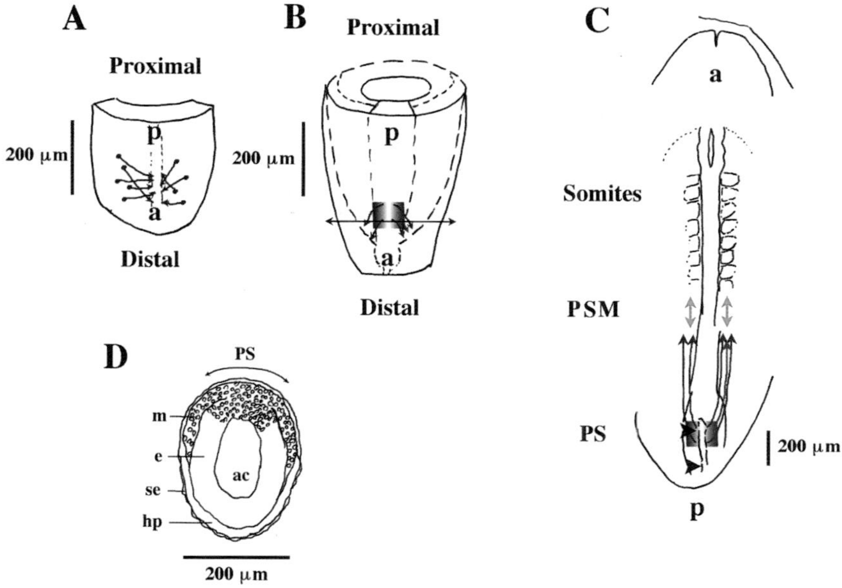

Figure 12 Elaboration of the cellular organization of a segment (a reconstruction integrating this clonal analysis and information from the literature). (A) Recruitment of the founder cells in the primitive streak. The ancestral cells of the myotomal segments occupy lateral regions in the early streak embryo. They converge toward the primitive streak. Very little of their relative order will be conserved in the mature segment. (B) Production of the paraxial mesoderm from the primitive streak. The rostrocaudal (anterior–posterior) order in the primitive streak is converted to a mediolateral order in the presomitic mesoderm. There is also a lateromedial order in the primitive streak: the more lateral cells contribute to only one side of the presomitic mesoderm (the red arrow). This characteristic decreases gradually in the medial direction (the graded blue). Some of the organization of the segment prefigures in the primitive streak. Arrowheads indicate the primitive streak. (C) The first myogenic cells at the nine somite stage. The mouse embryo has been flattened. The primitive streak (PS) has produced the presomitic mesoderm, PSM, and nine somites. The mediolateral order in the presomitic mesoderm still corresponds to the anteroposterior order in the PS (the blue arrows) and in the somites. The ancestral cells of a segment in the presomitic plaque participate in only one to four (in general two) adjacent segments (the green arrows, see the plurisegmented unilateral clones in blue in Fig. 7). (D) Drawing of an histological section through the primitive streak at the level indicated by the arrow in B. Each small circle represents the nucleus of a cell. Only the nuclei of mesodermal cells and cells in the region of the primitive streak have been indicated. Note the relative continuity in the organization of the epiblast, primitive streak, and mesoderm. a, Anterior; p, posterior; PMS, presomitic mesoderm; PS, primitive streak; m, mesoderm; e, epiblast; se, squamous endoderm; hp, head process; ac, amniotic cavity.

The proposition of a stem cell system at the origin of the paraxial mesoderm is reminiscent of observations regarding the persistence of cells between E7 and E7.5 in the anterior part of the primitive streak (Tam and Beddington, 1987; Lawson and Pedersen, 1992) and the node (Beddington, 1994). The retrospective clonal analysis described here has amplified these findings by permitting long-term studies

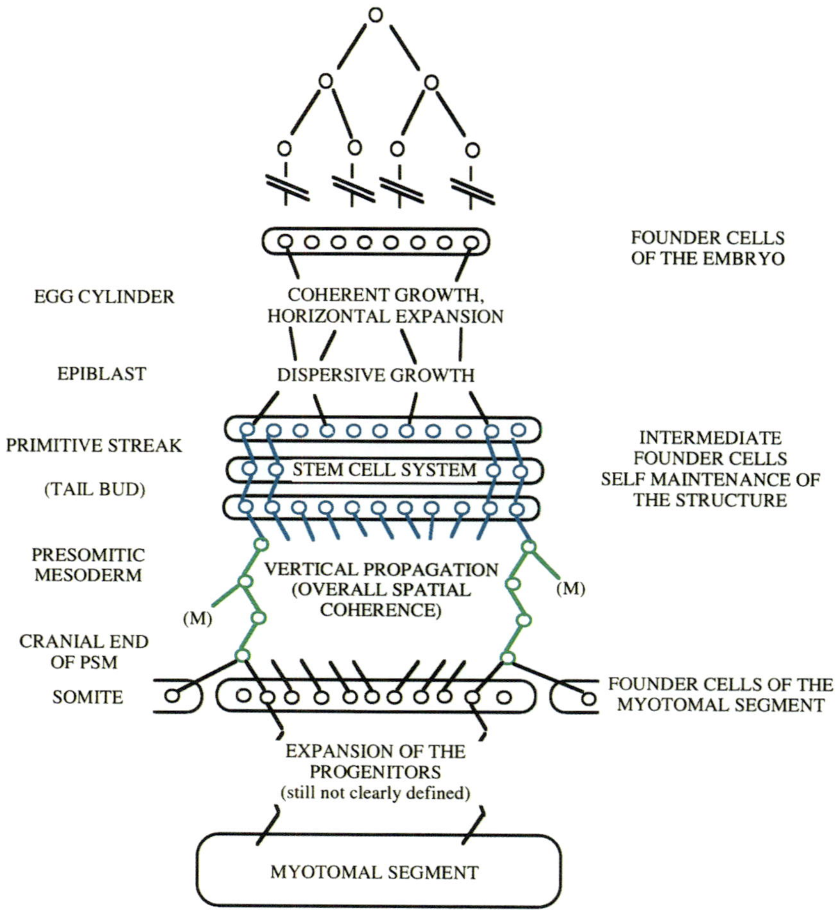

Figure 13 The genealogical tree of the ancestral cells for a myotomal segment. For the myotome, the oldest ancestral cells derive from the primitive streak through the formation of a pool of intermediary founder cells. These cells enter a period of vertical propagation in the primitive streak and then in the tail bud for the stem cells and in the presomitic mesoderm and then in the somites for their descendants. Nonclonal growth during this period of vertical propagation is due to the particularity of the stem cell system. There is acquisition of positional information during this period of vertical propagation. The founder cells of the myotomal segment produce the progenitors of the myotome. They have been instructed during their period of vertical propagation. The affiliated cells of their ancestral cells contribute to other myotomal segments. The first period of horizontal expansion is hypothetically placed during the period of coherent clonal growth of the cells of the embryo before gastrulation and after the allocation of the founder cells of the embryo proper.

and, therefore, an evaluation of the consequences of very early events on a differentiated structure. Our analysis reveals the persistence of stem cells throughout axiogenesis. This fact is remarkable, because it shows that the same cell system persists after regression of the primitive streak, when somites form from the tail bud, a structure with little resemblance to the primitive streak. In fact, the tail bud corresponds to a mass of morphologically uniform cells in continuity with axial and paraxial structures and in which the germ layers cannot be distinguished (Griffith *et al.,* 1992) and this contrasts with the epithelial mesenchymal interface of the primitive streak until E9 (Kaufman, 1998). However, in spite of these obvious morphological differences, it is not possible to detect major changes in cell behavior between the two structures, as there are stem cells producing bilateral progenitors in both. This similarity parallels the maintenance of the organizer function in these structures (Gont *et al.,* 1993; Tucker and Slack, 1995; Gofflot *et al.,* 1997; Knezevic *et al.,* 1998). Therefore, the two structures exhibit both cellular and functional continuity.

The proposition of a coherent cell organization in the primitive streak has not been clearly formulated for mammals. However, clonal labeling of epiblast cells at E6.5 and at E7.5 has demonstrated that the rostrocaudal order of prospective mesodermal cells within the primitive streak is maintained during ingression, such that more medial mesodermal regions are derived from more rostral streak levels and more lateral mesodermal regions are derived from more caudal streak levels (Tam and Beddington, 1987; Lawson *et al.,* 1991; Wilson and Beddington, 1996; J. L. Smith *et al.,* 1994). Our results considerably reinforce this proposition by showing that the organization also involves very small territories of the paraxial mesoderm and this coherence is also maintained in a differentiated structure such as the myotome.

Many experiments have been performed in the chick, and they also suggest a relationship between the rostrocaudal order in the primitive streak and the mediolateral order in the mesoderm (Ooi *et al.,* 1986; Selleck and Stern, 1991; Schoenwolf *et al.,* 1992; Psychoyos and Stern, 1996; Wilson and Beddington, 1996). In addition, a more precise analysis of presomitic mesoderm has shown the existence of cells contributing to either lateral or medial sectors, at least at certain stages of development (Selleck and Stern, 1991; Psychoyos and Stern, 1996). The migratory routes of primitive streak cells have also been studied, and these cells appear to follow defined pathways (Psychoyos and Stern, 1996), very reminiscent of what we propose (Fig. 12). Altogether (Tam and Beddington, 1987; Lawson *et al.,* 1991; J. L. Smith *et al.,* 1994; Wilson and Beddington, 1996), it is now clear that there is conservation of migratory routes between the chick and mouse embryos, which provides an additional argument in favor of their importance in development.

The questions these conclusions raise are, why a stem cell system and why such an early pattern of coherent growth (i.e., restricted movements in the primitive streak and paraxial mesoderm)? The stem cell system could provide a simple solution to the problem of producing a series of elements by a unique group of cells in the absence of massive external recruitment, and also helps to maintain the structure through self-renewal. The long bilateral clones suggest that the myotomal segment is produced from a closed cellular territory (compare Fig. 9D, c and d).

Furthermore, the permanent residence of the stem cells in the same structure, and the temporal reiterative production of the segments, may constitute a cellular basis for the temporal activation of important developmental genes, such as the *Hox* genes. Clearly, *Hox* genes are expressed as gastrulation proceeds (Burke *et al.,* 1995) and specification by position has occurred before somite formation begins. The transition between the stem cells and their nonstem cell progeny is an intriguing stage in the development of the mesoderm, because it is concomitant with the morphological transition from epithelium to mesenchyme. Along with this transition is the acquisition of prospective intrinsic properties, the most spectacular being the rhythmic expression of the mammalian homologues, *Fringe* and *Hairy* (Palmeirim *et al.,* 1997; McGrew *et al.,* 1998), and the acquisition of an intrinsic metameric pattern in only the mesenchymal paraxial mesoderm, acquired at the time of its birth in Hensen's node in the chick (Menkes and Sandor, 1977; Bellairs, 1979).

The spatial coherence could result from the constraints imposed by the migratory movements during gastrulation, but it could also be a condition required for the sequential expression of different sets of genes during patterning: cell movements and cell dispersion would be tolerated only within the limited sphere of activity of the patterning systems. Since the transition between the primitive streak and paraxial mesoderm may coincide with instructive genetic programs within cells (from the expression of genes, such as *Hox, Hairy,* and *Fringe*), the acquisition of coherent growth patterns at precisely this stage is not surprising and, perhaps, necessary. Indeed, the maintenance of the same coherent pattern of growth during axiogenesis, somitogenesis, and dermomyotome formation argues in favor of this hypothesis, and for the importance of this early patterning in development of the myotome. This coherence would, therefore, correspond to a "prepatterning" or coordinating event that would rely on the subsequent action of genes, such as *Hox* (longitudinal organization) or *Pax* (dorsoventral organization). The whole system (that is, stem cells and coherent growth) has probably evolved as a compromise between cell dispersion, due to the massive requirement of cells for tissue production, and the need to immobilize cells for the purpose of sequential patterning. It may be because of the difficulty in achieving this balance that the fundamental aspects of the system have been conserved in amniotes.

The elaboration of the cellular organization of a segment, as one could envisage it at the present time, is illustrated in Fig. 12 and can be summarized as follows. In the epiblast at E6–E6.5, there is a region that contains the ancestors to the founder cells in the primitive streak (Fig. 12A) (Tam and Beddington, 1987; Lawson *et al.,* 1991); however, none of the spatial organization seen in this epiblast region is maintained in the segment: neither its bilaterality, which is reshaped during mesoderm production from the primitive streak, nor its mediolateral organization (as far as we know), which cell convergence toward the primitive streak obliterates (the arrows in Fig. 12A) (Lawson *et al.,* 1991; Gardner and Cockroft, 1998). In contrast, the mediolateral organization of the future segment is already represented in the rostrocaudal organization of the primitive streak, and the future bilaterality of the segment is also partially represented (Fig. 12B). The most lateral cells are restricted to one side of the embryo, but other cells are only statistically restricted, with an equal contribution to both sides by, presumably, the most axial cells. Therefore, the future bilaterality is represented by a partial mediolateral organization (Fig. 12B, the gradient of color), and this is reminiscent of a system based on either two lateral influences or an axial symetrical system. Whatever the case may be, a cellular organization along the mediolateral axis clearly exists at the early stages of development, and this raises the interesting possibility that either the most lateral cells or the very axial cells play a special role in the process.

Finally, in the presomitic mesoderm, a longitudinal organization is added onto the newly acquired bilaterality and mediolateral organization, inherited from previous events: the participation of cells in the muscle system is restricted to only one to four (in general two) of these adjacent segments (Fig. 12C, the green line).

The ancestral cells arriving in the somites, from which the founder cells of the dermomyotome derive, have already been "instructed" during their period of vertical propagation. This period is not neutral for the cells, as it was described in the schematic model at the beginning of this chapter, since the founder cells of a structure represent a pool of already partially instructed cells (Fig. 13).

It would be premature to relate the cell behavior, just described, and the expression patterns of the known developmental genes, but there is a strong indication here that a significant fraction of cells follow a coherent pattern of growth and coherent routes of migration, and that the longitudinal organization may be contained in the way mesodermal cells are produced. This raises the problem of whether there is pre-existing prepatterning in the dermomyotome. If such a prepatterning, indeed, exists, its basis must be first investigated in the cellular and molecular interactions that occur between the primitive streak and the node.

In conclusion, there is still much to be learned concerning many of the questions relating to muscle formation. For instance, the mode of production of muscle cells by the dermomyotome is not clearly defined. The possibility that the observations made for the myotome apply to other somite derivatives has not yet been fully investigated; however, these questions can be approached by the available methodology of clonal analysis. A description of development at the cellular level is clearly the only way to have a profound understanding of this complex process.

Acknowledgments

We thank Suzanne Capgras and Valérie Guyot for their help during the screen of the 6000 embryos and Françoise Kamel for secretarial assistance. We thank Joan Shellard for careful reading of this version of the manuscript. This work has been financially supported by grants from the CNRS (URA 1947), Association française contre les Myopathies, AFM, and Association pour la Recherche sur le Cancer, ARC. J.F.N. is from the Institut National de la Santé et de la Recherche Médicale.

References

Balakier, H., and Pedersen, R. A. Allocation of cells to inner cell mass and trophectoderm lineages in preimplantation mouse embryos. *Dev. Biol.* **90,** 352–362.

Beddington, R. S. (1994). Induction of a second neural axis by the mouse node. *Development* **120,** 613–620.

Beddington, R., and Lawson, K. A. (1990). Clonal analysis of cell lineages. *In* "Postimplantation Mammalian Embryos—A Practical Approach" (A. J. Copp and D. L. Cockroft, Eds.), pp. 267–291. Oxford Univ. Press, Oxford.

Bellairs, R. (1979). The mechanism of somite segmentation in the chick embryo. *J. Embryol. Exp. Morphol.* **51,** 227–243.

Bellairs, R. (1985). A new theory about somite formation in the chick. *Prog. Clin. Biol. Res.* **171,** 25–44.

Bonnerot, C., and Nicolas, J.-F. (1993a). Application of LacZ gene fusions to postimplantation development. *In* "Methods in Enzymology: Guide to Techniques in Mouse Development" (P. M. Wassarman and M. L. DePamphilis, Eds.), Vol. 225, pp. 451–469. Academic Press, San Diego.

Bonnerot, C., and Nicolas, J. F. (1993b). Clonal analysis in the intact mouse embryo by intragenic homologous recombination. *C. R. Acad. Sci.* **316,** 1207–1217.

Bonnerot, C., Rocancourt, D., Briand, P., Grimber, G., and Nicolas, J. F. (1987). A β-galactosidase hybrid protein targeted to nuclei as a marker for developmental studies. *Proc. Natl. Acad. Sci. U.S.A.* **84,** 6795–6699.

Borycki, A. G., Mendham, L., and Emerson, C. P., Jr. (1998). Control of somite patterning by Sonic hedgehog and its downstream signal response genes. *Development* **125,** 777–790.

Brand-Saberi, B., Ebensperger, C., Wilting, J., Balling, R., and Christ, B. (1993). The ventralizing effect of the notochord on somite differentiation in chick embryos. *Anat. Embryol.* **188,** 239–245.

Burke, A. C., Nelson, C. E., Morgan, B. A., and Tabin, C. (1995). *Hox* genes and the evolution of vertebrate axial morphology. *Development* **121,** 333–346.

Cepko, C. L., Ryder, E. F., Austin, C. P., Walsh, C., and Fekete, D. M. (1993). Lineage analysis using retrovirus vectors. *In* "Methods in Enzymology: Guide to Techniques in Mouse Development" (P. M. Wasserman and M. L. DePamphilis, Eds.), Vol. 225, pp. 933–960. Academic Press, San Diego.

Christ, B., and Ordahl, C. P. (1995). Early stages of chick somite development. *Anat. Embryol.* **191,** 381–396.

Christ, B., Jacob, H. J., and Jacob, M. (1978). On the formation of the myotomes in avian embryos. An experimental and scanning electron microscopic study. *Experientia* **34,** 514–516.

Christ, B., Brand-Saberi, B., Jacob, H. J., Jacob, M., and Seifert, R. (1990). Principles of early muscle development. *In* "The Avian Model in Developmental Biology: From Organism to Genes." (N. Le Douarin, F. Dieterlen-Lièvre and J. Smith, Eds.), pp. 139–151. Editions du CNRS, Paris.

Christ, B., Schmidt, C., Huang, R., Wilting, J., and Brand-Saberi, B. (1998). Segmentation of the vertebrate body. *Anat. Embryol.* **197,** 1–8.

Cossu, G., Tajbakhsh, S., and Buckingham, M. (1996). How is myogenesis initiated in the embryo? *Trends Genet.* **12,** 218–223.

Denetclaw, W. F., Jr., Christ, B., and Ordahl, C. P. (1997). Location and growth of epaxial myotome precursor cells. *Development* **124,** 1601–1610.

Dietrich, S., Schubert, F. R., Healy, C., Sharpe, P. T., and Lumsden, A. (1998). Specification of the hypaxial musculature. *Development* **125,** 2235–2249.

Eloy-Trinquet, S., and Nicolas, J.-F. (1999). In preparation.

Gardner, R. L., and Cockroft, D. L. (1998). Complete dissipation of coherent clonal growth occurs before gastrulation in mouse epiblast. *Development* **125,** 2397–2402.

Gardner, R. L., and Lawrence, P. A. (1985). Single cell marking and cell lineage in animal development. *Philos. Trans. R. Soc. London* **B312,** 1–187.

Gardner, R. L., and Lyon, M. F. (1971). X chromosome inactivation studied by injection of a single cell into the mouse blastocyst. *Nature* **231,** 385–386.

Garner, W., and McLaren, A. (1974). Cell distribution in chimaeric mouse embryos before implantation. *J. Embryol. Exp. Morphol.* **32,** 495–503.

Gearhart, J. D., and Mintz, B. (1972). Clonal origins of somites and their muscle derivatives: Evidence from allophenic mice. *Dev. Biol.* **29,** 27–37.

Gofflot, F., Hall, M., and Morriss-Kay, G. M. (1997). Genetic patterning of the developing mouse tail at the time of posterior neuropore closure. *Dev. Dyn.* **210,** 431–445.

Gont, L. K., Steinbeisser, H., Blumberg, B., and DeRobertis, E. M. (1993). Tail formation as a continuation of gastrulation: The multiple cell populations of the *Xenopus* tailbud derive from the late blastopore lip. *Development* **119,** 991–1004.

Goulding, M., Lumsden, A., and Paquette, A. J. (1994). Regulation of *Pax-3* expression in the dermomyotome and its role in muscle development. *Development* **120,** 957–971.

Griffith, C. M., Wiley, M. J., and Sanders, E. J. (1992). The vertebrate tail bud: Three germ layers from one tissue. *Anat. Embryol.* **185,** 101–113.

Hadjantonakis, A. K., Gertsenstein, M., Ikawa, M., Okabe, M., and Nagy, A. (1998). Generating green fluorescent mice by germline transmission of green fluorescent ES cells. *Mech. Dev.* **76,** 79–90.

Hatada, Y., and Stern, C. D. (1994). A fate map of the epiblast of the early chick embryo. *Development* **120,** 2879–2889.

Kahane, N., Cinnamon, Y., and Kalcheim, C. (1998a). The origin and fate of pioneer myotomal cells in the avian embryo. *Mech. Dev.* **74,** 59–73.

Kahane, N., Cinnamon, Y., and Kalcheim, C. (1998b). The cellular mechanism by which the dermomyotome contributes to the second wave of myotome development. *Development* **125,** 4259–4271.

Kato, N., and Aoyama, H. (1998). Dermomyotomal origin of the ribs as revealed by extirpation and transplantation experiments in chick and quail embryos. *Development* **125,** 3437–3443.

Kaufman, M. H. (1998). "The Atlas of Mouse Development 1." Harcourt Brace & Company, London.

Kessel, M., and Gruss, P. (1990). Murine developmental control genes. *Science* **249,** 374–379.

Knezevic, V., De Santo, R., and Mackem, S. (1998). Continuing organizer function during chick tail development. *Development* **125,** 1791–1801.

Krumlauf, R. (1992). Evolution of the vertebrate *Hox* homeobox genes. *BioEssays* **14,** 245–252.

Lawrence, P. A. (1992). "The Making of a Fly. The Genetics of Animal Design 1." Blackwell Scientific, Oxford.

Lawson, K. A., and Pedersen, R. A. (1992). Early mesoderm formation in the mouse embryo. *In* "Formation and Differentiation of Early Embryonic Mesoderm" (R. Bellairs, Ed.), pp. 33–46. Plenum, New York.

Lawson, K. A., Meneses, J. J., and Pedersen, R. A. (1991). Clonal analysis of epiblast fate during germ layer formation in the mouse embryo. *Development* **113,** 891–911.

Lewis, J. H., Summerbell, D., and Wolpert, L. (1972). Chimaeras and cell lineage in development. *Nature* **239,** 276–279.

Lyons, G. E., and Buckingham, M. E. (1992). Developmental regulation of myogenesis in the mouse. *Semin. Dev. Biol.* **3,** 243–253.

McGrew, M., Dale, J. K., Fraboulet, S., and Pourquié, O. (1998). The *lunatic Fringe* gene is a target of the molecular clock linked to somite segmentation in avian embryos. *Curr. Biol.* **8,** 979–982.

McLaren, A. (1972). Numerology of development. *Nature* **239,** 274–276.

McMahon, J. A., Takada, S., Zimmerman, L. B., Fan, C. M., Harland, R. M., and McMahon, A. P. (1998). Noggin-mediated antagonism of BMP signaling is required for growth and patterning of the neural tube and somite. *Genes Dev.* **12,** 1438–1452.

Mathis, L., and Nicolas, J.-F. (1997). Analyse clonale rétrospective chez les vertébrés: Méthodes, concepts et résultats. *Ann. Inst. Pasteur/Actualités* **8,** 3–17.

Mathis, L., and Nicolas, J. F. (1998). Autonomous cell labelling using *LaacZ* reporter transgenes to produce genetic mosaics during development. *In* "Microinjections and Transgenesis. Strategies and Protocols" (A. Cid-Arrequi and A. Garcia-Carranca, Eds.), pp. 439–458. Springer-Verlag.

Mathis, L., Bonnerot, C., Puelles, L., and Nicolas, J. F. (1997). Retrospective clonal analysis of the cerebellum using genetic laacz/lacz mouse mosaics. *Development* **124,** 4089–4104.

Menkes, B., and Sandor, S. (1977). Somitogenesis: Regulation potencies, sequence determination and primordial interactions. *In* "Vertebrate Limb and Somite Morphogenesis" (J. R. Hinchliffe, Ed.), pp. 405–419. Cambridge Univ. Press, Cambridge.

Mintz, B. (1965). Genetic mosaicism in adult mice of quadriparental lineage. *Science* **148,** 1232–1233.

Mintz, B. (1967). Gene control of mammalian pigmentary differentiation. I. Clonal origin of melanocytes. *Proc. Natl. Acad. Sci. U.S.A.* **58,** 344–351.

Münsterberg, A. E., Kitajewski, J., Bumcrot, D. A., McMahon, A. P., and Lassar, A. B. (1995). Combinatorial signaling by Sonic hedgehog and Wnt family members induces myogenic bHLH gene expression in the somite. *Genes Dev.* **9,** 2911–2922.

Nicolas, J. F., and Bonnerot, C. (1988). Recombinant retrovirus, cell lineage and gene expression in the mouse embryo. *Prog. Clin. Biol. Res.* **284,** 125–145.

Nicolas, J. F., Mathis, L., and Bonnerot, C. (1996). Evidence in the mouse for self-renewing stem cells in the formation of a segmented longitudinal structure, the myotome. *Development* **122,** 2933–2946.

Nüsslein-Volhard, C., and Wieschaus, E. (1980). Mutations affecting segment number and polarity in *Drosophila. Nature* **287,** 795–801.

Ooi, V. E. C., Sanders, E. J., and Bellairs, R. (1986). The contribution of the primitive streak to the somites in the avian embryo. *J. Embryol. Exp. Morphol.* **92,** 193–206.

Ordahl, C. P., and Le Douarin, N. M. (1992). Two myogenic lineages within the developing somite. *Development* **114,** 339–353.

Ott, M. O., Bober, E., Lyons, G., Arnold, H., and Buckingham, M. (1991). Early expression of the myogenic regulatory gene, *myf-5,* in precursor cells of skeletal muscle in the mouse embryo. *Development* **111,** 1097–1107.

Palmeirim, I., Henrique, D., Ish-Horowicz, D., and Pourquie, O. (1997). Avian hairy gene expression identifies a molecular clock linked to vertebrate segmentation and somitogenesis. *Cell* **91,** 639–648.

Patapoutian, A., Yoon, J. K., Miner, J. H., Wang, S., Stark, K., and Wold, B. (1995). Disruption of the mouse *MRF4* gene identifies multiple waves of myogenesis in the myotome. *Development* **121,** 3347–3358.

Pourquié, O., Coltey, M., Teillet, M. A., Ordahl, C., and LeDouarin, N. M. (1993). Control of dorsoventral patterning of somitic derivatives by notochord and floor plate. *Proc. Natl. Acad. Sci. U.S.A.* **90,** 5242–5246.

Pourquié, O., Coltey, M., Bréant, C., and LeDouarin, N. M. (1995). Control of somite patterning by signals from the lateral plate. *Proc. Natl. Acad. Sci. U.S.A.* **92,** 3219–3223.

Pourquié, O., Fan, C. M., Coltey, M., Hirsinger, E., Watanabe, Y., Bréant, C., Francis-West, P., Brickell, P., Tessier-Lavigne, M., and LeDouarin, N. M. (1996). Lateral and axial signals involved in avian somite patterning: A role for BMP4. *Cell* **84,** 461–471.

Pownall, M. E., and Emerson, C. P. (1992). Molecular and embryological studies of avian myogenesis. *Semin. Dev. Biol.* **3,** 229–241.

Psychoyos, D., and Stern, C. D. (1996). Fates and migratory routes of primitive streak cells in the chick embryo. *Development* **122,** 1523–1534.

Rawls, A., Valdez, M. R., Zhang, W., Richardson, J., Klein, W. H., and Olson, E. N. (1998). Overlapping functions of the myogenic bHLH genes *MRF4* and *MyoD* revealed in double mutant mice. *Development* **125,** 2349–2358.

Rossant, J. (1984). Somatic cell lineages in mammalian chimeras. "Chimeras in Developmental Biology" (N. Le Douarin and A. McLaren, Eds.), pp. 89–109. Academic Press, New York.

Rossant, J. (1987). Cell lineage analysis in mammalian embryogenesis. *Curr. Top. Dev. Biol.* **23,** 115–146.

Rudnicki, M. A., Schnegelsberg, P. N., Stead, R. H., Braun, T., Arnold, H. H., and Jaenisch, R. (1993). MyoD or Myf-5 is required for the formation of skeletal muscle. *Cell* **75,** 1351–1359.

Russell, L. B. (1978). "Genetic Mosaics and Chimeras in Mammals 1." Plenum, New York.

Sanes, J., Rubenstein, J., and Nicolas, J. F. (1986). Use of a recombinant retrovirus to study post-implantation cell lineage in mouse embryos. *EMBO J.* **5,** 3133–3142.

Schoenwolf, G. C., Garcia-Martinez, V., and Dias, M. S. (1992). Mesoderm movement and fate during avian gastrulation and neurulation. *Dev. Dyn.* **193,** 235–248.

Scott, M. P. (1987). Complex loci of *Drosophila. Annu. Rev. Biochem.* **56,** 195–227.

Scott, M. P., and Carroll, S. B. (1987). The segmentation and homeotic gene network in early *Drosophila* development. *Cell* **51,** 689–698.

Selleck, M. A. J., and Stern, C. D. (1991). Fate mapping and cell lineage analysis of Hensen's node in the chick embryo. *Development* **112,** 615–626.

Smith, J. L., Gesteland, K. M., and Schoenwolf, G. C. (1994). Prospective fate map of the mouse primitive streak at 7.5 days of gestation. *Dev. Dyn.* **201,** 279–289.

Smith, T. H., Block, N. E., Rhodes, S. J., Konieczny, S. F., and Miller, J. B. (1993). A unique pattern of expression of the four muscle regulatory factor proteins distinguishes somitic from embryonic, fetal and newborn mouse myogenic cells. *Development* **117,** 1125–1133.

Smith, T. H., Kachinsky, A. M., and Miller, J. B. (1994). Somite subdomains, muscle cell origins, and the four muscle regulatory factor proteins. *J. Cell Biol.* **127,** 95–105.

Soriano, P., and Jaenisch, R. (1986). Retroviruses as probes for mammalian development: Allocation of cells to the somatic and germ cell lineages. *Cell* **46,** 19–29.

Spörle, R., Gunther, T., Struwe, M., and Schughart, K. (1996). Severe defects in the formation of epaxial musculature in open brain (opb) mutant mouse embryos. *Development* **122,** 79–86.

Stern, H. M., Brown, A. M., and Hauschka, S. D. (1995). Myogenesis in paraxial mesoderm: Preferential induction by dorsal neural tube and by cells expressing Wnt-1. *Development* **121,** 3675–3686.

Tajbakhsh, S., and Buckingham, M. E. (1994). Mouse limb muscle is determined in the absence of the earliest myogenic factor myf-5. *Proc. Natl. Acad. Sci. U.S.A.* **91,** 747–751.

Tam, P. P. L. (1981). The control of somitogenesis in mouse embryos. *J. Embryol. Exp. Morphol.* **65**(Suppl.), 103–128.

Tam, P. P. L. (1988). The allocation of cells in the presomitic mesoderm during somite segmentation in the mouse embryo. *Development* **103,** 379–390.

Tam, P. P. L., and Beddington, R. S. P. (1987). The formation of mesodermal tissues in the mouse embryo during gastrulation and early organogenesis. *Development* **99,** 109–126.

Tam, P. P. L., and Behringer, R. R. (1997). Mouse gastrulation: The formation of a mammalian body plan. *Mech. Dev.* **68,** 3–25.

Tam, P. P. L., and Trainor, P. A. (1994). Specification and segmentation of the paraxial mesoderm. *Anat. Embryol.* **189,** 275–305.

Tarkowski, A. K. (1963). Studies on mouse chimaeras developed from eggs fused *in vitro. Natl. Cancer Inst. Monogr.* **2,** 51–71.

Theiler, K. (1989). Development and normal stages from fertilization to 4 weeks of age. *In* "The House Mouse." Springer-Verlag, Berlin.

Tucker, A. S., and Slack, J. M. W. (1995). The *Xenopus laevis* tail-forming region. *Development* **121,** 249–262.

Vogel, M. W., and Herrup, K. (1993). A theoretical and experimental examination of cell lineage relationship among cerebellar Purkinje cells in the mouse. *Dev. Biol.* **156,** 49–68.

Wachtler, F., and Christ, B. (1992). The basic embryology of skeletal muscle formation in vertebrates: The avian model. *Semin. Dev. Biol.* **3,** 217–227.

West, J. D. (1978). Clonal growth versus cell mingling. *In* "Genetic Mosaics and Chimeras in Mammals" (L. B. Russell, Ed.), pp. 435–444. Plenum, New York.

Williams, B. A., and Ordahl, C. P. (1994). *Pax-3* expression in segmental mesoderm marks early stages in myogenic cell specification. *Development* **120,** 785–796.

Williams, B. A., and Ordahl, C. P. (1997). Emergence of determined myotome precursor cells in the somite. *Development* **124,** 4983–4997.

Wilson, V., and Beddington, R. S. (1996). Cell fate and morphogenetic movement in the late mouse primitive streak. *Mech. Dev.* **55,** 79–89.

Yoon, J. K., Olson, E. N., Arnold, H. H., and Wold, B. J. (1997). Different MRF4 knockout alleles differentially disrupt Myf-5 expression: cis-regulatory interactions at the MRF4/Myf-5 locus. *Dev. Biol.* **188,** 349–362.

Yoon, K., Thiede, M. A., and Rodan, G. A. (1988). Alkaline phosphatase as a reporter enzyme. *Gene* **66,** 11–17.

Zernicka-Goetz, M. (1999). Green fluorescent protein: A new approach to understanding spatial patterning and cell fate in early mammalian development. *In* "Cell Lineage and Fate Determination" (S. A. Moody, Ed.), pp. 521–526. Academic Press, San Diego.

Zinyk, D. L., Mercer, E. H., Harris, E., Anderson, D. J., and Joyner, A. L. (1998). Fate mapping of the mouse midbrain–hindbrain constriction using a site-specific recombination system. *Curr. Biol.* **8,** 665–668.

3

Segmentation of the Paraxial Mesoderm and Vertebrate Somitogenesis

Olivier Pourquié
Laboratoire de Génétique et de Physiologie du Développement
Developmental Biology Institute of Marseille
CNRS-INSERM-Université de la Méditerranée-AP de Marseille
Campus de Luminy
13288 Marseille Cedex 09, France

Somites are the most obviously segmented features of the vertebrate embryo. Although the way segmentation is achieved in the fly is now well described, little was known about the molecular mechanisms underlying vertebrate somitogenesis. Through the recent identification of genes important for vertebrate somitogenesis and the analysis of their function, several theoretical models accounting for somitogenesis such as the clock and wavefront model, which have been proposed over the past 20 years, are now starting to receive experimental support. A molecular clock linked to somitogenesis has been identified which might act as a periodicity generator in the presomitic cells. This temporal periodicity is then translated into a tightly controlled spatial periodicity which is revealed by the expression of several genes. Analysis of mouse mutants in the Notch–Delta pathway suggest that this signaling mechanism might play an important

Current Topics in Developmental Biology, Vol. 47

0070-2153/00 $30.00

role at this level. The final step of the cascade is to translate these genetically specified segments into morphological units: the somites. Importantly, these studies have helped in dissociating the segmentation and the somitogenesis processes in vertebrates. In addition, although segmentation was classically thought to have arisen independently in protostomes and deuterostomes, recent evidence suggests that part of the segmentation machinery might actually have been conserved. The conservation of segmentation mechanisms reported in the fly such as the pair-rule pattern, however, remain a subject of controversy.

I. Definition of the Mesodermal Segment

A. Segmentation in the Animal Kingdom

The segmented aspect of both vertebrate and invertebrate embryos is a morphological feature which has long fascinated embryologists. Cooke and Zeeman (1976) have defined segmentation as "A regular spatial alternation between a few modes of cell behavior, such that successive populations of cells become separated." In invertebrates external segmentation of the exoskeleton, the cuticle (of ectodermal origin), is most evident, whereas in vertebrates it is the endoskeleton, the vertebral column (of mesodermal origin), that is the most prominently segmented structure. An important consequence of segmentation is the articulation and therefore movement of hard tissues like bones or cuticle.

Two different modes of segmentation are classically recognized in animal development. The first is observed both in the development of long germ band insects such as *Drosophila* and in formation of the vertebrate hindbrain. It involves the subdivision of a preexisting territory (i.e., the blastoderm or the rhombencephalon) into repeated units corresponding to the prospective segments (Lawrence, 1992; Lumsden and Krumlauf, 1996). The second, and by far the more frequently observed, is that found, for example, in more primitive insects like the grasshopper, in other classes of arthropods, as well as in annelids and in vertebrates (Weisblat *et al.,* 1994). This mode involves a progressive generation of segments from a terminal growth zone of the embryo.

From an evolutionary point of view it is classically thought that segmentation arose independently in the different classes of protostomes such as annelids and arthropods, and in the segmented deuterostomes like vertebrates. Therefore, from a conventional point of view, segmentation of annelids or of short germ band insects and vertebrate somitogenesis are analogous rather than homologous processes. This view is in marked contrast with the extreme conservation of other major patterning systems used during the development of insects and of vertebrates. Recent data suggests, however, that some molecular aspects of the segmentation machinery might have been conserved between insects and chordates (Kimmel, 1996; De Robertis, 1997). Within the chordate phylum, the cephalochordate amphioxus and vertebrates exhibit clear segmentation evidenced by the presence of somites. No segmented somites are found in the other chordate classes such as tunicates (Satoh, 1994).

B. Somites and the Mesodermal Segments of Vertebrates

The obvious segmented structures of the vertebrate mesoderm are the somites and many of their derivatives, that is, subsets of striated muscles (epaxial and intercostal), and the vertebral column, including the vertebrae, intervertebral disks, and ribs (Tam and Trainor, 1994; Christ and Ordahl, 1995). Other segmented mesodermal structures include muscles of the gills in lower vertebrates and intermediate mesoderm and its early derivatives. Segmentation of the lateral plate mesoderm remains unclear. A subset of blood vessels, called the intersegmental arteries develop according to a segmented pattern as a consequence of somite segmentation. Epidermal structures such as scales and feathers, which can include somatic derivatives from the dermis (in the back of the embryo) and an ectodermal component are also segmented. The mechanisms used to achieve this repetitive pattern of feathers and scales might, however, be different from that employed during somitogenesis and could be more closely related to the spacing mechanisms used during the development of cuticular appendages in insects (Cooke, 1981).

C. Characteristics of Somite Formation

In most vertebrate species, the somites are formed from bilateral rods of mesenchyme, the paraxial mesoderm which is initially produced at the level of the primitive streak and Hensen's node, and, later at the level of the tail bud. In most vertebrate species, individual pairs of somites, located symetrically on either side of the neural tube emerge from the rostral end of the presomitic mesoderm (PSM) in a coordinated fashion and at a defined pace, while new mesenchymal cells enter the caudal paraxial mesoderm, as a consequence of gastrulation. The speed at which somite formation occurs is precisely defined, and somite number is often used as a measure for the developmental age of embryos in several species.

In cold-blooded animals, the pace at which somites are produced varies depending on the species and the breeding temperature. In salmons, it has been shown that this pace is a direct function of temperature and is different in various genetic strains (Gorodilov, 1992). Remarkably, hybrid salmon embryos retain the pace of somite formation characteristic of the male parent's strain. At a given temperature, this pace is species-dependent even within the same animal class. For instance, in amphibians raised at 15°C, it can vary from 2 hr 20 min in *Rana temporaria* to 4 hr 30 min in *Bufo vulgaris* (Davidson, 1988). Somite formation takes 20 min in zebrafish (Morin-Kensicki and Eisen, 1997) at 28°C and 40 min in *Xenopus* at room temperature (Cooke and Zeeman, 1976). In amniotes such as the chick or the mouse in which development occurs at 37°–40°C, the time to generate a somite is around 1 hr 30 min (Tam and Trainor, 1994; Palmeirim *et al.,* 1997). Soon after they segment from the presomitic mesoderm, somites become patterned in response to local inducing signals produced by adjacent structures.

In both arthropods and vertebrates, the number of body segments is usually precisely defined within a species, whereas it can vary greatly between different species. In vertebrates, most species have less than 100 somite pairs but some reptiles like snakes and some fishes can develop several hundred. In most vertebrates, the number of head and trunk somites is tightly controlled, whereas in the tail the somite number is slightly more variable. Virtually nothing is known about somite number control though it seems that the cessation of somitogenesis has to be actively defined and does not merely reflect a physiological constraint on a cellular process because some species are able to produce 10 times more somites than others.

The mechanism of somite number control is independent of cell number because, amphibian haploid embryos possess half the number of cells that diploid embryos have, yet they develop a normal number of somites (Fankhauser, 1945). Further experiments aimed at studying this question have been performed in which abnormally small amphibian embryos have been produced by removing parts of the blastula (Cooke, 1975). This produced animals in which the number of segments remains unaffected while the number of cells allocated to each segment is significantly reduced at least for the anterior segments.

D. The Head Somites, the Somitomeres, and the Initiation of Segmentation

In the head region of cephalochordates and elasmobranchs, somites are formed by pinching off from the anterior gut, a process termed enterocoely. These somites extend rostrally to the anterior tip of the animal and they are thought to essentially give rise to head muscles (Gilland and Baker, 1993; Holland *et al.,* 1997). Amphioxus has eight such somites while only five are found in shark embryos (three preotic and two metotic). In teleosts and tetrapods, the first somite lies at the level of the otic vesicle. Anterior to the first somite is found a stripe of paraxial mesoderm called cephalic mesoderm which gives rise to head muscles and bones of the base of the skull but which is never seen to overtly segment (Noden, 1983; Couly *et al.,* 1993).

Based on scanning electron microscopy studies, Meier reported the existence of concentric organizations of cells repeated along the anteroposterior axis in the cephalic paraxial mesoderm of the chick embryo (Meier, 1984; Jacobson, 1988). He termed these condensations somitomeres postulating that they reflected a covert segmentation of the head mesoderm similar to that found in lower vertebrates. The number of head somitomeres was found to vary between species from four in the frog to seven in the chick or the Medaka fish (Jacobson, 1988).

In amniotes, the regression of the primitive streak first results in head formation and then progressively produces the body axis. Therefore, most of the head (including the cephalic mesoderm) is laid down prior to the actual onset of somitogenesis. An attractive consequence of the existence of head somitomeres is the coupling of the start of paraxial mesoderm segmentation (i.e., when the first somitomere is formed in the cephalic mesoderm) to the onset of primitive streak regression in amniotes (Fig. 1). According to this idea somitogenesis would correspond to the continuation of head somitomere segmentation.

Meier also observed similar condensations in the presomitic mesoderm and argued that they prefigured the somites to be formed (Meier, 1984). However, a difference between the trunk and head somitomeres is that the latter never become epithelial somites. Since these original observations in the chick, similar structures have been reported in the presomitic mesoderm of species ranging from teleosts, amphibians, reptiles, birds, and mammals (Jacobson, 1988). In the chick embryo, up to 12 somitomeres were observed in the presomitic mesoderm and this was shown to correspond to the number of prospective somites in the presomitic mesoderm (Packard, 1976). As in the head, the number of somitomeres varies between different classes of animals, for instance only six somitomeres are found in the mouse presomitic mesoderm (Tam *et al.,* 1982; Jacobson, 1988).

Therefore, the somitomere theory implies that the paraxial mesoderm becomes segmented all along the anteroposterior axis as soon as it is produced by gastrulation, although the segments are initially covert. However, no additional evidence for the existence of somitomeres has been obtained. Moreover, it is particularly suprising that among the impressive number of genes now identified, which are expressed in a segmented fashion in the paraxial mesoderm, none are expressed according to the proposed somitomeric pattern.

E. Resegmentation and Sclerotome Compartmentalization

Somites are transient embryonic structures. As differentiation proceeds, the ventral part of the epithelial somite becomes mesenchymal and forms the sclerotome which yields the axial skeleton whereas the still epithelial dorsal part of the somite, the dermomyotome, then gives rise to dermis and muscle. In the adult vertebrate body, all these derivatives are obviously segmented structures. Based on the analysis of histological sections of developing embryos, Remak proposed that somites and vertebrae are not in register. He postulated that a vertebra is derived from the caudal part of one somite and the rostral part of the next somite. He called this phenomenon "Neugliederung" or "resegmentation" (Remak, 1850). Although

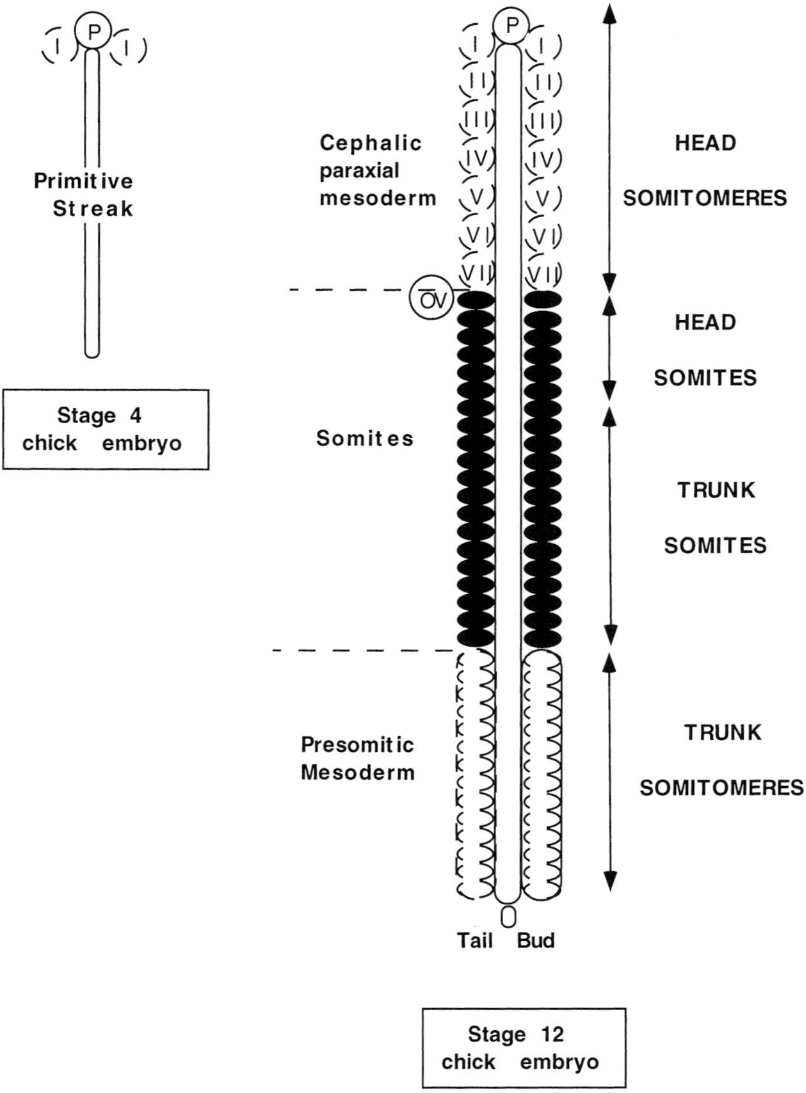

Figure 1 Schematic formation of the somitomeres. Formation of head somitomeres begins as soon as the primitive streak starts its regression, at stage 4 in the chick embryo. Seven somitomeres will be formed in the chick head mesoderm before formation of the first true somite immediately caudal to the otic vesicle (OV). The trunk somitomeres were proposed to correspond to a prepattern of segmentation prefiguring the prospective somites in the presomitic mesoderm. Their number was proposed to be characteristic of the species considered (12 in the chick). (P) Prechordal mesoderm.

this hypothesis has been challenged by several workers on the basis of histological observations (Verbout, 1976, 1985), convincing experimental evidence in favor of this theory was recently obtained using the quail–chick chimeras technique (Bagnall *et al.,* 1988; Aoyama and Asamoto, 1988). In these experiments, a chick somite or a series of somites was substituted by its stage-matched quail counterpart. When a somite was grafted alone, it was found to contribute to the caudal part of one vertebra and to the rostral part of the caudally adjacent vertebra (Bagnall *et al.,* 1988, 1989). In the case of three grafted somites, an analysis of the graft borders revealed a contribution restricted to the caudal part of one vertebra at the anterior level and to the rostral part of the vertebra in the posterior domain of the graft (Aoyama and Asamoto, 1988). Additional evidence for resegmentation came from replacing a series of three to four somites in a chick host by reconstituted somites formed only of caudal or rostral halves of quail somites (Goldstein and Kalcheim, 1992). Somites composed of caudal halves but not those composed of rostral halves gave rise to a structure characteristic of the rostral vertebra, the vertebral pedicles. In these experiments, intervertebral disks were found to be entirely derived from the rostral somitic halves. Another line of evidence in favor of resegmentation came from fate-mapping experiments performed by injecting β-gal-expressing defective retroviruses into the somitocoele of individual somites (Ewan and Everett, 1992). Descendants of the infected somite cells were observed in the caudal part of one vertebra and in the rostral part of the consecutive one. There seems now to be a general consensus on the fact that a vertebra arises from two consecutive somites (Christ and Wilting, 1992; see also Chap. 9 by Brand-Saberi and Christ, this volume).

The resegmentation hypothesis implies the existence of an intrasomitic boundary defining a rostral and a caudal somitic half which contribute to two consecutive vertebrae. Such a boundary has been identified in the ventral somitic compartment, the sclerotome, which is fated to give rise to the skeletal derivatives. This boundary is morphologically identifiable in amniotes and is called the Von Ebner's fissure (Solursh *et al.,* 1979; Stern and Keynes, 1987; Keynes and Stern, 1988). It demarcates the anterior part of the sclerotome from the posterior. Microsurgical experiments in the chick embryo have demonstrated that rostral and caudal parts of the sclerotome exhibit different functional properties (Stern and Keynes, 1987; see Chap. 7 by Bronner-Fraser, this volume). Neural crest cells and axonal migration are impaired in the caudal sclerotome and the cells and axons are therefore channeled into the rostral sclerotome (Keynes and Stern, 1984; Rickmann *et al.,* 1985). This anteroposterior identity of the sclerotome provides the basis for the segmentation of the peripheral nervous system in amniotes (spinal nerves and dorsal root sensory ganglia). Molecular evidence for this anteroposterior subdivision of the sclerotome has also been observed. Molecules such as T-cadherin, Pax1 and 9, and epitopes recognized by lectins are differentially distributed in the rostral or the caudal sclerotome (Stern *et al.,* 1986; Ranscht and Bronner-Fraser, 1991; Balling *et al.,* 1996). In addition, the cellular arrangements are different between the rostral and caudal sclerotome: caudal sclerotomal cells are more densely packed than rostral ones.

In anamniotes like zebrafish and *Xenopus,* the sclerotome does not represent a major somitic derivative and its differentiation has been poorly studied. An anteroposterior subdivision of the sclerotome was described in the zebrafish in which both anterior and posterior compartments follow different migration pathways and exhibit different cell fates. Cells of the anterior compartment give rise essentially to sclerotomal derivatives while cells from the posterior compartment yield both sclerotomal derivatives and muscle (Morin-Kensicki and Eisen, 1997). Surprisingly, ablation of the sclerotome in the zebrafish does not affect segmentation of the peripheral nervous system. This could be due to the intrinsic segmentation of the spinal cord in this species evidenced by the distribution of the primary motoneurons (Kimmel *et al.,* 1988). Therefore the role of the anteroposterior subdivision of the sclerotome in patterning the nervous system is not obvious outside the amniotes. Such a subdivision has not been reported in *Xenopus,* in which most of the somite is comprised of the myotome (Hamilton, 1969; Woo Youn and Malacinski, 1981).

Since sclerotome is only found in differentiated somites [from somite V onward in the chick embryo (Christ and Ordahl, 1995)] it can be argued that this anteroposterior subdivision of the somite is a late event. However, inversion of the presomitic mesoderm or of the newly formed somites along the anteroposterior axis in the chick embryo has established that this subdivision is determined in the segmental plate, prior to somite formation (Aoyama and Asamoto, 1988; Bronner-Fraser and Stern, 1991). More recently, molecular markers have been identified that are expressed selectively in the rostral or caudal aspect of the newly formed somite in which the sclerotome has not yet formed. These include genes such as *Notch-1, delta-1/Dll1,* and *chairy-1* in the caudal somite (Reaume *et al.,* 1992; Henrique *et al.,* 1995; Bettenhausen *et al.,* 1995; Palmeirim *et al.,* 1997), and *c-ret, Dll3,* and *FGFR1* in the rostral somite (Yamaguchi *et al.,* 1994; Robertson and Mason, 1995; Dunwoodie *et al.,* 1997). Therefore, somitic subdivision into rostral and caudal is defined much earlier than sclerotome formation, that is, at the epithelial somite stage or even earlier in the rostral presomitic mesoderm. This suggests that the specification of the anterior and posterior compartment of the somite is likely to occur prior to somitogenesis itself in the amniotes. Surprisingly, it might be different in amphibians since anteroposterior inversion of the PSM does not result in an inversion of the anteroposterior orientation of the myotomal chevrons (Deuchar and Burgess, 1967), suggesting that anteroposterior polarity of the somites might not be determined in the PSM.

In contrast to the sclerotomal derivatives, the early segmented axial myotubes that form the early myotome arising from the dorsal part of the somite (the dermomyotome) are in register with the somite (Kaehn *et al.,* 1988; Denetclaw *et al.,* 1997). The reason for this uncoupling can be easily understood from a functional

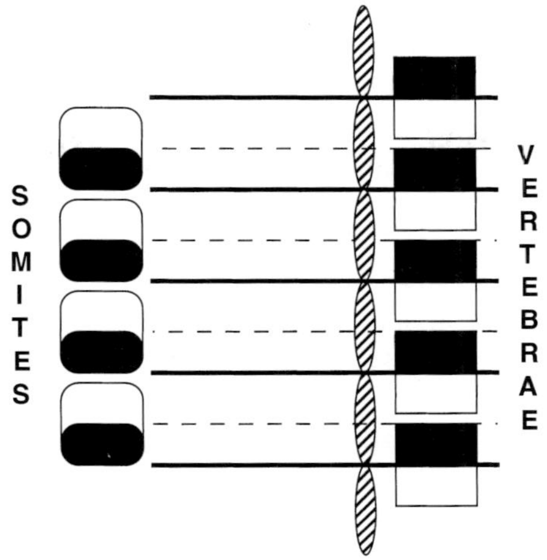

Figure 2 Schematic correspondence between the somites and the muscles and vertebrae they produce. The axial muscles are in register with the somites while the vertebrae derive from two adjacent somites. The rostral halves of vertebrae are composed of caudal somite parts (in black) and halves of caudal vertebrae derive from rostral somite cells (in white). This shift in segmental periodicity is known as the resegmentation process.

point of view since these muscles are responsible for vertebral column movements and thus attach in the middle of each vertebra (Fig. 2). However, in the chick, this early segmental restriction only holds true for the early stages of myotome differentiation. In older embryos, axial muscles keep their original segmental arrangement but receive a contribution from several adjacent somites (Bagnall *et al.,* 1989).

Therefore, different metameric units of the paraxial mesoderm can be defined and include: the somite, the half somite, the vertebra, the myotome, and the axial muscle. This defines, not only one, but several segmental periodicities along the body axis. The periodicity of the half-somite segmentation which underlies vertebral formation is double that of myotome segmentation. When segmentation in the presomitic mesoderm, and thus the nature of the basic segmental unit of the paraxial mesoderm first appears, remains a controversial issue and will be discussed below.

II. Models for Somite Formation

A. Early Somitogenesis Models

Experiments in the chick embryo performed in the late 1950s have led to different types of models to account for somitogenesis. Following transection experiments of the blastoderm in chick, Spratt (1955) proposed the existence of somite centers located on both sides of Hensen's node. Following ingression in the primitive streak and Hensen's node, prospective paraxial mesoderm cells would acquire a somitogenic potential having passed via the somite centers. The somite centers therefore corresponded more to a geographical zone of the embryo than to a particular population of cells. This hypothesis was challenged by a series of experiments in which somites were obtained after ablation of the region corresponding to the proposed somite centers (Bellairs, 1963).

Another kind of model proposed that somites are induced by adjacent structures such as Hensen's node or the neural tube and that segmentation is simply transferred from these structures to the presomitic mesoderm (Fraser, 1963). However, in addition to being of little explanatory value, these models were difficult to reconcile with the fact that somites can be obtained in absence of the surrounding structures (with the exception of the ectoderm, see below).

B. Reaction–Diffusion and Positional Information Based Models

A reaction–diffusion based model proposed that somite formation results from the confrontation of cells with different identities (such as anterior and posterior) which cannot mix (Meinhardt, 1986; Keynes and Stern, 1988). Cells of the rostral PSM would flip-flop between these different states as a result of an anteroposterior gradient of positional information and reaction–diffusion mechanisms. Cell state stabilization would lead to the confrontation of cells with incompatible identities, resulting in the segregation of alternate domains corresponding to the prospective anterior and posterior compartment to form a somite. Such a model would, however, not differentiate between anteroposterior boundary in the middle of a somite and an intersomitic boundary. To explain somite formation based on this twofold periodicity of alternating states, Meinhardt postulated an additional state corresponding to the somite boundary.

This model is attractive particularly since, as discussed above, formation of the anterior and posterior compartments actually precedes somite formation and intermingling does not occur between cells of these two somitic compartments (Stern and Keynes, 1987). However, the fact that somitogenesis continues to proceed autonomously even when small pieces of presomitic mesoderm are isolated tends to argue against reaction–diffusion mechanisms being responsible for the oscillatory mechanisms (Menkes and Sandor, 1977; Palmeirim *et al.*, 1997).

C. The Somitogenic Cluster Model

This model is based on a series of experiments involving transections of primitive streak stage chick embryos (Bellairs and Veini, 1984). It postulates that as early as stage four, a population of "somitogenic cells" are set aside and become distributed in an anteroposterior fashion in the primitive streak, reflecting their future distribution along the embryonic axis. While the primitive streak and Hensen's node regress, these cells are progressively released from the primitive streak and enter the presomitic mesoderm where they associate with PSM resident cells to form the somites. Thus segmentation reflects the progressive release of these somitogenic cells. Similar kinds of ideas involving somite founder cells have also been proposed for mouse embryos (Gearhart and Mintz, 1972; Tam, 1981).

D. Stem Cell Models

Other models propose the existence of a pool of stem cells resident in the node that give rise to all somitic cells. Such a population of stem cells was identified both by single cell injection of fluorescent tracers in the chick (Stern *et al.,* 1992) and by retrospective clonal analysis in the mouse using the LaacZ reporter gene (Nicolas *et al.,* 1996; see also Chap. 17 by Eloy-Trinquet *et al.,* this volume). The latter technique relies on the use of a transgenic mouse strain which carries a LaacZ construct. This construct includes a muscle specific promoter which drives expression of a modified *LacZ* gene in which a short internal sequence duplication interrupts the translational reading frame and encodes an inactive protein. This construct is linked to a muscle specific promoter. Internal homologous recombination which restores the LacZ coding frame occurs at very low frequency (less than one event per embryo). Sometimes this event happens in myotomal precursors, resulting in formation of a myotomal clone expressing the *LacZ* gene. By compiling observations on large numbers of clones, the genealogical history of myotome progenitor cells has been studied. The result of such an analysis shows that the myotome derives from a population of 100–150 stem cells (S cells) which then divide asymetrically to generate an S cell and a progenitor cell (P cells), P cell descendants are found to disperse within a maximum of six consecutive myotomal segments. Not all somitic derivatives (i.e., sclerotomal and dermal ones) have been examined in this study because of the tissue specificity of the promoter. However, fate mapping studies of Hensen's node and PSM have established that prior to somite formation, cells remain able to give rise to sclerotome, dermatome, and myotome derivatives and no evidence for specific progenitors for the myotome was ever obtained (Stern *et al.,* 1988). Therefore this pool of myotomal stem cells is likely to correspond to the pool of somite stem cells.

Based on these observations, it is clear that mesoderm segmentation cannot arise from a simple mechanism such as the one observed in the leech. In this annelid, oriented asymmetric division of a pool of stem cells called the teloblasts yields a progeny arranged in a linear array in which each cell prefigures a segment or a half segment depending on the lineage considered (Weisblat *et al.,* 1994). The periodicity of segmentation is thus directly linked to that of cell division.

Fate-mapping studies in vertebrates show that in many instances the progeny of marked clones is distributed along the anteroposterior axis in a periodic fashion, which is variable even for clones of the same lineage and usually corresponds to several somites (Kimmel *et al.,* 1988; Stern *et al.,* 1992; Nicolas *et al.,* 1996). Therefore unlike in annelids, in vertebrates, the periodic distribution of the descendants of a clone more likely reflects a consequence of the segmentation process rather than the driving force of this mechanism.

E. Clock-Based Models

1. The Clock and Wavefront Model

The clock and wavefront model was originally aimed at explaining somite number regulation observed after experimental reduction of embryonic size (see above). This model proposes the existence of an intracellular oscillator or clock inside the somitic cells that is responsible for the generation of the periodicity (Cooke and Zeeman, 1976). In the clock and wavefront model, presomitic cells oscillate synchronously according to an internal clock, and then halt their oscillation when they are reached by a posteriorly progressing wavefront while they are in the appropriate (somitogenic) phase of the clock cycle. The wavefront corresponds to the anteroposterior gradient of maturation which sweeps slowly back along the embryo. Thus the clock segments in response to the progress of the wavefront.

The wavefront was proposed to be either of propagatory or of "kinematic" nature. A kinematic wave is the consequence of an inherent graded property of the cells, it does not depend on the propagation of a signal and is not stopped by a cut across its path (Cooke and Zeeman, 1976). The kinematic nature of the somitogenic wave has been endorsed experimentally in amphibians (Deuchar and Burgess, 1967) and chick embryos (Menkes and Sandor, 1977) in which somitogenesis is able to jump across a gap experimentally produced in the presomitic mesoderm.

Experimental support for this model was obtained by provoking short heat shocks during amphibian development (Davidson, 1988). In the treated animals, a localized segmentation defect was reproducibly observed, corresponding to somite disruption at the level of cells located in the presomitic mesoderm at the time of heat shock. These experiments suggested that cells of a given anteroposterior level of the presomitic mesoderm were in a similarly receptive state at the time of heat shock, that is, in the same phase of the clock cycle. Similar kinds of results were observed in the zebrafish (Kimmel *et al.,* 1988). Heat shock treatment in amniotes led to slightly different results (see below). In agreement with this model, molecular evidence for the existence of a clock linked to segmentation was recently obtained in the chick embryo and will be discussed below (Palmeirim *et al.,* 1997).

2. The Cell Cycle Based Model

This model is based on the result of heat shock experiments in avian embryos (Primmett *et al.,* 1988; Stern *et al.,* 1988; Primmett *et al.,* 1989). As was observed in amphibians, this treatment generated segmentation defects in cells that were located in the presomitic mesoderm at the time of heat shock (Davidson, 1988). Segmentation defects were repeated along the anteroposterior axis with a periodicity corresponding to six to seven somites. Since a synchrony of the cell cycle exists in the PSM and because the time interval between two segmentation defects corresponds approximately to the duration of a cell cycle in the PSM, it was proposed that the cell cycle acts as an internal clock underlying segmentation (Stern *et al.,* 1988). Thus, it was proposed that cells of the rostral segmental plate which are in a similar phase of the cell cycle synchronously increase their adhesive properties and segregate to form a new somite. According to this model, the behavior of the cells fated to form part of the same somite is entirely autonomous and controlled by the cell cycle. A further modification of the model was proposed in which the wavefront was reintegrated as a kinematic wave of maturation progressing caudally along the anteroposterior axis (Polezhaev, 1992).

The six to seven somite interval corresponding to the cell-cycle duration in the PSM may be of biological relevance for it generally corresponds to the regional vertebral modules. For instance, most amniotes have 6 occipital vertebrae, and mammals have 7 cervical vertebrae and 12 (twice 6) thoracic, whereas chicks have 14 (twice 7) cervical vertebrae and 7 thoracic ones. In addition, fate mapping of the myotome progenitor cells in the mouse uncovered two types of clones from which the myotome originates: large clones extending from a given anteroposterior level to the tail of the embryo and smaller clones which never span more than six somites, that is, a regional module (Nicolas *et al.,* 1996). It remains unclear how the cell cycle could play a role in defining the somite periodicity but it could be involved in defining regional specification which is later superimposed on the somite periodicity.

III. Molecular Aspects of Vertebrate Somitogenesis

A. Expression of Avian *hairy1* Gene Defines a Molecular Clock Linked to Vertebrate Segmentation

An avian homolog of the *Drosophila* pair-rule gene, *hairy, c-hairy1* has recently been shown to be strongly expressed in the presomitic mesoderm where its mRNA exhibits cyclic waves of expression whose temporal periodicity corresponds to the formation time of one somite (Palmeirim *et al.,* 1997). The apparent posterior to anterior movement of these waves is not due to massive cell displacement along the anteroposterior axis, but arises from pulses of *c-hairy1* expression that are coordinated in time and space. Interestingly, the *c-hairy1* wave resolves into the prospective somite caudal half where it is maintained at least for a few hours after somite formation. Analysis of *in vitro* cultures of isolated presomitic mesoderm demonstrates that rhythmic *c-hairy1* mRNA production and degradation is an autonomous property of the paraxial mesoderm and does not result from caudal-to-rostral propagation of an activating signal.

c-hairy1 mRNA oscillations start as soon as cells exit the gastrulation site and acquire their paraxial mesoderm identity. Since approximately 12 prospective somites are contained within the chick PSM, cells of the PSM will undergo 12 *c-hairy1* expression cycles between their exit from the gastrulation site and their incorporation into a somite (Fig. 3). These rhythmic oscillations provide molecular evidence for a developmental clock linked to segmentation and somitogenesis such as that proposed in the clock and wavefront model.

The clock mechanisms best understood are the circadian clocks which oscillate over a 24-hr period. These clocks were shown to rely on transcriptional mechanisms involving negative autoregulatory feedback loops (Sassone-Corsi, 1994; Dunlap, 1996). Since *c-hairy1* belongs to a family of transcriptional repressors, it is conceivable that the rhythmic production of *c-hairy1* mRNA is driven by negative autoregulation of the gene on its own promoter. However, blocking protein synthesis does not alter the kinetics of *c-hairy1* expression, indicating that negative autoregulation of *c-hairy1* expression is unlikely to control its dynamic expression. Therefore *c-hairy1* mRNA oscillations are more likely to reflect an output of the clock than a crucial component of the clock itself.

B. Implication of the Vertebrate Homologs of the Fly Neurogenic Genes in Somitogenesis in the Control of the Segmentation Process

Some of the major contributions to the understanding of the molecular mechanisms underlying somitogenesis have come from mouse genetics. Unexpectedly, the identification of vertebrate counterparts of the so-called neurogenic genes of *Drosophila* revealed that homologs of the Notch receptor and its ligand Delta are prominently expressed in the presomitic mesoderm (Reaume *et al.,* 1992; Henrique *et al.,* 1995; Bettenhausen *et al.,* 1995). Accordingly, their knockout in the mouse leads to a major disruption of somitic segmentation (Conlon *et al.,* 1995; Hrabe de Angelis *et al.,* 1997).

Null mutant mice for the *Notch* gene die at around 10 days, that is, before somitogenesis is fully completed. In these mutants, somite formation is not prevented but its coordination is affected so that somites are often not aligned across the midline (Conlon *et al.,* 1995). In embryos mutant for *Delta1* (also called *Deltalike1* or *DLL1*), a similar phenotype is observed at the trunk level while in the tail no somites are formed (Hrabe de Angelis *et al.,* 1997). Expression of *Delta1* in the

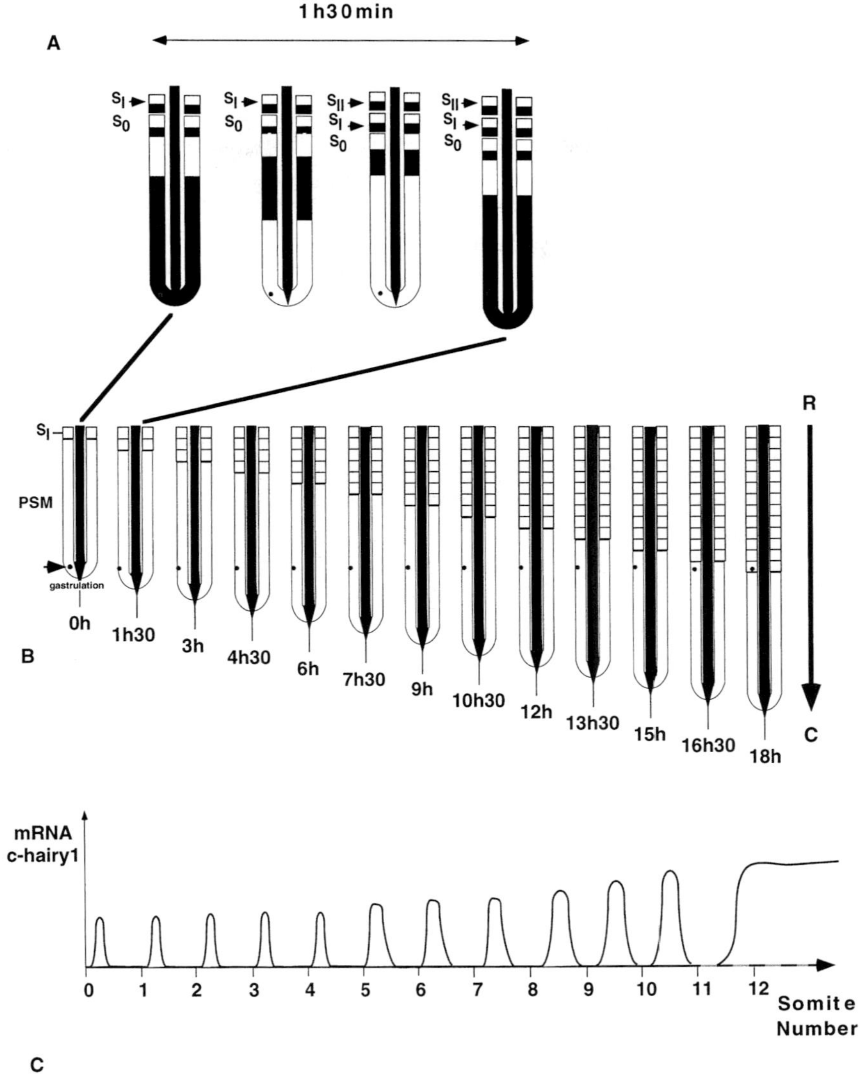

Figure 3 Schematic representation of *c-hairy1* mRNA expression in the presomitic mesoderm during one somite formation. (A) *c-hairy1* expression appears as a wave arising in a broad domain from the posterior region of the embryo and progressing in a rostral direction as a narrower domain. The *c-hairy1* expression becomes finally stabilized in the caudal somitic domain in which its expression remains. S_0, Forming somite; S_I, Newly formed somite; S_{II}, Last but one somite. (B) History of a presomitic mesoderm (PSM) cell (black spot) between the moment it exits gastrulation (0 hr) and the moment it becomes itself incorporated into a somite (18 hr). This time interval corresponds to the formation of 12 somites, which is the number of presumptive somites found in the presomitic mesoderm. (C) Oscillations of the *c-hairy1* mRNA expression in the cell figured in B during the time it spends in the PSM. At the twelfth cycle, c-hairy expression remains on if the cell is in the caudal somitic compartment or turns off if it is in the rostral compartment. These oscillations of the *c-hairy1* mRNA occur in every cell of the PSM and define a clock linked to somite segmentation.

caudal part of the somite suggested that it could play a role in the establishment and/or the maintenance of the anteroposterior subdivision of the somite. Analysis of anteroposterior subdivision of the sclerotome in the mutant shows an increase of the anterior compartment at the expense of the posterior one. Consequently, the peripheral nervous system becomes improperly segmented in these mutants. Interestingly in both *Notch1* and *Delta1* mutants, while somite formation is affected, the basic segmentation process is still operating since a periodicity of paraxial mesodermal derivatives is still observed, though this is irregular. Thus these genes are unlikely to be involved in defining the basic segmental pattern but rather act in coordinating and refining the somite pattern. Nevertheless, since several Notch receptors and ligands have been identified in vertebrates, some of which are expressed the PSM (Weinmaster *et al.,* 1992; Saga *et al.,* 1997; Dunwoodie *et al.,* 1997), the possibility remains that the full phenotype of these mutations may be hidden by gene redundancy.

The involvement of Notch–Delta signaling during somitogenesis has also been confirmed by a series of overexpression studies in *Xenopus.* In the frog, several constituents of the Notch–Delta pathway that are expressed in the presomitic mesoderm have been identified. These include *X-Notch1, X-Delta1,* and *X-Delta2* (Coffman *et al.,* 1990, 1993; Jen *et al.,* 1997). mRNA injection of wild-type *X-Delta2,* and of a dominant negative form of *X-delta2* leads to a disruption of the segmented pattern of the myotomes (Jen *et al.,* 1997). In addition, the striped pattern of genes like *X-delta2* or *X-hairy2a* in the PSM is completely perturbed. However, muscle differentiation is not prevented. In contrast to the mouse mutants, it is not clear in these experiments whether the global segmental pattern is maintained. Surprisingly, no such phenotypes were reported following injection of the constitutively activated form of *X-Notch1,* or of *X-delta1,* the homolog of mouse *delta1* or indeed of the dominant negative version of *X-delta1* (Coffman *et al.,* 1993; Chitnis *et al.,*1995).

Based on studies in the fly, a number of models might explain how Notch–Delta signaling could be functioning in somitogenesis. First, in *Drosophila, Notch* and *Delta* genes have been implicated in the control of the timing of cell differentiation by preventing the cells in which Notch is activated to respond to surrounding inducing signals (Artavanis-Tsakonas *et al.,* 1995; Cossu *et al.,* 1996). Such a role is not obvious for the mouse *Notch1,* and *Delta1/Dll1* genes since differentiation of the paraxial mesoderm into dermomyotome and sclerotome and their later derivatives occurs fairly normally in the mouse mutants (Conlon *et al.,* 1995; Hrabe de Angelis *et al.,* 1997). Accordingly, overexpression studies in the frog of wild-type or dominant–negative versions of these genes lead to disruption of the myotome segmentation but do not affect muscle differentiation (Jen *et al.,* 1997). An alternative possibility inspired by the functioning of Notch–Delta signaling via lateral inhibition during neurogenesis in *Drosophila* has been proposed to account for their role during somitogenesis (Conlon *et al.,* 1995). Finally, these genes might be involved in the definition of the paraxial somitic mesoderm boundaries much as they are implicated in defining the dorsoventral boundary in the fly developing wing disk (Blair, 1997).

The mutation of mouse homologs of other genes involved in the Notch signaling pathway such as *RBPJϰ* or *Presenilin1* (*PS1*) also give rise to somitogenesis defects (Oka *et al.,* 1995; Wong *et al.,* 1997). *RBPJϰ* is a vertebrate homolog of the fly neurogenic gene *Suppressor of Hairless* which codes for a transcription factor acting as a crucial mediator of the Notch response (Fortini and Artavanis-Tsakonas, 1994; Tamura *et al.,* 1995; Jarriault *et al.,* 1995). Homozygous mutants of this gene die around the same age as the *Notch* mutants and form irregularly shaped somites. However, differentiation of the paraxial mesoderm stops earlier than in the *Notch1* and *Delta1* mutants since myogenin expression is never detected. These results suggest that the Notch–Delta pathway may not only play a role in coordinating the somitogenesis process, but is also likely to be involved in somite differentiation. In the frog, however, injection of a dominant negative version of *suppressor of hairless* does not prevent muscle differentiation though it does disrupt segmentation (Jen *et al.,* 1997).

The *Presenilin1* gene (*PS1*) codes for a multipass transmembrane protein implicated in the genesis of Alzheimer's disease and whose homolog in *Caenorhabditis elegans, SEL12,* has been shown to interact with the neurogenic genes pathway. The mouse mutant of this gene also shows a somitogenesis defect (Wong *et al.,* 1997). Its phenotype is very similar to that of the *delta1 −/−* mice including misaligned somites. Similarly, analysis of later embryos shows that the posterior sclerotome identity has been lost. Accordingly, *Notch1* and *Delta1* gene expression in the PSM is strongly downregulated indicating that *PS1* may act upstream of these genes.

Another category of recently identified genes important for somitogenesis are the product of the *Mesp* genes. They code for basic helix–loop–helix (bHLH) transcription factors that share a common expression domain in the rostral PSM (Saga *et al.,* 1996, 1997). Mutation of the *Mesp2* gene, which is exclusively expressed as a stripe in the rostral PSM, leads to abnormal somitogenesis, particularly at the level of caudal somites (Saga *et al.,* 1997). An opposite effect on sclerotomal polarity to that exhibited by *Delta1* mutants is observed in *Mesp2−/−* mutants since the anterior somitic identity is lost. Interestingly, the global segmental pattern is not fully disrupted as evidenced by the segmented arrangement of ribs and vertebrae and the distribution of myogenin expressing cells. Somite patterning occurs fairly normally since both sclerotomal and myotomal derivatives are observed. However, examination of genes expressed in the presomitic mesoderm indicates that *Mesp2* acts upstream of *Notch1* and *2* as well as *FGFR1* since their expression is strongly downregulated in the mutant, while Delta1 expression is not affected.

Therefore none of these mutants lose the basic segmental pattern. Rather, the patterning of the segments which is characterized by the extreme regularity and coordination of somite production is altered. Thus, these analyses suggest that this group of genes plays a role in the fine tuning of the segmentation process or that gene redundancy occurs attenuating the null phenotypes.

C. Role of the Basic Helix–Loop–Helix *Paraxis* and Somite Epithelialization

The *Paraxis* gene encodes a bHLH transcription factor selectively expressed in the paraxial mesoderm (Burgess *et al.,* 1995; Quertermous *et al.,* 1994; Blanar *et al.,* 1995). *Paraxis* is expressed in the rostral PSM and in the newly formed epithelial somite. Later, its expression becomes restricted to the epithelial dorsal compartment, the dermomyotome. Mutation of this gene in the mouse leads to an embryo in which somitogenesis is abolished while the segmented pattern of paraxial mesoderm derivatives and the peripheral nervous system is retained (Burgess *et al.,* 1996). This result clearly dissociates the segmentation process of mesodermal derivatives from the formation of epithelial somite. Experiments on *Paraxis* regulation in the chick embryo have shown that its expression correlates with epithelialization of the PSM and that these events are controlled by factors derived from the ectoderm and from the neural tube (Sosic *et al.,* 1997). In addition, experiments using antisense oligonucleotides blocking *Paraxis* expression in chick result in an impairment of somite epithelization (Barnes *et al.,*1997).

D. Other Categories of Genes Affecting Vertebrate Segmentation and Somitogenesis

Other genes have been implicated in somitogenesis. For instance, mutations in *Wnt3A* or *FGFR1* (Takada *et al.,* 1994; Yamaguchi *et al.,* 1994) in mice lead to axis truncation or production of axial tissue instead of paraxial mesoderm, respectively. Therefore, these genes are likely to act very early when PSM cells become determined. Thus, their role in segmentation may be rather indirect.

FGFR1 is also expressed in the rostralmost part of the PSM and might play a role in the definition of the prospective anterior somitic compartment (Yamaguchi *et al.,* 1994). In addition, in the *Mesp2* mutant, *Notch1* downregulation is observed in territories that do not express *Mesp2* indicating a non cell-autonomous effect of the mutation. The FGF pathway could be an interesting candidate for mediating this effect as its interaction with the Notch–Delta pathway has been observed in other systems (Mitsiadis *et al.,* 1997).

IV. A Model for Vertebrate Segmentation and Somitogenesis

One important lesson from the mouse genetic studies is that the process of periodicity generation can be dissociated from that of somitogenesis and its coordination. Identification of the *c-hairy1* oscillations in the avian embryo strongly suggests the existence of a developmental clock, which could act as the periodicity generator in the PSM (Fig. 4). A second role assigned to this clock could be to measure time in the PSM to determine the precise moment of segment formation which occurs synchronously on both sides of the PSM. It is likely that such a mechanism has been conserved among vertebrates and will be identified in the mouse. Most of the genes identified thus far mainly by mouse genetics are likely to play a role downstream of this clock mechanism since segmental periodicity is retained in these mutants. It is likely that a mutation in a crucial component of the clock would result in a dramatic disruption of somitogenesis starting from the first somite resulting in a very early phenotype which would be difficult to interpret.

A second important process after acquisition of this temporally rhythmic behavior demonstrated by *c-hairy1* expression is its translation into a spatially periodic pattern. Several lines of evidence indicate that segmental identity is established in the rostral PSM prior to somitogenesis (Burgess *et al.,* 1996; Saga *et al.,* 1997). In particular, the fact that in the paraxis null mutant, segmentation proceeds fairly normally while no epithelial somites are formed suggests that the epithelial somite is not required to achieve a segmented arrangement of mesodermal derivatives. Accordingly, in *Xenopus,* no comparable somite epithelialization occurs and somite formation corresponds essentially to myotome formation (Hamilton, 1969). However, other amphibians like the axolotl do form somites exhibiting the epithelial rosette aspect of the amniote somite. Some specific variations in the epithelialization process, therefore, exist among the different classes of vertebrates but the resulting patterning of the paraxial mesoderm is very similar.

Segmentation of the rostral PSM can only be visualized using molecular markers such as genes of the Notch–Delta pathway or *Mesp2* (Henrique *et al.,* 1995; Bettenhausen *et al.,* 1995; Dunwoodie *et al.,* 1997; Saga *et al.,* 1997). Surprisingly, the vast majority of these genes are not expressed in domains corresponding to prospective somites as the somitomere theory would predict. Rather, their expression occurs according to a twofold periodicity in the rostral PSM. Since this periodicity is retained in the newly formed somite where it demarcates the anterior and posterior somite compartments, it is often assumed that expression of these genes marks these prospective compartments in the rostral PSM. This is, however, far from clear and is not true for genes like *c-hairy1* which sweeps along the PSM prior to its expression becoming resolved in the caudal somite. This situation is to some extent analogous to what is observed in the fly in which establishment of gene activities in alternate expression domains by the pair-rule genes corresponds to the first hint of metamerism but does not necessarily match the definitive segments.

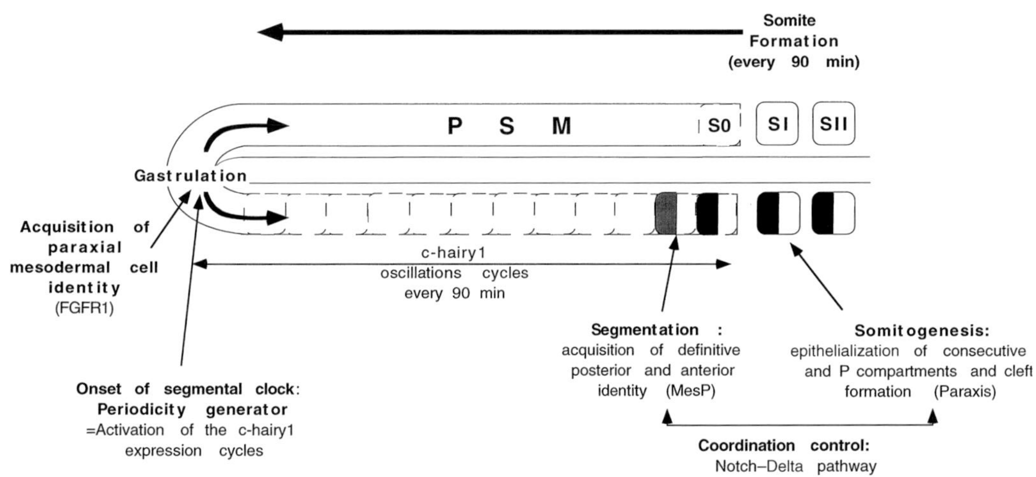

Figure 4 Recapitulation of the major events leading to somite segmentation (see text).

In the fly, pair-rule genes are expressed with a periodicity of two segments, whereas in the PSM of vertebrates the periodicity appears rather of two half-segments, when considering the somite as the basic segmental unit. Nevertheless, there remains a striking similarity between the pair-rule patterns of the fly and PSM segmentation. In zebrafish, it has been reported that a homolog of the fly *hairy* gene called *Her1* follows a true pair-rule expression pattern since it was found to be expressed in alternate primordia of presumptive somites (Muller *et al.*, 1996). Such a systematic two somite periodicity was, however, not observed with the avian *c-hairy1* gene to which *Her1* is only distantly related (Palmeirim *et al.*, 1997). *c-hairy1* shows an alternate stripe periodicity in the rostral PSM and the newly formed somites but the stripes correspond to prospective half somites. In invertebrates, pair-rule expression patterns during segmentation have not been demonstrated outside the evolved insects like dipterans or coleopterans. Surprisingly, homologs of several fly pair-rule genes such as *even-skipped* do not exhibit such a pattern in short germ band insects whose body segmentation closely resembles somitogenesis (Patel *et al.,* 1992). Therefore the issue of whether this pattern of segmentation is ancestral remains unclear.

The issue of whether the pair-rule patterning system in segmentation has been conserved within and outside the arthropod phylum remains a controversial one. New evidence is accumulating in favor of a common origin of segmentation in protostomes and deuterostomes. It would thus appear that almost all the major basic patterning systems used in invertebrates have been conserved in vertebrates, highlighting the amazing conservation of developmental processes in the animal kingdom.

Acknowledgments

The author thanks Kim Dale, Estelle Hirsinger, David Ish-Horowicz, Mike McGrew, and Isabel Palmeirim for their critical reading and helpful comments on the manuscript.

References

Aoyama, H., and Asamoto, K. (1988). Determination of somite cells: Independence of cell differentiation and morphogenesis. *Development* **104,** 15–28.

Artavanis-Tsakonas, S., Matsuno, K., and Fortini, M. E. (1995). Notch signaling. *Science* **268,** 225–232.

Bagnall, K. M., Higgins, S. J., and Sanders, E. J. (1988). The contribution made by a single somite to the vertebral column: Experimental evidence in support of resegmentation using the chick–quail chimaera model. *Development* **103,** 69–85.

Bagnall, K. M., Higgins, S. J., and Sanders, E. J. (1989). The contribution made by cells from a single somite to tissues within a body segment and assessment of their integration with similar cells from adjacent segments. *Development* **107,** 931–943.

Balling, R., Helwig, U., Nadeau, J., Neubuser, A., Schmahl, W., and Imai, K. (1996). Pax genes and skeletal development. *Ann. N. Y. Acad. Sci.* **785,** 27–33.

Barnes, G. L., Alexander, P. G., Hsu, C. W., Mariani, B. D., and Tuan, R. S. (1997). Cloning and characterization of chicken Paraxis: A regulator of paraxial mesoderm development and somite formation. *Dev. Biol.* **189,** 95–111.

Bellairs, R. (1963). The development of somites in the chick embryo. *J. Embryol. Exp. Morphol.* **11,** 697–714.

Bellairs, R., and Veini, M. (1984). Experimental analysis of control mechanisms in somite segmentation in avian embryos. II. Reduction of material in the gastrula stages of the chick. *J. Embryol. Exp. Morphol.* **79,** 183–200.

Bettenhausen, B., Hrabe de Angelis, M., Simon, D., Guenet, J. L., and Gossler, A. (1995). Transient and restricted expression during mouse embryogenesis of Dll1, a murine gene closely related to *Drosophila* Delta. *Development* **121,** 2407–2418.

Blair, S. S. (1997). Limb development: Marginal fringe benefits. *Curr. Biol.* **7,** 686–690.

Blanar, M. A., Crossley, P. H., Peters, K. G., Steingrimsson, E., Copeland, N. G., Jenkins, N. A., Martin, G. R., and Rutter, W. J. (1995). Meso1, a basic-helix-loop-helix protein involved in mammalian presomitic mesoderm development. *Proc. Natl. Acad. Sci. U.S.A.* **92,** 5870–5874.

Bronner-Fraser, M., and Stern, C. (1991). Effects of mesodermal tissues on avian neural crest cell migration. *Dev. Biol.* **143,** 213–217.

Burgess, R., Cserjesi, P., Ligon, K. L., and Olson, E. N. (1995). Paraxis: A basic helix-loop-helix protein expressed in paraxial mesoderm and developing somites. *Dev. Biol.* **168,** 296–306.

Burgess, R., Rawls, A., Brown, D., Bradley, A., and Olson, E. N. (1996). Requirement of the paraxis gene for somite formation and musculoskeletal patterning. *Nature* **384,** 570–573.

Chitnis, A., Henrique, D., Lewis, J., Ish-Horowicz, D., and Kintner, C. (1995). Primary neurogenesis in *Xenopus* embryos regulated by a homologue of the *Drosophila* neurogenic gene Delta [see comments]. *Nature* **375,** 761–766.

Christ, B., and Ordahl, C. P. (1995). Early stages of chick somite development. *Anat. Embryol.* **191,** 381–396.

Christ, B., and Wilting, J. (1992). From somites to vertebral column. *Anat. Anz.* **174,** 23–32.

Coffman, C., Harris, W., and Kintner, C. (1990). Xotch, the *Xenopus* homolog of *Drosophila* notch. *Science* **249,** 1438–1441.

Coffman, C. R., Skoglund, P., Harris, W. A., and Kintner, C. R. (1993). Expression of an extracellular deletion of Xotch diverts cell fate in *Xenopus* embryos. *Cell* **73,** 659–671.

Conlon, R. A., Reaume, A. G., and Rossant, J. (1995). Notch1 is required for the coordinate segmentation of somites. *Development* **121,** 1533–1545.

Cooke, J. (1975). Control of somite number during morphogenesis of a vertebrate, *Xenopus laevis. Nature* **254,** 196–199.

Cooke, J. (1981). The problem of periodic patterns in embryos. *Philos. Trans. R. Soc. London B Biol. Sci.* **295,** 509–524.

Cooke, J., and Zeeman, E. C. (1976). A clock and wavefront model for control of the number of repeated structures during animal morphogenesis. *J. Theor. Biol.* **58,** 455–476.

Cossu, G., Tajbakhsh, S., and Buckingham, M. (1996). How is myogenesis initiated in the embryo? *Trends Genet.* **12,** 218–223.

Couly, G., Coltey, P., and Douarin, N. (1993). The triple origin of skull in higher vertebrates: A study in quail–chick chimeras. *Dev. Suppl.* **117,** 409–429.

Davidson, D. (1988). Segmentation in frogs. *Development* **104,** 221–230.

De Robertis, E. M. (1997). Evolutionary biology. The ancestry of segmentation. *Nature* **387,** 25–26.

Denetclaw, W. F. J., Christ, B., and Ordahl, C. P. (1997). Location and growth of epaxial myotome precursor cells. *Development* **124,** 1601–1610.

Deuchar, E., and Burgess, A. M. C. (1967). Somite segmentation in amphibian embryos: Is there a transmitted control mechanism? *J. Embryol. Exp. Morphol.* **17,** 349–358.

Dunlap, J. C. (1996). Genetics and molecular analysis of circadian rhythms. *Annu. Rev. Genet.* **30,** 579–601.

Dunwoodie, S. L., Henrique, D., Harrison, S. M., and Beddington, R. S. P. (1997). Mouse Dll3: A novel divergent Delta gene which may complement the function of other Delta homologues during early pattern formation in the mouse embryo. *Development* **124,** 3065–3076.

Ewan, K. B., and Everett, A. W. (1992). Evidence for resegmentation in the formation of the vertebral column using the novel approach of retroviral-mediated gene transfer. *Exp. Cell Res.* **198,** 315–320.

Fankhauser, G. (1945). The effect of change in chromosome number on amphibian development. *Q. Rev. Biol.* **20,** 20–78.

Fortini, M. E., and Artavanis-Tsakonas, S. (1994). The suppressor of hairless protein participates in notch receptor signaling. *Cell* **79,** 273–282.

Fraser, R. C. (1963). Somite genesis in the chick. III. The role of induction. *J. Exp. Zool.* **145,** 151–167.

Gearhart, J. D., and Mintz, B. (1972). Clonal origins of somites and their muscle derivatives: Evidence from allophenic mice. *Dev. Biol.* **29,** 27–37.

Gilland, E., and Baker, R. (1993). Conservation of neuroepithelial and mesodermal segments in the embryonic vertebrate head [published erratum appears in *Acta Anat.* (1994) 149(2), 164]. *Acta Anat.* **148,** 110–123.

Goldstein, R. S., and Kalcheim, C. (1992). Determination of epithelial half-somites in skeletal morphogenesis. *Development* **116,** 441–445.

Gorodilov, Y. N. (1992). Rhythmic processes in lower vertebrates embryogenesis and their role for developmental control. *Zool. Sci.* **9,** 1101–1111.

Hamilton, L. (1969). The formation of somites in *Xenopus. J. Embryol. Exp. Morphol.* **22**(2), 253–264.

Henrique, D., Adam, J., Myat, A., Chitnis, A., Lewis, J., and Ish-Horowicz, D. (1995). Expression of a Delta homologue in prospective neurons in the chick. *Nature* **375,** 787–790.

Holland, L. Z., Kene, M., Williams, N. A., and Holland, N. D. (1997). Sequence and embryonic expression of the amphioxus engrailed gene (AmphiEn): The metameric pattern of transcription resembles that of its segment-polarity homolog in *Drosophila. Development* **124,** 1723–1732.

Hrabe de Angelis, M., McIntyre, J., and Gossler, A. (1997). Maintenance of somite borders in mice requires the Delta homologue DII1. *Nature* **386,** 717–721.

Jacobson, A. G. (1988). Somitomeres: Mesodermal segments of vertebrate embryos. *Development* **104**(Suppl.), 209–220.

Jarriault, S., Brou, C., Logeat, F., Schroeter, E. H., Kopan, R., and Israel, A. (1995). Signalling downstream of activated mammalian Notch. *Nature* **377,** 355–358.

Jen, W. C., Wettstein, D., Chitnis, D., and Kintner, C. (1997). The Notch ligand, X-Delta-2, mediates segmentation of the paraxial mesoderm in *Xenopus* embryos. *Dev. Suppl.* **124,** 1169–1178.

Kaehn, K., Jacob, H. J., Christ, B., Hinrichsen, K., and Poelmann, R. E. (1988). The onset of myotome formation in the chick. *Anat. Embryol.* **177,** 191–201.

Keynes, R. J., and Stern, C. D. (1984). Segmentation in the vertebrate nervous system. *Nature* **310,** 786–789.

Keynes, R. J., and Stern, C. D. (1988). Mechanisms of vertebrate segmentation. *Development* **103,** 413–429.

Kimmel, C. B. (1996). Was Urbilateria segmented? *Trends. Genet.* **12,** 329–331.

Kimmel, C. B., Sepich, D. S., and Trevarrow, B. (1988). Development of segmentation in zebrafish. *Development* **104**(Suppl.), 197–207.

Lawrence, P. A. (1992). "The Making of a Fly." Blackwell Science, Oxford.

Lumsden, A., and Krumlauf, R. (1996). Patterning the vertebrate neuraxis. *Science* **274,** 1109–1115.

Meier, S. (1984). Somite formation and its relationship to metameric patterning of the mesoderm. *Cell Differ.* **14,** 235–243.

Meinhardt, H. (1986). Models of segmentation. *In* "Somites in Developing Embryos" (R. Bellairs, D. A. Ede, and J. W. Lash, Eds.), pp. 179–191. Plenum, New York and London.

Menkes, B., and Sandor, S. (1977). Somitogenesis, regulation potencies, sequence determination and primordial interactions. *In* "Vertebrate Limb and Somite Morphogenesis" (D. A. Ede, D. A. Hinchcliffe, and M. Balls, Eds.), British Society Developmental Biology Symposium 3, pp. 405–419. Cambridge Univ. Press, Cambridge.

Mitsiadis, T. A., Henrique, D., Thesleff, I., and Lendahl, U. (1997). Mouse Serrate-1 (Jagged-1): Expression in the developing tooth is regulated by epithelial–mesenchymal interactions and fibroblast growth factor-4. *Development* **124,** 1473–1483.

Morin-Kensicki, E. M., and Eisen, J. S. (1997). Sclerotome development and peripheral nervous system segmentation in embryonic zebrafish. *Development* **124,** 159–167.

Muller, M., Weizsacker, E., and Campos-Ortega, J. A. (1996). Expression domains of a zebrafish homologue of the *Drosophila* pair-rule gene hairy correspond to primordia of alternating somites. *Development* **122,** 2071–2078.

Nicolas, J. F., Mathis, L., Bonnerot, C., and Saurin, W. (1996). Evidence in the mouse for self-renewing stem cells in the formation of a segmented longitudinal structure, the myotome. *Development* **122,** 2933–2946.

Noden, D. M. (1983). The embryonic origins of avian cephalic and cervical muscles and associated connective tissues. *Am. J. Anat.* **168,** 257–276.

Oka, C., Nakano, T., Wakeham, A., de la Pompa, J. L., Mori, C., Sakai, T., Okazaki, S., Kawaichi, M., Shiota, K., Mak, T. W., and Honjo, T. (1995). Disruption of the mouse RBP-J kappa gene results in early embryonic death. *Development* **121,** 3291–3301.

Packard, D. S. J. (1976). The influence of axial structures on chick somite formation. *Dev. Biol.* **53,** 36–48.

Palmeirim, I., Henrique, D., Ish-Horowicz, D., and Pourquie, O. (1997). Avian hairy gene expression identifies a molecular clock linked to vertebrate segmentation and somitogenesis. *Cell* **91,** 639–648.

Patel, N. H., Ball, E. E., and Goodman, C. S. (1992). Changing role of even-skipped during the evolution of insect pattern formation. *Nature* **357,** 339–342.

Polezhaev, A. A. (1992). A mathematical model of the mechanism of vertebrate somitic segmentation. *J. Theor. Biol.* **156,** 169–181.

Primmett, D. R., Stern, C. D., and Keynes, R. J. (1988). Heat shock causes repeated segmental anomalies in the chick embryo. *Development* **104,** 331–339.

Primmett, D. R., Norris, W. E., Carlson, G. J., Keynes, R. J., and Stern, C. D. (1989). Periodic segmental anomalies induced by heat shock in the chick embryo are associated with the cell cycle. *Development* **105,** 119–130.

Quertermous, E. E., Hidai, H., Blanar, M. A., and Quertermous, T. (1994). Cloning and characterization of a basic helix-loop-helix protein expressed in early mesoderm and the developing somites. *Proc. Natl. Acad. Sci. U.S.A.* **91,** 7066–7070.

Ranscht, B., and Bronner-Fraser, M. (1991). T-cadherin expression alternates with migrating neural crest cells in the trunk of the avian embryo. *Development* **111,** 15–22.

Reaume, A. G., Conlon, R. A., Zirngibl, R., Yamaguchi, T. P., and Rossant, J. (1992). Expression analysis of a Notch homologue in the mouse embryo. *Dev. Biol.* **154,** 377–387.

Remak, R. (1850). "Untersuchungen über die entwicklung der Wirbeltiere." Reimer, Berlin.

Rickmann, M., Fawcett, J. W., and Keynes, R. J. (1985). The migration of neural crest cells and the growth of motor axons through the rostral half of the chick somite. *J. Embryol. Exp. Morphol.* **90,** 437–455.

Robertson, K., and Mason, I. (1995). Expression of ret in the chicken embryo suggests roles in regionalisation of the vagal neural tube and somites and in development of multiple neural crest and placodal lineages. *Mech. Dev.* **53,** 329–344.

Saga, Y., Hata, N., Kobayashi, S., Magnuson, T., Seldin, M. F., and Taketo, M. M. (1996). MesP1:

A novel basic helix-loop-helix protein expressed in the nascent mesodermal cells during mouse gastrulation. *Development* **122,** 2769–2778.

Saga, Y., Hata, N., Koseki, H., and Taketo, M. M. (1997). Mesp2: A novel mouse gene expressed in the presegmented mesoderm and essential for segmentation initiation. *Genes Dev.* **11,** 1827–1839.

Sassone-Corsi, P. (1994). Rhythmic transcription and autoregulatory loops: Winding up the biological clock. *Cell* **78,** 361–364.

Satoh, N. (1994). "Developmental Biology of Ascidians." Cambridge Univ. Press, New York.

Solursh, M., Fisher, M., Meier, S., and Singley, C. T. (1979). The role of extracellular matrix in the formation of the sclerotome. *J. Embryol. Exp. Morphol.* **54,** 75–98.

Sosic, D., Brand-Saberi, B., Schmidt, C., Christ, B., and Olson, E. N. (1997). Regulation of paraxis expression and somite formation by ectoderm- and neural tube-derived signals. *Dev. Biol.* **185,** 229–243.

Spratt, N. T. (1955). Analysis of the organizer center in the early chick embryo. I Localisation of prospective notochord and somite cells. *J. Exp. Zool.* **128,** 121–162.

Stern, C. D., and Keynes, R. J. (1987). Interactions between somite cells: The formation and maintenance of segment boundaries in the chick embryo. *Development* **99,** 261–272.

Stern, C. D., Sisodiya, S. M., and Keynes, R. J. (1986). Interactions between neurites and somite cells: Inhibition and stimulation of nerve growth in the chick embryo. *J. Embryol. Exp. Morphol.* **91,** 209–226.

Stern, C. D., Fraser, S. E., Keynes, R. J., and Primmett, D. R. (1988). A cell lineage analysis of segmentation in the chick embryo. *Development* **104,** 231–244.

Stern, C. D., Hatada, Y., Selleck, M. A., and Storey, K. G. (1992). Relationships between mesoderm induction and the embryonic axes in chick and frog embryos. *Dev. Suppl.,* 151–156.

Takada, S., Stark, K. L., Shea, M. J., Vassileva, G., McMahon, J. A., and McMahon, A. P. (1994). Wnt-3a regulates somite and tailbud formation in the mouse embryo. *Genes Dev.* **8,** 174–189.

Tam, P. P. (1981). The control of somitogenesis in mouse embryos. *J. Embryol. Exp. Morphol.* **65**(Suppl.), 103–128.

Tam, P. P., and Trainor, P. A. (1994). Specification and segmentation of the paraxial mesoderm. *Anat. Embryol.* **189,** 275–305.

Tam, P. P., Meier, S., and Jacobson, A. G. (1982). Differentiation of the metameric pattern in the embryonic axis of the mouse. II. Somitomeric organization of the presomitic mesoderm. *Differentiation* **21,** 109–122.

Tamura, K., Taniguchi, Y., Minoguchi, S., Sakai, T., Tun, T., Furukawa, T., and Honjo, T. (1995). Physical interaction between a novel domain of the receptor Notch and the transcription factor RBP-J kappa/Su(H). *Curr. Biol.* **5,** 1416–1423.

Verbout, A. J. (1976). A critical review of the 'neugliederung' concept in relation to the development of the vertebral column. *Acta Biotheor.* **25,** 219–258.

Verbout, A. J. (1985). The development of the vertebral column. *Adv. Anat. Embryol. Cell Biol.* **90,** 1–122.

Weinmaster, G., Roberts, V. J., and Lemke, G. (1992). Notch2: A second mammalian Notch gene. *Development* **116,** 931–941.

Weisblat, D. A., Wedeen, C. J., and Kostriken, R. G. (1994). Evolution of developmental mechanisms: Spatial and temporal modes of rostrocaudal patterning. *Curr. Top. Dev. Biol.* **29,** 101–134.

Wong, P. C., Zheng, H., Chen, H., Becher, M. W., Sirinathsinghji, D. J., Trumbauer, M. E., Chen, H. Y., Price, D. L., Van der Ploeg, L. H., and Sisodia, S. S. (1997). Presenilin 1 is required for Notch1 and DII1 expression in the paraxial mesoderm. *Nature* **387,** 288–292.

Woo Youn, B., and Malacinski, G. M. (1981). Somitogenesis in the amphibian *Xenopus laevis:* Scanning electron microscopic analysis of intrasomitic cellular arrangements during somite rotation. *J. Embryol. Exp. Morphol.* **64,** 23–43.

Yamaguchi, T. P., Harpal, K., Henkemeyer, M., and Rossant, J. (1994). fgfr-1 is required for embryonic growth and mesodermal patterning during mouse gastrulation. *Genes Dev.* **8,** 3032–3044.

4

Segmentation: A View from the Border

Claudio D. Stern and Daniel Vasiliauskas
Department of Genetics and Development
College of Physicians and Surgeons of Columbia University
New York, New York 10032

Segmentation, or metamerism, consists of the subdivision of the body into discrete units that subsequently acquire regional specializations. In vertebrates, the most obvious manifestation of this phenomenon is seen during the formation of the mesodermal somites and their derivatives. This review surveys three different models for how somites form, and how they relate to recent molecular data suggesting the involvement of transcription factors and cell surface molecules. A new model (the "Morse code" model) is proposed to convey positional information to somitogenic cells. Finally, the molecular events of boundary formation (during the initial epithelialization of somites) and boundary maintenance (between adjacent somite halves as well as in resegmentation) are discussed.

I. Introduction

Segmentation, or metamerism, is a fundamental feature of the organization of the body plan of higher animals, in which a basic structure is repeated a number of times along the length of the body axis. Each segment (or metamere) can diverge from the prototype, acquiring specialized functions and organs. Metamerism

Current Topics in Developmental Biology, Vol. 47

0070-2153/00 $30.00

brings with it the advantage that the animal can have sufficient rigidity to contain the internal organs while still allowing mobility at segment boundaries, irrespective of whether the segments and skeletal elements are external (as in insects) or internal (as in vertebrates). This basic architecture has been so successful that it is displayed by many invertebrate phyla as well as throughout the chordates. However, there has been considerable controversy about whether this body plan evolved independently in long germ band insects and chordates or whether there is a common segmented ancestor ("Urbilateria") for both groups (see De Robertis, 1997, for a recent discussion).

There appear to be two distinct mechanisms for generating metamerism during embryonic development (Fig. 1; see also Keynes and Stern, 1988; Sander, 1988; Stern, 1990). The first ("segmentation by subdivision") is exemplified by long germ band insects (e.g., *Drosophila*) and is also seen in the hindbrain and diencephalon of higher vertebrates. In this mode, the future segmented domain (the whole body in the case of long germ band insects) becomes gradually subdivided into domains that become progressively smaller by the appearance of boundaries or of further domains of gene expression. The second mode ("sequential segmentation") is displayed by all other segmented animals, but there are three minor variations of this theme. In annelids, short germ band insects, and other arthropods, the metameres and the borders between them are added sequentially, from a posteriorly located growth zone that may contain stem cells. In amniote vertebrates (reptiles, birds, and mammals), there is a similar growth zone (the primitive streak and node), which continuously generates cells that enter the segmental plate, within which they transit toward the head. Segmentation is delayed, because only some time later do groups of cells within the segmental plate adhere to each other to form an epithelial sphere (the somite); later still, each somite breaks down into a ventromedial, mesenchymal sclerotome, and a dorsolateral, still epithelial, dermomyotome. In most anuran amphibians and some teleost fishes (including the "model" organisms *Xenopus* and zebrafish), segmentation also proceeds sequentially, but there appears to be no growth zone with stem cells. In these organisms, where there is little or no increase in volume of the embryo at early stages, the paraxial mesoderm becomes segmented through coordinate turning of blocks of cells, such that the length of one extended cell in the block initially spans the entire length of the metamere and later these blocks change shape to form diagonal "chevrons." In these animals virtually the entire somite has a myotome-like structure, and there is little or no sclerotome or dermomyotome at the time of somite formation. However, unlike the amniote case, the myotome contributes to both epaxial and hypaxial muscles. Urodele amphibians and some fishes are intermediate between *Xenopus* and amniotes: as in anurans, there appears to be no obvious growth zone but as in the amniotes, somites form epithelial spheres and subsequently become subdivided into dermomyotome and sclerotome.

In long germ band insects, many of the gene products involved in the generation and maintenance of segmental organization have been identified, as have

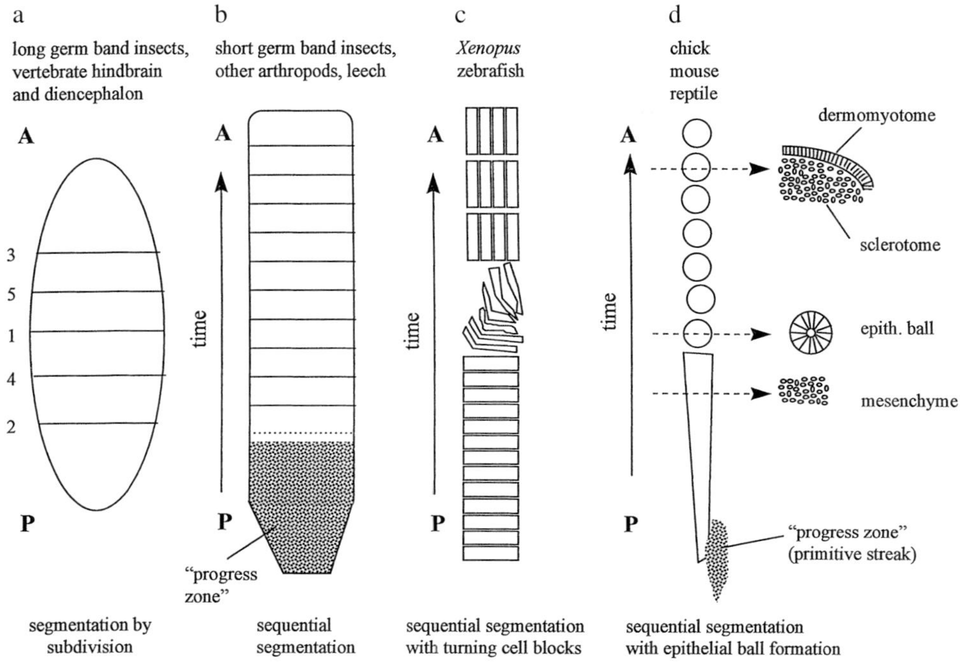

Figure 1 Different modes of segmentation in different animals. (a) In long germ band insects, as well as in the hindbrain and diencephalon of vertebrates, segment boundaries appear so as to subdivide an initially fixed domain into progressively smaller units. In the remaining animals segmental units and the boundaries between them are added sequentially from the posterior end of the animal. (b) In short germ band insects, other arthropods, and segmented worms, a posteriorly located growth zone ("progress zone," by analogy to the limb) adds material derived from the progeny of stem-cell-like precursors. As these cells differentiate they become organized metamerically. (c) In anuran amphibians and some teleost fishes, there is no posterior growth zone and no increase in the tissue mass that will contribute to mesodermal segments. An initial column of cells acquires segmental organization by coordinate turning of cell blocks and give rises predominantly to myotome. The origin and position of the cells that contribute to the vertebrate column in these animals is unknown, as the sclerotome is very small or absent at these stages of development. (d) Amniotes possess a posterior growth zone, as in long germ band insects, but segmentation is delayed by a period during which somitogenic cells reside in a mesenchymal segmental plate. Somite formation consists of the formation of an epithelial sphere. Later, each somite becomes subdivided into a dorsolateral dermomyotome (which retains its epithelial character for some time) and a ventromedial sclerotome (which becomes mesenchymal again). Urodeles and some other fishes display an intermediate type of segmentation, including formation of epithelial spheres and subsequent formation of sclerotome and dermomyotome, but appear to lack a posterior growth zone. A, Anterior; P, posterior. In a, the numbers on the left indicate the approximate order of formation of the boundaries. In d, the dashed arrows show the axial level through which the schematic sections shown on the right are obtained.

those that confer individual identity to specific metameres. We are still far from this level of understanding in other animals. During the last few years, the remarkable finding has been made that at least some of the underlying molecular players are strikingly conserved (see De Robertis, 1997). Despite our deeper knowledge in *Drosophila,* however, many of the genes identified to play critical roles in establishing metamerism in this organism are transcription factors, and most of the remaining ones are involved in signaling pathways that establish or refine segment polarity; cell-surface molecules likely to restrict cell mixing between adjacent metameres have not yet been identified. Indeed, there is still very little information about the mechanisms that define and subsequently maintain the boundaries between metameres in any animal, and it is probably true that here more rapid progress is being made in vertebrates than in *Drosophila.*

Irrespective of how they form, and of whether they are in the interior of the animal or at the outer surface, metameres are always defined by the borders that separate them from their neighbors. These boundaries are not only essential during initial formation of a metamere, but at least in some cases they are required to maintain the segmental organization if cells become motile at some later stage in development. For example, in amniotes, boundaries initially delimit the groups of cells that epithelialize together to form a somite sphere. However, some hours later, when the ventromedial aspect of the somite loses its epithelial structure, some mechanism must exist to prevent the mesenchymal sclerotome cells from mixing across metameres, destroying the original segmental pattern. Relatively little attention has been paid to somite boundaries and their subsequent development. Until now, each of the models proposed for somite formation has been heavily influenced by data from one or another group of animals, and one of the challenges that lies ahead is to find a unifying mechanism (if such a thing exists) to accommodate different modes of metamerization (at least for the trunk of different vertebrates). It is only then that we will begin to understand how segmentation may have evolved.

II. Segments and Parasegments

Martínez-Arias and Lawrence (1985; see also Lawrence, 1988) proposed that the fundamental unit of metamerism in long germ band insects is not the segment, but a unit (the "parasegment") out of phase with it. This is a familiar concept to vertebrate embryologists, because of the idea put forward by Remak (1855; see also Chap. 9 by Brand-Saberi and Christ, this volume) that vertebrae develop out of phase with the original somites such that each vertebra arises from a fusion of the anterior half of one sclerotome and the posterior half of the preceeding sclerotome. But in those embryos (including urodele amphibians, some fishes, and all amniotes) in which somite formation is accompanied by epithelialization of a somitic sphere, an earlier event is also out of phase with the somites: The formation of each successive border *simultaneously* defines the caudal part of the somite that is currently becoming epithelial and the rostral edge of the segmental plate. Although it is likely that the cellular events that lead to the formation of intersomitic borders is a more gradual process, beginning several prospective somites before overt epithelialization, scanning electron microscopy of the rostral tip of the segmental plate reveals that cells seem to acquire epithelial continuity at about the same time as the newly formed somite undergoes the same process (see Bellairs, 1979; Tam, 1986; Tam and Trainor, 1994). The question arises as to which is the segment and which the parasegment, and whether epithelialization, boundary formation, or the specification of rostral and caudal half-segments constitute the key events that define individual metameres.

In addition to the alignment of earlier and later structures, there are two critical features that define metameric parasegments: adjacent parasegments must be lineally independent, and must be accompanied by patterns of gene expression consistent with the position of (para)segmental boundaries (García-Bellido *et al.,* 1973; Morata and Lawrence, 1975; Lawrence, 1988). In amniotes, both criteria are fulfilled at both times in development when the unit of segmentation appears to change. Lineage restrictions to the spread of clones first arise at about the same time as segmentation becomes visible (Stern *et al.,* 1988). From this time, the progeny of individual cells no longer spreads into adjacent somites, but also respects the intrasomite boundary ("von Ebner's fissure"). By the time of sclerotome formation, the failure of "unlike" sclerotome cells to mix (see Section VII and Stern and Keynes, 1987) is probably responsible for the maintenance of these lineage restrictions, but the mechanisms underlying the restriction that appears at the time of somite formation are still unclear. It is also around the time of segmentation that somite- and half-somite-specific gene expression appears (see below and Nieto *et al.,* 1992; Palmeirim *et al.,* 1997; Biben *et al.,* 1998). As development proceeds, more genes appear whose expression is restricted at either an intersomite or at an intrasomite boundary. It will be particularly interesting to determine accurately whether the rostral boundaries of *Hox* gene expression in some somite derivatives respect the original somite boundaries, whether the expression boundaries are out of phase in the sclerotome with respect to the blocks of epaxial muscle, and at what time these boundaries become restricted to lineally defined groups of cells. It is surprising that this analysis has not yet been done in sufficient detail.

III. Three Models for the Control of Somite Formation

The above summary shows that many fundamental questions about the cellular and molecular biology of somite development are still unanswered. During the last two decades, several models have been put forward to identify the mechanisms that determine the position of segmental boundaries. Three of these will be discussed below: the "clock and wavefront model" of Cooke and Zeeman (1976) (Fig. 2), Meinhardt's (1982, 1986) model (Fig. 3), and the "cell cycle model" (Stern *et al.,* 1988; Primmett *et al.,* 1989) (Fig. 4).

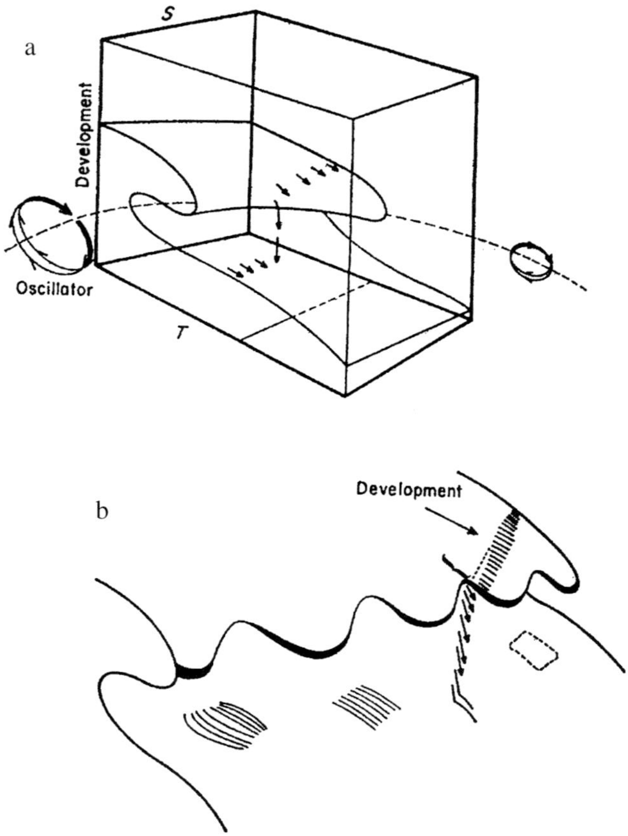

Figure 2 The clock and wavefront model (from Cooke and Zeeman, 1976). The "oscillator" is some cyclic event intrinsic to the cells, and the folded surface shown in (a) and (b) represents a condition with two stable states (upper and lower). *S* represents spatial position and *T* time. As cells approach the edge of the "catastrophe" surface, the forward phase of the oscillator takes them over the edge. The oscillator together with the surface partitions groups of cells that will segment together. From Cooke and Zeeman (1976), reproduced with kind permission of Dr. Jonathan Cooke and Academic Press.

A. Clock and Wavefront Model

Cooke and Zeeman (1976) proposed this model (Fig. 2) to explain the observation (Cooke, 1975) that amphibian embryos of reduced size appear to form the correct number of somites, each containing a smaller number of cells. Despite its name, the model was proposed to have three elements: (a) a smooth gradient extending over the entire length of the embryo (or somite-forming region); (b) a slow-passing wave, progressing from anterior to posterior regions of the embryo whose speed is controlled by the slope of the gradient (a), and (c) an intracellular oscillator which gates the wave into discrete blocks of cells that will segment together. In the original model, the "forward" direction of the oscillator takes cells to a state that triggers a sudden change in adhesive or motile properties, while the "reverse" direction of the oscillator has no visible consequence (Fig. 2). The sudden change is necessary to account for the fact that no cells exist between somites, and that in amphibians somite formation can be represented as a coordinated change in cell orientation along the body axis. Cooke and Zeeman (1976) used the topological analogy of catastrophe theory proposed by René Thom (1975), where the sudden change is represented by the edge of the catastrophe surface, and the oscillator controls the position of this edge within the surface.

Cooke and Zeeman (1976) state that their model applies equally well to amniotes, because here too somite formation occurs as a wave, progressing from head to tail. However, some modifications have to be introduced to account for a number of features of somitogenesis specific to vertebrates other than anuran amphibians and some fishes (see Keynes and Stern, 1988). First, amniote somites form by epithelialization of cells to form a sphere (see Fig. 1). Therefore, each new border simultaneously defines the posterior half of the forming somite and the anterior portion of the next somite to form (see Section II). In other words, in *Xenopus,* somite formation occurs "segmentally," while in amniotes it appears to be "parasegmental." The clock and wavefront model does not obviously explain why some cells decide to form part of one somite while others join the next one. Second, unlike lower vertebrates, in amniotes there does not exist a specific time in development when all prospective somitic cells are aligned along the rostrocaudal axis such that a whole-body gradient can relay information about the size of the body to the population of somitogenic cells and hence control the speed of the wave. Instead, the early progeny of somitic precursor cells residing within the anterior part of the primitive streak will colonize rostral somites while later progeny from the same precursor cells will contribute to more caudal ones (see Selleck and Stern, 1991; Psychoyos and Stern, 1996; Chap. 1 by Tam *et al.,* this volume). Third, there is as yet no clear evidence that in amniotes the total number of somites is maintained when total body size is altered (in fact, a reduction in cell number in the mouse leads to abnormal numbers of vertebra and ribs; Gregg and Snow, 1983). One modification that can be made to the Cooke and Zeeman model and which may help overcome some of these problems is to postulate that the oscillator contains two states. In essence, this is the model proposed by Meinhardt (1982, 1986).

B. Meinhardt Model

Meinhardt (1982, 1986) proposed that the formation of each segment may be preceded by a number of oscillations between two states, "A" (anterior) and "P" (posterior) (Fig. 3). The period of the oscillator is equal to the time that it takes for one somite to form, and the number of oscillations between the genesis of a prospective segmenting cell and the time of segment formation is determined by a

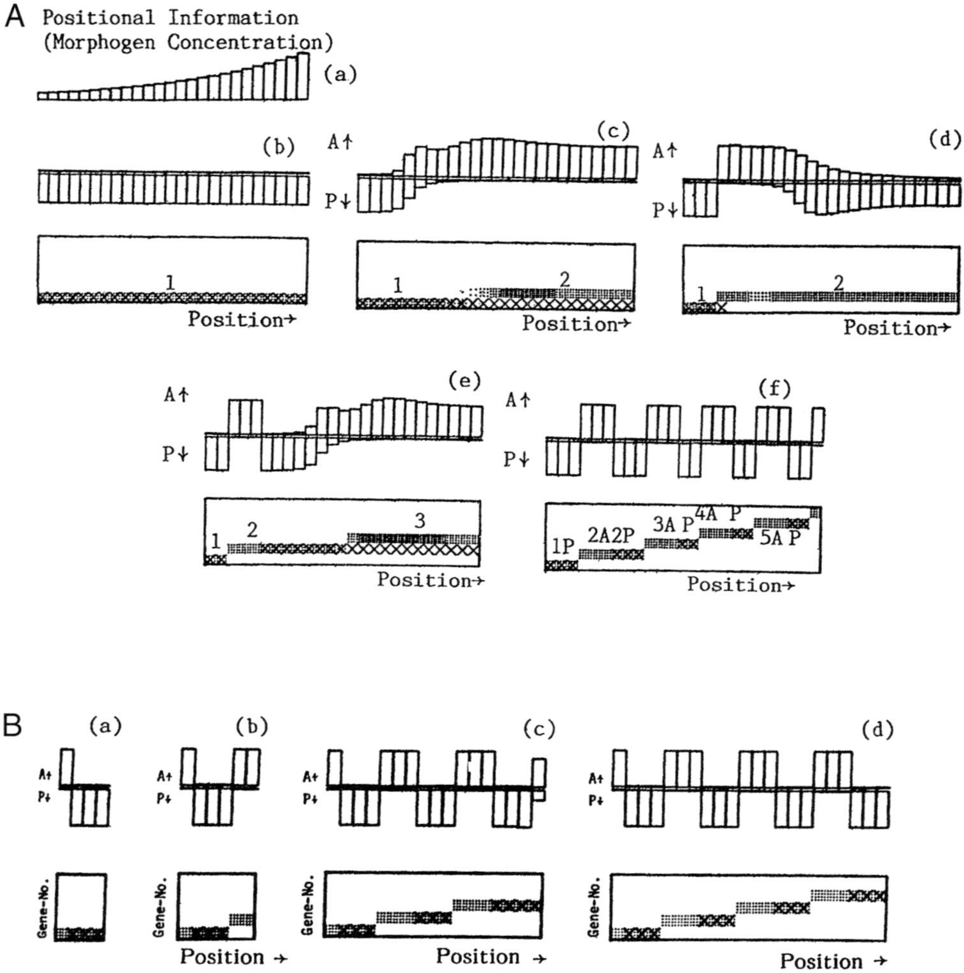

Figure 3 Two versions of the model proposed by Meinhardt (1982, 1986). (A) In cases where all prospective somitogenic cells are present in the embryo at the same time, a gradient of a morphogen initiates oscillations between "A" (anterior) and "P" (posterior) states, gradually generating a periodic pattern that converts temporal A/P oscillations into spatial (segmental) alternation of A and P states. (B) Modification of the model where cells are added sequentially from a growing zone. After each full cycle, a new gene is expressed, specifying the position of the segment along the axis. In all diagrams, posterior is to the right. From Meinhardt (1986), reproduced with kind permission of Dr. Hans Meinhardt and Plenum Press.

gradient. Meinhardt likened the process to a grandfather clock: the gradient is equivalent to the height to which the weight is initially raised, and this drives a pendulum oscillating between left and right (A/P). At each cycle, a hand of the clock advances by one unit. This advance would be equivalent to the acquisition by a somite of its rostrocaudal identity (somite 1, 2, 3, . . .). He predicted "this model would obtain strong support if the postulated oscillations in the mesoderm before somite formation could be detected. One full cycle of this oscillation should take precisely the same time as that required for the formation of one somite . . ." (Meinhardt, 1986, p. 188). As will be explained later, this prediction has been remarkably well fulfilled by the finding that expression of chick *hairy-1,* a marker for posterior somite tissue, oscillates in the segmental plate with a period equal to that required for the formation of a single somite (90 min; Palmeirim *et al.,* 1997). In fact, the profile of the oscillations depicted by Meinhardt (1986) is strikingly similar to those observed by Palmeirim *et al.* (1997) (compare Figs. 3A and 5).

Like Cooke and Zeeman (1976), Meinhardt also proposes a long-range gradient of a morphogen (Fig. 3A). He does so to accommodate the amphibian data on size regulation (although he does not distinguish explicitly between the two modes of somite formation) as well as to assign different positional values to each somite formed, generating regional identity. Unlike the clock and wavefront model, however, Meinhardt's conditions can generate a periodic pattern without a gradient when cells are added at one end of the field (the "tail bud," or primitive streak; Fig. 3B).

Meinhardt was also concerned with assigning a polarity to each somite, and observed that simple alternation between two (A/P) states is insufficient to determine unambiguous polarity. He therefore proposed a third state, equivalent to the segment border (S), and suggested that a physical boundary arises at the confrontation between P and S states. One reason why such polarity may be required is to explain how, although epithelialization of the posterior half of the forming somite appears to take place at the same time as epithelialization of the anterior tip of the segmental plate, the former cells join the forming somite while the latter remain attached to the segmental plate (see above). The proposal of a third state (in addition to A and P) also receives some support from recent molecular data. For example, the mouse *Delta*-like genes (*Dll1* and *Dll3*), which are required for the normal formation of segment boundaries, are expressed in a subset of somitic cells, smaller than a half-somite, just anterior (*Dll1*) or posterior (*Dll3*) to the somite boundary, even at the time of somite formation (Dunwoodie *et al.,* 1997; see also below). This finding is particularly remarkable because were it not for this expression pattern, one would not have predicted that an epithelial sphere (where each cell reaches both the lumen and the outer surface of the somite) can be patterned into more than two equal portions along one axis.

C. Cell Cycle Model

A single, brief heat shock applied to chick embryos can give rise to anomalies in somite formation repeated at about seven somite intervals along the axis (Primmett

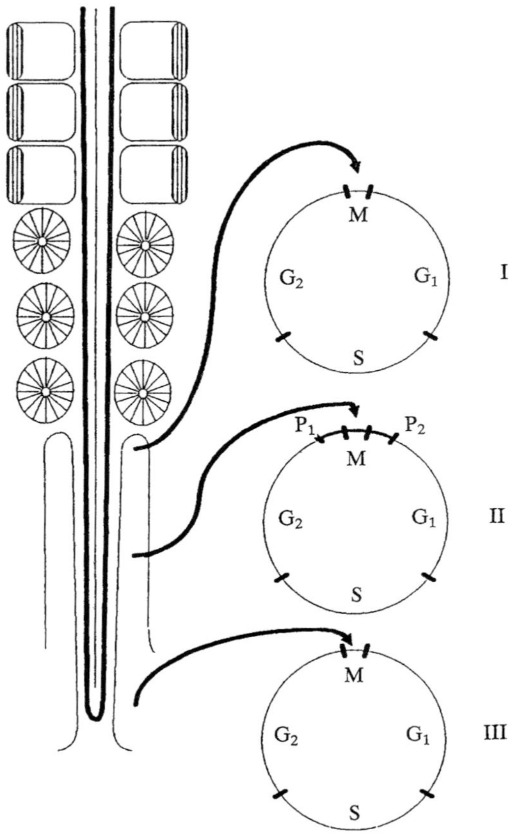

Figure 4 The cell cycle model (Stern *et al.*, 1988; Primmett *et al.*, 1989) in diagram form, correlating stages of the process of somite formation with stages of the cell division cycle. Three cell cycles are shown, labeled I, II, and III. The mitotic division of the first of these occurs at about the time of somite formation, while cells at cycle III will traverse two further full cycles before taking part in segmentation. Cells are proposed to become apportioned to a particular somite during cycle II, which precedes somite formation by one full cycle. This grouping of code into presumptive somites is suggested to occur by the existence of two special points, P_1 and P_2, which might be situated on either side of the M phase of cycle II. The time interval between these two points was proposed to be 90 min, equal to the time taken by one pair of somites to form. "Pioneer" cells arriving at P_2 would signal to other cells situated between P_1 and P_2, but not to those in other positions of the cell cycle, grouping those cells situated within this time window. From Primmett *et al.* (1989), reproduced with kind permission of the Company of Biologists Ltd.

et al., 1988). To explain this, Stern *et al.* (1988) and Primmett *et al.* (1989) proposed that cells destined to segment together are synchronous in their passage through the cell cycle (Fig. 4). Somite progenitor cells in the anterior primitive streak were suggested to have self-renewing ("stem cell") properties (Selleck and Stern, 1991, 1992). At each division of one of these progenitor cells, one daughter remains in the node/streak (self-renewing) while the other daughter enters the caudal segmental plate. An asynchronously dividing population of somite precursor cells should generate a continuous stream of cells, which enter the segmental plate at a constant position in the cell cycle (presumably G_1). The model predicts that the cell cycle duration in somitogenic cells should be about 10 hr (seven somites times ~100 min, the time taken for a single somite to form), and this has been confirmed experimentally (Primmett *et al.,* 1989). Consistent with this suggestion, accumulations of mitotic cells have been found at about seven prospective somite intervals in the segmental plate (Stern and Bellairs, 1984; Primmett *et al.,* 1989).

The cell cycle model proposes a gating mechanism to partition the continuous stream of cells into individual somites, and that the gate is coupled to the cell cycle. One such mechanism invokes the existence of two special points, P_1 and P_2, in a specific cell cycle (perhaps one full cell cycle before segmentation; Fig. 4). These two points should be 100 min apart. As cells pass P_2, they emit a local signal to which any cell between P_1 and P_2 can respond by activating gene(s) encoding cell adhesion molecules. The protein(s) they encode are inserted into the membrane one cell cycle later, resulting in adhesion ("segmentation") of those cells that passed synchronously through the P_1–P_2 gate. Cells just before P_1 in the critical cell cycle must wait until they reach P_2 (100 min later) to become part of a synchronous group of presomitic cells, and give rise to the next intersomitic border (Stern *et al.,* 1988; Primmett *et al.,* 1989).

Although this mechanism fits much of the experimental data available for amniotes, it also does not explain why about half of the cells adjacent to the forming border become part of the forming somite while the other half adhere to each other at the cranial tip of the segmental plate. To explain this, some additional complexity is required, such as ascribing polarity to each somite, as proposed by Meinhardt (1982, 1986). The discovery (Palmeirim *et al.,* 1997) that *c-hairy-1* expression oscillates between three phases, adding up to a period of about 90 min, could provide a starting point to modify the model, by suggesting a gating mechanism that can simultaneously assign polarity to the somite.

IV. A Molecular Clock

A recent paper by Palmeirim *et al.* (1997) reports the cloning of a chick homologue of *Drosophila hairy, c-hairy-1,* expressed in the posterior half of the somites. However, within the segmental plate, the appearance of the expression domain is variable even in embryos with the same number of somites (but each embryo is

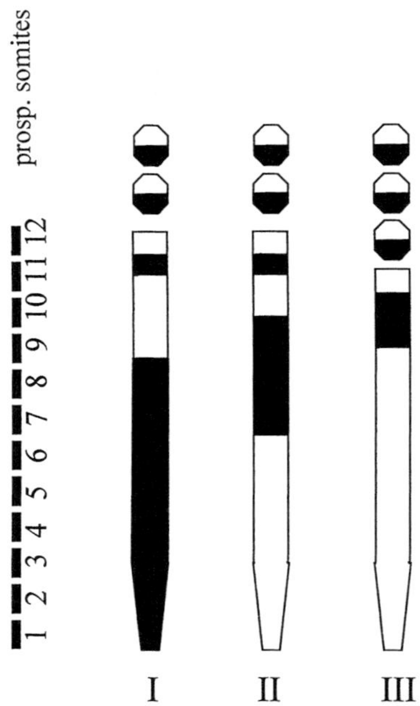

Figure 5 The three phases (I, II, III) of expression of *hairy-1* in the chick segmental plate. Two newly formed somites are also shown, with expression of *c-hairy-1* retained in the caudal half. On the left, the approximate position of each of the 12 or so prospective somites ("somitomeres") is also shown.

always perfectly bilaterally symmetrical). The patterns seen in the segmental plate could be classified into three types (Fig. 5). In a series of beautifully designed experiments, Palmeirim *et al.* (1997) were able to conclude that these three types of patterns follow one another in time, making up a 90-min cycle (one somite's worth). It therefore seems that presomitic cells oscillate between "A" (*c-hairy-1* "off") and "P" (*c-hairy-1* "on") 10 or 11 times before segmenting, and that segmentation is accompanied by the stable acquisition of *c-hairy-1* expression by half of the somitic cells (P-half) and no expression by the other (A-half). These oscillations in *c-hairy-1* expression are therefore strikingly reminiscent of the oscillations proposed by Meinhardt (see above).

V. Two Interpretations of the *c-hairy-1* Clock

Two alternative interpretations of the meaning of this expression pattern can be postulated. One view, apparently favored by Palmeirim *et al.* (1997), is that the three types of patterns represent snapshots of what is in reality a smooth, continuous wave of *c-hairy-1* expression. A second possibility is that these are three discrete patterns that follow each other more abruptly. The implications of each of these interpretations to the models presented above will now be discussed briefly.

A. A Smooth Wave of *c-hairy-1* Expression

Although *c-hairy-1* transcripts appear to oscillate between three states in the whole of the segmental plate of the chick, with a regular period of 90 min or so, Palmeirim *et al.* (1997) are careful to point out that it is possible to generate a cyclic alternation between the three phases (numbered I, II, and III; see Fig. 5) in a smooth way, by postulating that a wave of expression spreads from posterior to anterior regions of the segmental plate and slows down gradually (Fig. 6; see also http://www.cell.com/supplemental/91/5/639/ for an animated simulation of this wave prepared by Dr. J. H. Lewis). This view predicts that cells transiting along the segmental plate would experience oscillations of *c-hairy-1* expression that gradually slow down (Fig. 6; see also Figure 8 of Palmeirim *et al.*, 1997). It is interesting

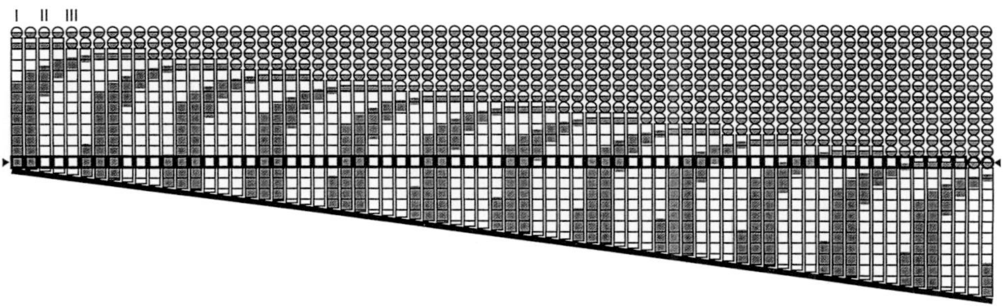

Figure 6 Schematic diagram showing *c-hairy-1* expression pattern in the segmental plate over an 18-hr period (from the time a cell enters the segmental plate to the time it becomes incorporated in the newly formed somite), as a wave progressing from caudal to rostral in the segmental plate. During the 18-hr period, a total of 12 somites is produced. Each column represents the mesoderm of the same embryo at 15-min intervals. Somites are shown as circles, and the segmental plate is subdivided into squares, each representing a prospective somite. Somites form every 90 min (six columns) from the anterior (top) end of the segmental plate. At the same time, cells leave the primitive streak (shown as a thick line along the bottom of the figure) and are continuously added to the caudal (bottom) end of the segmental plate. *c-hairy-1* is expressed (gray shading) in the posterior halves of the somites and as a wave that progresses anteriorly through the segmental plate. One row of squares is highlighted in bold; following this row from left to right, the history of this particular prospective somite can be followed. Cells in the upper part of this territory undergo 10 pulses of *c-hairy-1* expression and eventually become part of the anterior part of the somite, whereas cells in the lower part undergo 11 pulses of expression and the last pulse is maintained as cells become incorporated in the posterior half of the same somite. Note that cells enter the segmental plane at the correct phase of the *c-hairy-1* expression cycle. The Roman numerals indicate the three types of patterns used by Palmeirim *et al.* (1997) to classify different stages of *c-hairy-1* expression. From Stern and Vasiliauskas (1998), reproduced with kind permission of Wiley.

to reconsider the three models in the light of this interpretation of *c-hairy-1* expression. All three incorporate a wavelike event, but in both the clock and wavefront and Meinhardt's models, the wave progresses from anterior to posterior regions, while in the cell cycle model the wave (corresponding to successive stages in the cell cycle) passes from posterior to anterior, as does *c-hairy-1* in this representation.

B. Discrete Phases of *c-hairy-1* Expression: A Morse Code of Gene Expression?

If the three phases alternate abruptly, the oscillations in *c-hairy-1* expression experienced by cells are not as regular throughout the cells' sojourn in the segmental plate. As cells progress along the segmental plate toward its cranial end, *c-hairy-1* expression might spell out their position as a "Morse code" (Fig. 7). The entire history of a cell in the segmental plate could be represented somewhat like this:

□□□□□ (rostral)
●□□●□□●□□●□□●□□●□□●●□●●□□●●□□●●●□□ ↕segmentation
●●●●●● (caudal)

where time advances from left to right starting from the caudal end of the segmental plate, and each symbol represents 30 min: "●" is a 30 min pulse of *c-hairy-1*, and "□" is a 30 min period with no expression.

This notation suggests that the transit of a cell through the segmental plate is accompanied by regular oscillations of *c-hairy-1* expression lasting about 9 hr (the first 18 characters in the representation above). After this a more irregular pattern may start, containing longer pulses of *c-hairy-1* expression, and this continues for about 7 hr. Finally, the expression stabilizes: cells destined to contribute to the caudal half of the somite regain expression and maintain it thereafter, while those destined for the rostral half no longer express *c-hairy-1*. Based on the cell cycle measurements of Primmett *et al.* (1989), the initial 9 hr period should correspond to the entire first cell cycle in the segmental plate, the next 7 hr to the first portion (G_1 + S) of the second cell cycle, and the last 2 hr to the last portion of the second cycle (G_2 + M) (Fig. 6). The irregularity of the 7-hr period at the start of the second cell cycle could be used by synchronously cycling cells to give additional resolution to their assessment of their position within the segmental plate, and perhaps contribute to the decision of whether they belong to the posterior half of one somite or to the anterior half of the next. The stable expression of *c-hairy-1* that accompanies segmentation could reinforce such a decision, as cells that maintain their expression of this gene will always form part of the posterior half.

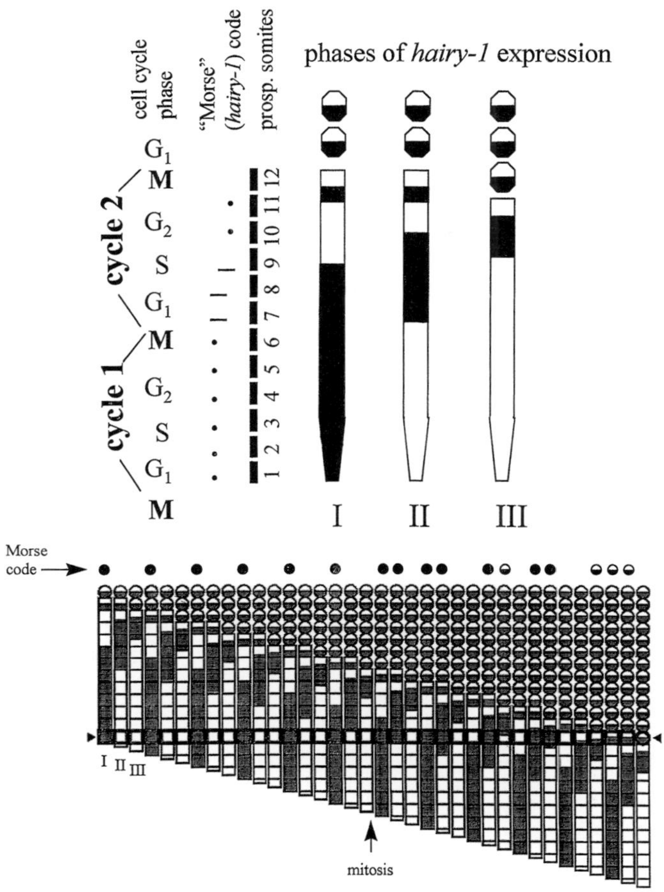

Figure 7 An alternative interpretation of the changing patterns of *c-hairy-1* expression in the chick, based on data by Palmeirim *et al.* (1997). In the upper part of the figure, the three types of patterns of expression are aligned with the prospective somites ("somitomeres") (see Fig. 5). To the left of this, a "Morse code" representation of the *c-hairy-1* expression for each somitomere is shown, where a dot represents brief (30 min) expression, a dash represents longer (60 min) expression, and a space represents no expression; every prospective somite represents 90 min. The approximate position of each phase of the cell cycle progression of cells in the segmental plate corresponding to the 12 somitomeres are shown to the far left, based on results of Primmett *et al.* (1989). In the lower part of the figure, this interpretation is depicted using the scheme introduced in Fig. 6 (except that each column represents 30 min, rather than 15, and that in this representation the three phases of expression follow each other abruptly), showing how the "Morse code" arises. Below this, the position of the "critical" mitosis proposed by the cell cycle model is indicated.

This Morse code is based on a rough estimate of the regions of segmental plate that express *c-hairy-1,* without direct reference to the position of expressing cells in the cell cycle or to the spread of any one clone of cells in the segmental plate.

It is therefore premature to propose a more rigorous model for the relationship between the gating mechanism and the cell cycle and/or *c-hairy-1* expression, or for the establishment of somite polarity. However, such a connection would be more likely if other genes were also found to change their expression in a similar, but not identical way to *c-hairy-1*. Recent data are starting to suggest that several such genes may indeed exist (A. Aulehla and R. Johnson, personal communication 1998; O. Pourquié, personal communication 1998).

VI. The Molecular Basis of Boundary Formation

The expression of *c-hairy-1,* a transcription factor, may indeed be involved in specifying which cells will segment together, or perhaps in ascribing polarity to future somites. However, it cannot be the effector of boundary formation. Other recent results strongly implicate the Notch/Delta Serrate/RBP-Jϰ pathway as direct effectors in this event. A *Delta* homolog, *C-Delta-1/Dll1,* is expressed in a subset of cells in the caudal half of somites (including newly formed ones), closest to the somite boundary. Before segmentation, it seems to be expressed throughout the presomitic mesoderm (albeit with regional differences in amount of transcripts; Dunwoodie *et al.,* 1997) up to almost the cranial end of the segmental plate, where it becomes excluded from the anterior part of the next somite to form (Henrique *et al.,* 1995; Myat *et al.,* 1996; Dunwoodie *et al.,* 1997). Another *Delta* homolog, *Dll3,* is also expressed throughout the paraxial mesoderm. As the somite forms, it is downregulated from the posterior part and remains expressed weakly at its anterior border; one somite later, transcripts are no longer detectable in either half (Dunwoodie *et al.,* 1997). The expression patterns of these two genes do not correspond precisely to half-somites, but rather demarcate a subset in each half, adjacent to the nearest somitic border. This suggests that rather than being involved in the specification of rostral and caudal identity, *Delta*-related genes may play a role in the formation and/or maintenance of segmental boundaries. This conclusion is strongly supported by the phenotype of mice lacking *Dll1* (Hrabě de Angelis *et al.,* 1997), its probable ligand *Notch-1* (Swiatek *et al.,* 1994; Conlon *et al.,* 1995), or *RBP-Jκ,* a component of the Notch signaling pathway (Oka *et al.,* 1995), all of which fail to form normal intersomitic boundaries.

However, it was recently reported that the mouse mutation *pudgy,* which disrupts the normal alternation of rostral and caudal sclerotome (Johnson, 1986), is due to loss of function of the *Dll3* gene (Kusumi *et al.,* 1998), suggesting that the molecular mechanisms that establish somite boundaries and those that ascribe polarity to the somite may be linked. An interesting additional question is whether Notch/Delta signaling might also control the boundary within the somite (separating anterior and posterior halves; "Von Ebner's fissure," Stern and Keynes, 1987). Strikingly, given their different mode of somite formation, *Xenopus* embryos also appear to use Notch/Delta interactions to establish somite borders: The *Xenopus* *X-Delta-2* gene is expressed in presomitic mesoderm in a series of stripes that correlate with the future position of segmental boundaries, and injection of antimorphic *X-Delta-2* or *X-Su(H)* suppresses segmentation but not the differentiation of somite derivatives (Jen *et al.,* 1997). This treatment also disrupts the banded expression of the *hairy* homolog *X-Hairy-2A,* suggesting that *hairy* is under the control of the Notch/Delta signaling pathway. In addition, members of the ephrin receptor/Eph ligand families are also expressed concomitantly with somite and boundary formation (see below).

Why are so many genes encoding cell surface molecules apparently involved in controlling somite and boundary formation? One possibility relates to the observations discussed above, that somite formation involves epithelialization of cells both within the forming somite and at the cranial tip of the segmental plate. A possible mechanism to subdivide the population of epithelializing cells is "differential affinity" (a concept first introduced by Townes and Holtfreter, 1955). Concomitant with the formation of a new border, cells just behind it (the cranial tip of the segmental plate, also future anterior half-somite cells) and those just in front of it (the forming somite, including future posterior half-somite cells joining cells previously at the cranial tip of the segmental plate) may differ in their adhesive properties. This difference could be encoded by overlapping expressions of different ligand–receptor pairs (e.g., Notch–Delta, ephrins–Eph receptors, etc.). Recent data reviewed above are starting to suggest the existence of two or more molecular clocks with different periods [*c-hairy-1* and *-2, Lunatic Fringe* (*LFng*)], and it is possible that each clock controls the expression of a specific set of cell surface adhesion molecules.

VII. The Molecular Basis of Boundary Maintenance

It is obvious that somite formation consists of the generation of boundaries that delimit discrete groups of cells in one somite from those of the next and previous somites. Once a somite has formed, it consists of an epithelial structure bound together both by strong intercellular junctions and also by a surrounding basement membrane (Bellairs, 1979). But some hours (about six somites' worth, or 10 hr) following somite formation, the cells in the ventromedial portion of the somite (sclerotome) lose their epithelial characteristics to become mesenchymal once again, at the same time as the basement membrane dissolves (Bellairs, 1979). Some mechanism must exist to maintain the boundaries in such a mesenchymal cell population. Stern and Keynes (1987) found that transplantations that result in juxtaposition of "like" sclerotome halves (e.g., anterior next to anterior) lead to loss of the boundary between them and extensive cell mixing, while juxtaposition of "unlike" halves generates a boundary between them. This result indicates that cells in the anterior and posterior halves of the sclerotome differ by some cell surface property that allows them to recognize each other. "Unlike" cells might repel each other by some mechanism, or alternatively, cells may have a greater affinity for "like" cells than for those of the opposite kind. Repulsion is probably the more likely mechanism, since caudal half-sclerotome cells repel the motor axons and neural crest cells (Keynes and Stern, 1984, 1988; Rickmann *et al.,* 1985; Stern *et al.,* 1986; Davies *et al.,* 1990).

Recent results have suggested a molecular mechanism that could underlie both the repulsion between "unlike" half-sclerotome cells and the repulsion of neural crest and motor axons by the posterior half-sclerotome. Krull *et al.* (1997) found that the tyrosine kinase receptor EphB3 is confined to the chick anterior half-sclerotome and to the neural crest cells that migrate through it, while one of its ligands, ephrin-B1, is expressed only in the caudal half-sclerotome. This raises the possibility that interaction of ephrin-B1 with its receptor EphB3 causes collapse of cell processes required for locomotion in the EphB3 receptor-expressing cells. In support of a role of this interaction in regulating neural crest migration, Krull *et al.* (1997) found that the addition of soluble ephrin-B1 caused migratory paralysis of neural crest cells. Similar findings were made by Wang and Anderson (1997) in the rat, but here the anterior half-sclerotome and neural crest cells express the related receptor EphB2 (which also binds to ephrin-B1). Therefore the interaction of the ephrin-B1 ligand and its tyrosine kinase receptors could underlie the repulsion of neural crest cells by the posterior half-sclerotome. In addition, this mechanism is also likely to account for the finding that anterior and posterior half-sclerotome cells fail to mix, but this has not yet been investigated directly.

It is interesting to note that a similar mechanism may account for the maintenance of boundaries in another segmented system, the rhombomeres of the hindbrain. Here, cells from "unlike" (odd- and even-numbered) rhombomeres also fail to mix (Guthrie and Lumsden, 1991), as in the sclerotome (Stern and Keynes 1986, 1987). In addition, neural crest cells derived from specific rhombomeres tend to populate specific branchial arches, generating peripheral neural segmentation in a manner analogous to the selective migration of neural crest cells and motor axons through the anterior half of the sclerotome. Although no appropriate ligand–receptor pair has yet been identified that could explain the failure of odd/even rhombomere combinations to mix, most of the members of this family are expressed in alternating rhombomeres in the hindbrain (e.g., EphA4 in rhombomeres three and five, ephrin-B2 in rhombomeres two, four, and six; see Nieto *et al.,* 1992; Becker *et al.,* 1994; Bergemann *et al.,* 1995; Smith *et al.,* 1997). When a truncated EphA4 receptor is misexpressed in zebrafish embryos, the normal alternation of odd/even rhombomere-specific gene expression is lost for several marker genes (Xu *et al.,* 1995), strongly suggesting that the maintenance of boundaries between rhombomeres requires ephrin ligand–Eph receptor interactions.

Interestingly, EphA4 is expressed in alternating stripes both in the hindbrain and at the anterior tip of the segmental plate and in the forming somite (Nieto *et al.,* 1992), hinting that a common mechanism may maintain and/or generate boundaries in segmented systems as different as the hindbrain and paraxial mesoderm. It is therefore surprising that with the exception of EphA4, none of the genes thus far implicated in the maintenance of boundaries in either the hindbrain or the sclerotome appear to be expressed in the segmental plate mesenchyme, raising the possibility that the mechanisms that initially establish the boundaries between adjacent somites may be different to those that maintain them. In addition, there is as yet no obvious molecular link between those genes (such as those belonging to the Notch/Delta/Serrate/RBP-Jκ pathway, *hairy-1*) that appear to be involved in boundary formation during initial epithelialization of the somite and those (such as the ephrin ligands and their Eph tyrosine kinase receptors) that might play a role in maintaining these boundaries later in development. However, it might be expected that the former set of genes should control the expression pattern of the latter group.

VIII. Determination of Somite Identity

An almost universal feature of metameric organization is that individual metameres become different from one another by acquiring functional specializations. For example, in *Drosophila,* each of the three thoracic segments is characterized by the appendages it carries: the most caudal segment (metathorax) develops a pair of halteres, the middle segment (mesothorax) develops a pair of wings, and the most rostral segment (prothorax) does not carry dorsal appendages. In vertebrates, each vertebra is morphologically recognizable. There is abundant evidence that the Hox gene clusters are responsible for regional specification in the mesoderm as well as in the central nervous system (Kessel and Gruss, 1991; Keynes and Krumlauf, 1994; Krumlauf, 1994; Tam and Trainor, 1994; Burke *et al.,* 1995), but how is their expression controlled? The first two of the three models summarized above incorporate a counter that could be used to measure the position of a cell in the rostrocaudal axis of the animal and thus control position-specific gene expression. Meinhardt suggested that somitogenic cells might count the number of oscillations between the A and P state, such that at each full cycle the next set of genes (e.g., more 5′ genes in each Hox cluster) is activated (Meinhardt, 1986), and a similar function could easily be attached to the oscillator of the clock and wavefront model (Cooke and Zeeman, 1976). In the cell cycle model, the time spent by progenitor cells in the node or primitive streak could be used as a measure of position along the axis (see Gaunt and Strachan, 1996). The note and anterior primitive streak are sites of retinoic acid synthesis (Hogan *et al.,* 1992; Chen *et al.,* 1992; Wagner *et al.,* 1990, 1992) and both concentration and time of exposure to this substance can control the expression of specific Hox genes (Boncinelli *et al.,* 1991; Kessel and Gruss, 1991; Lufkin, 1997). This provides a simple mechanism for translating temporal into spatial information.

The time spent in a growth zone such as the primitive streak is a viable mechanism for determining axial identity for those animals that possess such a growth zone (amniotes), but it is difficult to envisage how it might apply to the lower vertebrates. One possibility is that time is measured with respect to the moment that cell blocks start to turn to form epaxial muscle blocks (Fig. 1). However, by this time some other mechanism must have established the groups of cells that will turn together, and therefore some clocklike mechanism must operate before physical turning of the cells takes place. Meinhardt (1982, 1986) has provided the only attempt to date to account for the differences in regional specification in the two groups of animals (see Fig. 3).

It is clear, however, that the oscillations of *c-hairy-1* expression cannot be the mechanism by which cells estimate their position along the axis, because the number of full cycles of expression is the same for all somitogenic cells, regardless of the axial level they will come to occupy. Therefore, if a counter is involved in axial determination, as proposed by Meinhardt, this is unlikely to be mediated directly by *c-hairy-1* in amniotes. An interesting challenge for the next few years will be to compare the spatial and temporal properties of expression of different genes in different vertebrates, with reference to the differences as well as the similarities in their modes of development.

Acknowledgments

Our current research on somite development is supported by NIH Grant RO1-HD31942. We are grateful to Charlie Ordahl, Olivier Pourquié, and Andrea Streit for critical comments on the manuscript, and Hans Meinhardt for interesting discussions.

References

Becker, N., Seitanidou, T., Murphy, P., Mattei, M. G., Topilko, P., Nieto, M. A., Wilkinson, D. G., Charnay, P., and Gilardi-Hebenstreit, P. (1994). Several receptor tyrosine kinase genes of the Eph family are segmentally expressed in the developing hindbrain. *Mech. Dev.* **47,** 3–17.

Bellairs, R. (1979). The mechanism of somite segmentation in the chick embryo. *J. Embryol. Exp. Morphol.* **51,** 227–243.

Bergemann, A. D., Cheng, H. J., Brambilla, R., Klein, R., and Flanagan, J. G. (1995). ELF-2, a new member of the Eph-ligand family, is segmentally expressed in the region of the hindbrain and newly-formed somites. *Mol. Cell. Biol.* **15,** 4921–4929.

Biben, C., Stanley, E., Fabri, L., Kotecha, S., Rhinn, M., Drinkwater, C., Lah, M., Wang, C. C., Nash, A., Hilton, D., Ang, S. L., Mohun, T., and Harvey, R. P. (1998). Murine cerberus homologue mCer-1: A candidate anterior patterning molecule. *Dev. Biol.* **194,** 135–151.

Boncinelli, E., Simeone, A., Acampora, D., and Mavilio, F. (1991). Hox gene activation by retinoic acid. *Trends Genet.* **7,** 329–334.

Burke, A. C., Nelson, C. E., Morgan, B. A., and Tabin, C. (1995). Hox genes and the evolution of vertebrate axial morphology. *Development* **121,** 333–346.

Chen, Y., Huang, L., Russo, A. F., and Solursh, M. (1992). Retinoic acid is enriched in Hensen's node and is developmentally regulated in the early chicken embryo. *Proc. Natl. Acad. Sci. U.S.A.* **89,** 10056–10059.

Conlon, R. A., Reaume, A. G., and Rossant, J. (1995). *Notch 1* is required for the coordinate segmentation of somites. *Development* **121,** 1533–1545.

Cooke, J. (1975). Control of somite number during morphogenesis of a vertebrate, *Xenopus laevis*. *Nature* **254,** 196–199.

Cooke, J., and Zeeman, E. C. (1976). A clock and wavefront model for control of the number of repeated structures during animal morphogenesis. *J. Theor. Biol.* **58,** 455–476.

Davies, J. A., Cook, G. M. W., Stern, C. D., and Keynes, R. J. (1990). Isolation from chick somites of a glycoprotein fraction that causes collapse of dorsal root ganglion growth cones. *Neuron* **2,** 11–20.

De Robertis, E. M. (1997). The ancestry of segmentation. *Nature* **387,** 25–26.

Dunwoodie, S. L., Henrique, D., Harrison, S. M., and Beddington, R. S. P. (1997). Mouse *Dll3:* A novel divergent *Delta* gene which may complement the function of other *Delta* homologues during early pattern formation in the mouse embryo. *Development* **124,** 3065–3076.

García-Bellido, A., Ripoll, P., and Morata, G. (1973). Developmental compartmentalisation of the wing disk of *Drosophila. Nature New Biol.* **245,** 251–253.

Gaunt, S. J., and Strachan, L. (1996). Temporal colinearity in expression of anterior Hox genes in developing chick embryos. *Dev. Dyn.* **207,** 270–280.

Gregg, B. C., and Snow, M. H. (1983). Axial abnormalities following disturbed growth in mitomycin C-treated mouse embryos. *J. Embryol. Exp. Morphol.* **73,** 135–149.

Guthrie, S., and Lumsden, A. (1991). Formation and regeneration of rhombomere boundaries in the developing chick hindbrain. *Development* **112,** 221–229.

Henrique, D., Adam, J., Myat, A., Chitnis, A., Lewis, J., and Ish-Horowicz, D. (1995). Expression of a *Delta* homologue in prospective neurons in the chick. *Nature* **375,** 787–790.

Hogan, B. L., Thaller, C., and Eichele, G. (1992). Evidence that Hensen's node is a site of retinoic acid synthesis. *Nature* **359,** 237–241.

Hrabě de Angelis, M., McIntyre, J., and Gossler, A. (1997). Maintenance of somite borders in mice requires the *Delta* homologue *Dll1. Nature* **386,** 717–721.

Jen, W. C., Wettstein, D., Turner, D., Chitnis, A., and Kintner, C. (1997). The *Notch* ligand, *X-Delta-2,* mediates segmentation of the paraxial mesoderm in *Xenopus* embryos. *Development* **124,** 1169–1178.

Johnson, D. R. (1986). "The Genetics of the Skeleton: Animal Models of Skeletal Development." Oxford Univ. Press, Oxford.

Kessel, M., and Gruss, P. (1991). Homeotic transformations of murine vertebrae and concomitant alteration of Hox codes induced by retinoic acid. *Cell* **67,** 89–104.

Keynes, R. J., and Krumlauf, R. (1994). Hox genes and regionalization of the nervous system. *Annu. Rev. Neurosci.* **17,** 109–132.

Keynes, R. J., and Stern, C. D. (1984). Segmentation in the vertebrate nervous system. *Nature* **310,** 786–789.

Keynes, R. J., and Stern, C. D. (1988). Mechanisms of vertebrate segmentation. *Development* **103,** 413–429.

Krull, C. E., Lansford, R., Gale, N. W., Collazo, A., Marcelle, C., Yancopoulos, G. D., Fraser, S. E., and Bronner-Fraser, M. (1997). Interactions of Eph-related receptors and ligands confer rostrocaudal pattern to trunk neural crest migration. *Curr. Biol.* **7,** 571–580.

Krumlauf, R. (1994). Hox genes in vertebrate development. *Cell* **78,** 191–201.

Kusumi, K., Sun, E. S., Kerrebrock, A. W., Bronson, R. T., Chi, D. C., Bulotsky, M. S., Spencer, J. B., Birren, B. W., Frankel, W. N., and Lander, E. S. (1998). The mouse pudgy mutation disrupts Delta homologue Dll3 and initiation of early somite boundaries. *Nat. Genet.* **19,** 274–278.

Lawrence, P. A. (1988). The present status of the parasegment. *Development* **104**(Suppl.), 61–65.

Lufkin, T. (1997). Transcriptional regulation of vertebrate Hox genes during embryogenesis. *Crit. Rev. Eukaryotic Gene Expression* **7,** 195–213.

Martínez-Arias, A., and Lawrence, P. A. (1985). Parasegments and compartments in the insect embryo. *Nature* **313,** 639–642.

Meinhardt, H. (1982). "Models of Biological Pattern Formation." Academic Press, London.

Meinhardt, H. (1986). Models of segmentation. *In* "Somites in Developing Embryos" (R. Bellairs, D. A. Ede, and J. W. Lash, Eds.), NATO ASI Series 118, pp. 179–189. Plenum, New York.

Morata, G., and Lawrence, P. A. (1975). Control of compartment development by the engrailed gene in *Drosophila. Nature* **255,** 614–617.

Myat, A., Henrique, D., Ish-Horowicz, D., and Lewis, J. (1996). A chick homolog of *Serrate* and its relationship to *Notch* and *Delta* homologs during central neurogenesis. *Dev. Biol.* **174,** 233–247.

Nieto, M. A., Gilardi-Hebenstreit, P., Charnay, P., and Wilkinson, D. G. (1992). A receptor protein tyrosine kinase implicated in the segmental patterning of the hindbrain and mesoderm. *Development* **116,** 1137–1150.

Oka, C., Nakano, T., Wakeham, A., De la Pompa, J. L., Mori, C., Sakai, T., Okazaki, S., Kawaichi, M., Shiota, K., Mak, T. W., and Honjo, T. (1995). Disruption of the *RBP-Jκ* gene results in early embryonic death. *Development* **121,** 3291–3301.

Palmeirim, I., Henrique, D., Ish-Horowicz, D., and Pourquié, O. (1997). Avian *hairy* gene expression identifies a molecular clock linked to vertebrate segmentation and somitogenesis. *Cell* **91,** 639–648.

Primmett, D. R. N., Stern, C. D., and Keynes, R. J. (1988). Heat-shock causes repeated segmental anomalies in the chick embryo. *Development* **104,** 331–339.

Primmett, D. R. N., Norris, W. E., Carlson, G. J., Keynes, R. J., and Stern, C. D. (1989). Periodic segmental anomalies induced by heat-shock in the chick embryo are associated with the cell cycle. *Development* **105,** 119–130.

Psychoyos, D., and Stern, C. D. (1996). Fates and migratory routes of primitive streak cells in the chick embryo. *Development* **122,** 1523–1534.

Remak, R. (1855). "Untersuchungen über die Entwicklung der Wirbelthiere." Reimer, Berlin.

Rickmann, M., Fawcett, J. W., and Keynes, R. J. (1985). The migration of neural crest cells and the growth of motor axons through the rostral half of the chick somite. *J. Embryol. Exp. Morphol.* **90,** 437–455.

Sander, K. (1988). Studies in insect segmentation: From teratology to phenogenetics. *Development* **104**(Suppl.), 111–121.

Selleck, M. A. J., and Stern, C. D. (1991). Fate mapping and cell lineage analysis of Hansen's node in the chick embryo. *Development* **112,** 615–626.

Selleck, M. A. J., and Stern, C. D. (1992). Evidence for stem cells in the mesoderm of Hensen's node and their role in embryonic pattern formation. *In* "Formation and Differentiation of Early Embryonic Mesoderm" (J. W. Lash, R. Bellairs, and E. J. Sanders, Eds.), pp. 23–31. Plenum, New York.

Smith, A., Robinson, V., Patel, K., and Wilkinson, D. G. (1997). The EphA4 and EphB1 receptor tyrosine kinases and ephrin-B2 ligand regulate targeted migration of branchial neural crest cells. *Curr. Biol.* **7,** 561–570.

Stern, C. D. (1990). Two distinct mechanisms for segmentation? *Semin. Dev. Biol.* **1,** 109–116.

Stern, C. D., and Bellairs, R. (1984). Mitotic activity during somite segmentation in the chick embryo. *Anat. Embryol.* **169,** 97–102.

Stern, C. D., and Keynes, R. J. (1986). Cell lineage and the formation and maintenance of half-somites. *In* "Somites in Developing Embryos" (R. Bellairs, D. A. Ede, and J. W. Lash, Eds.), pp. 147–159. Plenum, London.

Stern, C. D., and Keynes, R. J. (1987). Interactions between somite cells: The formation and maintenance of segment boundaries in the chick embryo. *Development* **99,** 261–272.

Stern, C. D., and Vasiliauskas (1998). Clocked gene expression during chick somite formation. *BioEssays* **20,** 528–531.

Stern, C. D., Sisodiya, S. M., and Keynes, R. J. (1986). Interactions between neurites and somite cells: Inhibition and stimulation of nerve growth in the chick embryo. *J. Embryol. Exp. Morphol.* **91,** 209–226.

Stern, C. D., Fraser, S. E., Keynes, R. J., and Primmett, D. R. N. (1988). A cell lineage analysis of segmentation in the chick embryo. *Development* **104**(Suppl.), 231–244.

Swiatek, P. J., Lindsell, C. D., Del Amo, F. F., Weinmaster, G., and Gridley, T. (1994). *Notch 1* is essential for postimplantation development in mice. *Genes Dev.* **8,** 707–719.

Tam, P. P. L. (1986). A study of the pattern of prospective somites in the presomitic mesoderm of mouse embryos. *J. Embryol. Exp. Morphol.* **92,** 269–285.

Tam, P. P. L., and Trainor, P. A. (1994). Specification and segmentation of the paraxial mesoderm. *Anat. Embryol.* **189,** 275–305.

Thom R. (1975). "Structural Stability and Morphogenesis." Benjamin, Reading, U.K.

Townes, P. L., and Holtfreter, J. (1955). Directed movements and selective adhesion of embryonic amphibian cells. *J. Exp. Zool.* **128,** 53–120.

Wanger, M., Thaller, C., Jessell, T. M., and Eichele, G. (1990). Polarizing activity and retinoid synthesis in the floor plate of the neural tube. *Nature* **345,** 819–822.

Wagner, M., Han, B., and Jessell, T. M. (1992). Regional differences in retinoid release from embryonic neural tissue detected by an *in vitro* reporter assay. *Development* **116,** 55–66.

Wang, H. U., and Anderson, D. J. (1997). Eph family transmembrane ligands can mediate repulsive guidance of trunk neural crest migration and motor axon outgrowth. *Neuron* **18,** 383–396.

Xu, Q., Alldus, G., Holder, N., and Wilkinson, D. G. (1995). Expression of truncated Sek-1 receptor tyrosine kinase disrupts the segmental restriction of gene expression in the *Xenopus* and zebrafish hindbrain. *Development* **121,** 4005–4016.

5

Genetic Regulation of Somite Formation

Alan Rawls,[1,*] *Jeanne Wilson-Rawls,*[1] *and Eric N. Olson*[2]
[1]Department of Biology
Arizona State University
Tempe, Arizona 85287

[2]Department of Molecular Biology and Oncology
Hamon Center for Basic Cancer Research
The University of Texas Southwestern Medical Center at Dallas
Dallas, Texas 75235

Segmentation of the paraxial mesoderm into somites requires a strategy distinct from the division of a preexisting field of cells, as seen in the segmentation of the vertebrate hindbrain into rhombomeres and the formation of the body plan of invertebrates. Each new somite forms from the anterior end of the segmental plate; therefore, the conditions for establishing the anterior–posterior boundary must be re-created prior to the formation of the next somite. It has been established that regulation of this process is native to the anterior end of the segmental plate, however, the components of a genetic pathway are poorly understood. A growing library of candidate genes has been generated from hybridization screens and sequence homology searches, which include cell adhesion molecules, cell surface receptors, growth factors, and transcription factors. With the increasing accessibility of gene knockout technology, many of these genes have been tested for their role in regulating somitogenesis. In this chapter, we will review the significant advances in our understanding of segmentation based on these experiments.

* Author to whom correspondance should be addressed.

Current Topics in Developmental Biology, Vol. 47

0070-2153/00 $30.00

I. Introduction

A requisite step in the development of all vertebrate embryos is the establishment of a metameric body plan along the anterior–posterior (A/P) axis. This is evident in the segmental pattern of the axial skeleton (i.e., vertebrae and ribs), striated muscles, vasculature, and peripheral nerves. Segmentation occurs through the division of paraxial mesoderm on either side of the neural tube into pairs of epithelial balls, referred to as somites. Each somite in turn divides into three compartments, dermatome, myotome, and sclerotome, which are the anlagen for the dermis, skeletal muscle, and axial skeleton, respectively. The fidelity of somitogenesis is dependent on a highly coordinated set of changes in cell morphology, proliferation, migration, and gene expression. Understanding how these events are regulated at the gene level is a fundamental question in vertebrate embryo development. Several regulatory genes that control somitogenesis have been identified in the past decade, due in part to the identification of vertebrate homologs of members of the segmentation pathways established in *Drosophila.* Expression analysis of many of these genes predicts that they will play a similar role in vertebrate segmentation. In addition, advances in subtractive hybridization and directed functional screens of early embryo cDNA libraries have uncovered novel families of potential regulatory genes. Genetic manipulation of many of these genes is beginning to shed light on the underlying mechanisms of somitogenesis.

A. Vertebrate Somitogenesis

The process of somitogenesis begins with the recruitment of mesodermal cells from the late primitive streak into the caudal end of the segmental plates, which flank the neural tube and notochord (Tam and Trainor, 1994). As the embryo elongates along the A/P axis, the node and primitive streak move posteriorly, and the presomitic mesoderm of the segmental plate initiates a rostral to caudal segmentation into somite pairs. Cells from the primitive streak continue to migrate into the posterior end of the segmental plate until the point of posterior neuropore closure (Wilson and Beddington, 1996). Further posterior extension of the axis continues due to mesenchymal cells supplied solely from the tail bud. The length of the segmental plate remains relatively constant throughout somitogenesis, suggesting that the rate of somite formation is intimately linked to cell proliferation and cell migration.

The genesis of somites requires an ordered series of morphological events (Fig. 1). The paraxial mesoderm at the anterior end of the segmental plate undergoes an increase in cell number and density associated with a change in the extracellular matrix composition and cell shape, resulting in the formation of a simple epithelial structure (for review, see Keynes and Stern, 1988, and Tam and Trainor, 1994). The most newly formed somites are epithelial balls with a lumen contain-

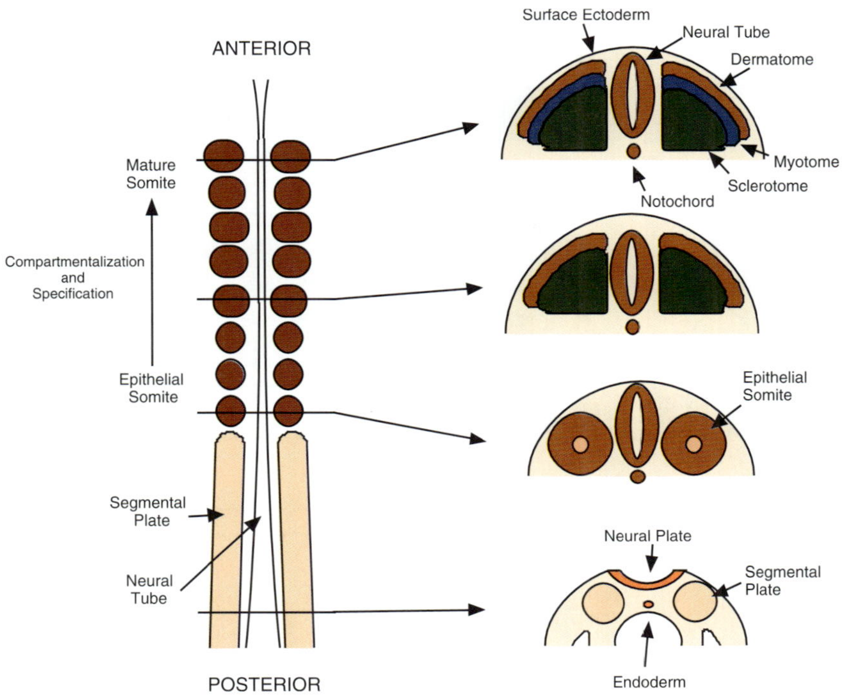

Figure 1 A schematic diagram of the morphological events associated with somitogenesis. Somites originate as an epithelial ball forms from the anterior end of the segmental plate. Somites mature through an increase in compartmentalization and cell lineage specification, resulting in the formation of the sclerotome (green), myotome (blue), and dermotome (orange).

ing mesenchymal cells, called the somiticeol. In response to signals emanating from the surrounding structures, the somite matures into three distinct compartments, sclerotome, myotome, and dermatome. Initially, the medial ventral half of the somite reacquires a mesenchymal phenotype, giving rise to the sclerotome, which is the anlage for the ribs and vertebrae (Wilting *et al.,* 1994). The myotome, which is the origin of most of the skeletal muscle, forms as cells from the dorsomedial region of the dermomyotome delaminate and migrate subjacent to the dermatome (Christ *et al.,* 1986). Finally, the dermatome undergoes an epithelial–mesenchymal transition and these cells migrate under the developing ectoderm to form the dermis.

How is a metameric pattern established from unsegmented tissue? Experimental evidence supports a model in which segmentation is prepatterned in the presomitic mesoderm. Morphologically distinct units, called somitomeres, can be observed by scanning electron microscopy as cellular swirls (Meier, 1979). It remains to be determined whether somitomeres correlate one to one with somites and what is their significance. The intrinsic periodicity of the segmental plate was

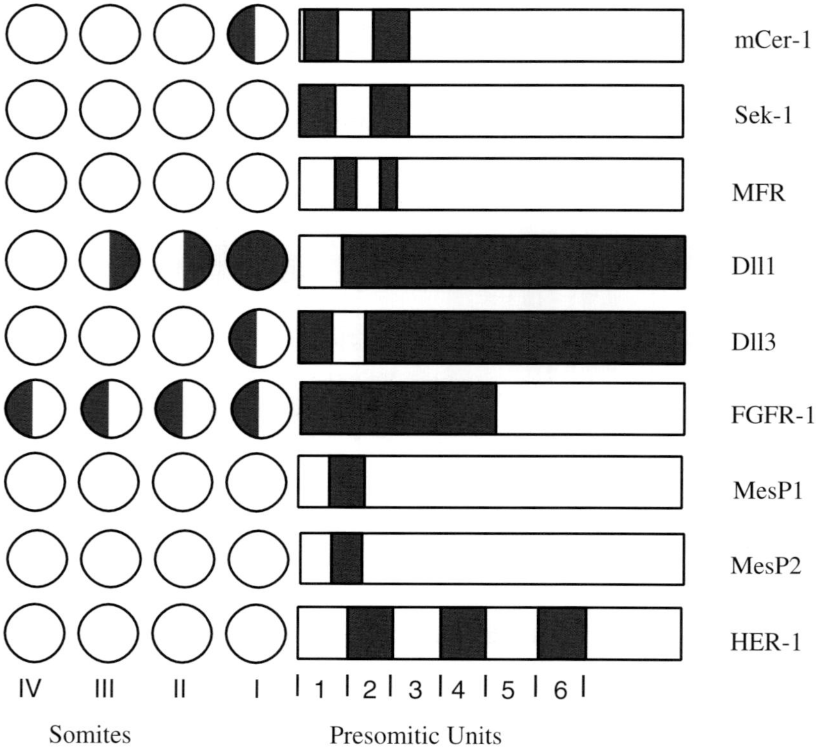

Figure 2 A schematic diagram of differential transcription of genes in the presomitic mesoderm and first four somites. *mCer-1* (Biben *et al.*, 1998), *Sek-1* (Nieto *et al.*, 1992), *MFR* (Johnston *et al.*, 1997), *Dll1* (Bettenhausen *et al.*, 1995; Dunwoodi *et al.*, 1997) and *Dll3*, *FGFR-1* (Fasel *et al.*, 1991), *MesP1* (Saga *et al.*, 1996), *MesP2* (Saga *et al.*, 1997), and *Her-1* (Müller *et al.*, 1996).

demonstrated by heat shock of *Xenopus* embryos (Elsdale *et al.*, 1976). Defects were not observed until the fourth and fifth somite pair formed after the heat shock treatment. This suggests that the mesoderm of the segmental plate possesses polarity and that events important for somitogenesis occur prior to cells migrating to the anterior end of the segmental plate. Segmentation in the presomitic mesoderm has also been described at the level of gene transcription (Fig. 2). As described below, characterization of the expression of these genes is beginning to provide clues as to how this process is regulated.

Segmentation of the paraxial mesoderm into somites requires a strategy distinct from the division of a preexisting field of cells, as seen in the segmentation of the vertebrate hindbrain into rhombomeres (Fraser *et al.*, 1990; Lumsden and Krumlauf, 1996), and the formation of the body plan of invertebrates. Each new somite forms from the anterior end of the segmental plate, therefore, the conditions for establishing the anterior–posterior boundary must be re-created prior to the formation of the next somite. Control of this process appears to be native to the anterior end of the segmental plate, since reorientation of the segmental plate along the A/P axis in the chick results in somites budding in the opposite direction. The list of candidate genes which might play a role in making the anterior end of the segmental plate competent for somitogenesis includes cell adhesion molecules, cell surface receptors, growth factors, and transcription factors (reviewed in Tam and Trainor, 1994; and this chapter). In this chapter, we will review significant advances in our understanding of how somitogenesis is regulated at the gene level.

II. The Role of Basic Helix–Loop–Helix Transcription Factors in Somitogenesis

Members of the basic helix–loop–helix family of transcription factors have been shown to play a role in cell-fate-specific gene expression in several cell types during embryo development (Jan and Jan, 1993). These proteins contain a basic helix–loop–helix motif (bHLH), which consists of a basic domain followed by two amphipathic α-helices, separated by a loop region. The HLH region mediates dimerization, which juxtaposes the basic regions to form a bipartite DNA binding domain that recognizes the consensus sequence CANNTG (E-box). In myogenesis and neurogenesis, a model of regulation has developed in which networks of bHLH factors are required for proper specification and differentiation. During somite formation, three distinct subfamilies of bHLH transcription factors have been identified, which are predicted to have a regulatory role based on gene transcription pattern, (1) paraxis (Burgess *et al.,* 1995), (2) Mesp2 (Saga *et al.,* 1997), and (3) members of the Hairy/enhancer of split family (Müller *et al.,* 1996; Palmeirim *et al.,* 1997).

A. Paraxis

Paraxis belongs to a subfamily of bHLH transcription factors that also includes scleraxis, which is expressed in chondrogenic precursor cells (Burgess *et al.,* 1996; Cserjesi *et al.,* 1995; Olson *et al.,* 1996). Paraxis transcripts are first detected in epiblast cells in a bilateral pattern, posterior to the node, at the onset of gastrulation; these cells are fated to become paraxial mesoderm. Paraxis is also expressed in the segmental plate and throughout newly formed somites. As somites mature, paraxis expression becomes restricted to the dermomyotome (Blanar *et al.,* 1995; Burgess *et al.,* 1995). The importance of paraxis is underscored by the conservation of the gene and its expression pattern in chick (c-paraxis; Sosic *et al.,* 1997; Barnes *et al.,* 1997), humans (bHLH-EC2; Quertermous *et al.,* 1994), hamsters (meso-1; Blanar *et al.,* 1995), and *Xenopus* (E. Olson, unpublished) (Fig. 3).

A targeted null mutation of *paraxis* revealed a deficiency in somite epithelial-

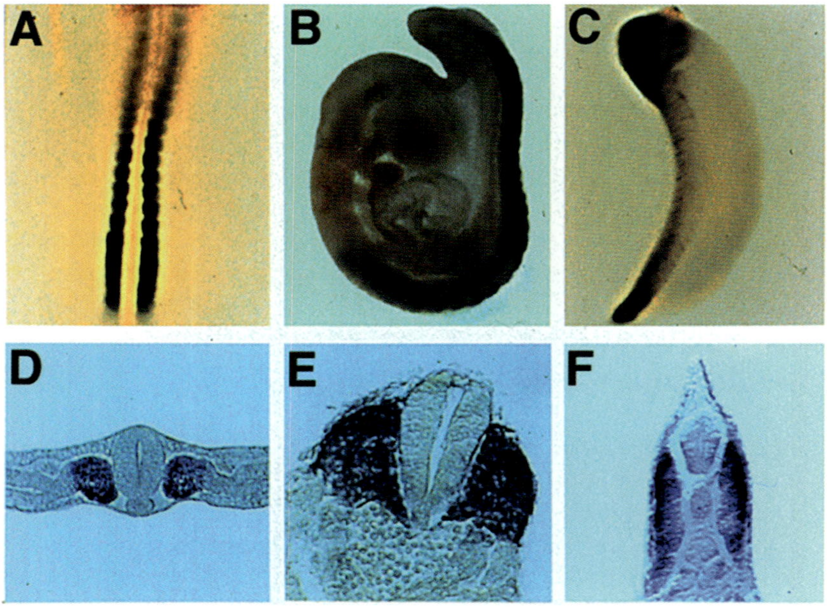

Figure 3 Conservation of paraxis expression in vertebrates. Paraxis mRNA transcripts were measured by whole mount *in situ* at equivalent stages of chicks (A), mice (B), and *Xenopus* (C). In all three embryos, paraxis is expressed in the segmental plate and all somites. Transverse thin sections (D–F) of these embryos, at the level of somite I, demonstrate that expression of paraxis is exclusive to paraxial mesoderm derivatives.

ization (Burgess *et al.,* 1996). In these mutants, the mesenchymal cells of the presomitic mesoderm segment at the same rate as in wild-type littermates, and with the appropriate midline registry, but with no evidence of epithelialization. Because a somite has been traditionally defined as an epithelial ball, it was concluded that somites fail to form in *paraxis*$^{-/-}$ embryos. However, the paraxial mesoderm segmented with normal fidelity, suggesting the existence of an independent regulatory pathway for this process. These data argue against a model of somitogenesis in which epithelialization drives segmentation (Conlon *et al.,* 1995), and are more consistent with a model in which segmentation is prepatterned in the segmental plate.

An analysis of cell-fate-specific markers revealed that paraxis is not required for commitment or differentiation of the axial skeleton, skeletal muscle, or dermis. At 13.5 days post coitus (dpc), *Pax-3,* expression was detectable in the dermatome, but at reduced levels in *paraxis*$^{-/-}$ embryos. The sclerotomal markers, *Pax-1, Pax-9,* and *scleraxis,* and the myotomal markers, *myf-5,* and *myogenin,* were expressed in the mutant at levels similar to wild-type littermates. It is important to note that the pattern of expression of these genes was altered in the absence of *paraxis* (Burgess *et al.,* 1996). This is clearest in the myotome, where the dorsal and ventral aspects are missing. The most obvious interpretation is that the epithelial characteristic of the dermatome is required for proper migration of the subjacent myotomal cells, either through instructive signals, or by acting as a scaffold. The role of the dermatome in the initiation of myotome formation has been well established (Denetclaw *et al.,* 1997). The phenotype of the *paraxis* mutant indicates that the dermatome controls the extent of migration as well.

The most dramatic consequence of the absence of *paraxis* is the radically shortened vertebral column (22 vertebrae). In mutant neonates, the caudal vertebrae are replaced by an unsegmented cartilaginous mass. It is not clear if this is due to a failure to form the complete number of somites, or to an apoptotic response. However, it does suggest that the requirement for *paraxis* differs in the cranial and caudal halves of the embryo. Several other mutations have been reported which display severe caudal defects, including mouse *Wnt3A* (Takada *et al.,* 1994), *MesP2* (Saga *et al.,* 1997), *Presenilin1* (Wong *et al.,* 1997; Shen *et al.,* 1997), and the zebrafish mutants *deadly seven, after eight,* and *white tail* (van Eeden *et al.,* 1996). The cervical and thoracic vertebrae are formed from cells migrating through the primitive streak, whereas the sacral and caudal vertebrae are exclusively derived from the tail bud. These mutations suggest that regulation of tail bud derived paraxial mesoderm is distinguishable from primitive streak-derived paraxial mesoderm.

A second consequence of the lack of epithelialization is morphological defects of the axial skeleton, which include fusion of the vertebrae and ribs, and underossification. Vertebrae normally form through resegmentation, a process whereby the sclerotome from the posterior half of one somite combines with the anterior half of the adjacent somite (Christ *et al.,* 1998). This results in a phase shift between the muscle and vertebrae. The anterior and posterior compartments of the sclerotome are distinguishable by the differential expression of surface antigens, cell compaction, and selective migration of neural crest and peripheral nerves (Keynes and Stern, 1988). Lateral fusion of vertebrae suggests a failure of the somites to form the intrasomitic anterior–posterior boundary. Targeted mutations with similar degrees of caudal truncation and vertebral fusions have been described for *MesP2* (Saga *et al.,* 1997), and *Presenilin1* (Wong *et al.,* 1997; Shen *et al.,* 1997). Further analysis is required to determine if these genes act in a common regulatory pathway determining anterior–posterior somitic boundaries.

The exact role of epithelialization in somitogenesis remains to be determined. The epithelial nature of the somite may function to prevent mixing of cells fated to different lineages. There is evidence that the intrasomitic A/P boundary is defined in the segmental plate prior to the formation of the somite. An inappropriate mixing of cells after they become specified could explain fusion of the vertebrae. Epithelialization may also play an important role in the onset of cell-fate-specific gene expression. A tight correlation exists between the transition from epithelium to mesenchyme, and the formation of the sclerotome and myotome. It is possible that the epithelial state prevents the early onset of sclerotome- and myotome-specific gene expression. Additionally, induction of mesenchyme by adjacent epithelium has been described in many developmental systems. It is possible that instructive signals from the epithelial dermomyotome or dermatome are required for the normal patterning of the myotome and sclerotome.

B. The MesP Family

MesP1 and MesP2 comprise a subfamily of bHLH transcription factors that is expressed early in gastrulation in the posterior mesoderm (Saga *et al.,* 1997). Expression within the paraxial mesoderm is restricted to overlapping bands at the anterior end of the segmental plate and is downregulated in newly formed somites. *MesP2*-deficient mice demonstrated a delay in segmentation of somites posterior to the cervicothoracic region (Saga *et al.*, 1997). However, the sclerotome and dermamyotome formed in the absence of segmentation. Later in development, these somites displayed segmentation in the dermatome, but not sclerotome. An analysis of compartment-specific genetic markers revealed that the myotome-specific gene, *myogenin,* and dermatome-specific gene, *paraxis,* were expressed in a metameric pattern in the mutant. In contrast, the sclerotome marker, *Pax-1,* lacked a segmental pattern of expression. Interestingly, this suggests that the segmentation of the sclerotome can be dissociated from that of the dermatome and myotome.

Disruptions of the sclerotome in *MesP2*$^{-/-}$ embryos result in defects in the patterning of the peripheral nerves and axial skeleton. Sensory nerves appear to inappropriately migrate through the center of the sclerotome of cervical somites and the dorsal root ganglia (DRG) fail to segment. In the thoracolumbar region, the axons failed to migrate ventrally. The axial skeleton of the *MesP2* mutant neonates bears a striking similarity to the *paraxis* mutants. Vertebrae were improperly formed with regions of underossification and lateral fusions, and the vertebral column was truncated at the sacral level. Segmentation of the proximal aspects of the ribs was weakly retained, but barely distinguishable from the ventral aspects of the vertebrae. These defects in the *Mesp2*$^{-/-}$ embryos suggest that the gene is required for establishing polarity and resegmentation of the sclerotome.

Despite phenotypic similarities, *MesP2* and *paraxis* appear to be regulated independently (Saga *et al.,* 1997; A. Rawls, unpublished), suggesting that they both may be required for activation of a downstream regulatory event in A/P polarity. It is more likely that MesP2 functions through the Notch and FGFR signaling pathways. In the *Mesp2*$^{-/-}$ embryos, neither *Notch1* or *FGFR1* are transcribed in the segmental plate or newly formed somites. These results are somewhat surprising due to the fact that expression pattern of *Notch1* and *FGFR1* is much broader and more posterior than *MesP2.* This raises the interesting possibility that the flow of information along the segmental plate is not strictly unidirectional (i.e., posterior to anterior).

C. Vertebrate Homologs of *Drosophila hairy/enhancer of split*

Members of a family of bHLH genes that are grouped by their homology to the *Drosophila hairy* and *enhancer of split* genes, have been implicated in segmentation of paraxial mesoderm. *hairy* belongs to the primary pair-rule class of genes that participate in the regulatory cascade that controls segmentation in seven stripes, alternating with nonexpressing cells during the blastoderm stage of *Drosophila* embryogenesis (Nüsslein-Volhard and Wieschaus, 1980; Pankratz and Jäckle, 1993). Hairy acts as a repressor of transcription when complexed with groucho (Paroush *et al.,* 1994; Fisher *et al.,* 1996). Interestingly, the chick homolog, *c-hairy1,* is expressed in cyclic waves in the segmental plate with a temporal periodicity corresponding to the time required to form one somite (Palmeirim *et al.,* 1997). The *c-hairy1* expression pattern can be divided into two components, a band corresponding to the caudal half of the next prospective somite, and a dynamic band, which initially covers the caudal 70% of the segmental plate and is subsequently restricted in a caudal to rostral direction. This oscillating pattern of expression does not appear to correspond to the migration of cells, nor does it require instructive signals from the surrounding axial structures. Instead, it has been proposed that *c-hairy1* is responsive to a kinematic wave of an uncharacterized transcriptional regulator(s), which is intrinsic to cells of the segmental plate. Such a pattern of expression is consistent with the clock and wavefront model of segmentation proposed by Cooke and Zeeman (1976). Prepatterning the segmental plate in this model is dependent on the interpretation of a kinematic wave of somitogenic cell determination by cells oscillating synchronously between a segmentation-competent and -incompetent state. Overt segmentation will occur only after cells in close proximity have become determined, and oscillate to the on state. This model is based on heat shock experiments in *Xenopus.* The expression pattern of *c-hairy1* represents the first direct evidence for the existence of such a periodic wave. To understand the molecular nature of this wave, it will be important to identify factors which regulate *c-hairy1* expression.

The expression pattern of the zebrafish gene, *her1* (Müller *et al.,* 1996) suggests an additional layer of regulation in segmental prepatterning. *her1* has homology to members of the *Drosophila E(Spl)*-complex. Expression of *her1* is restricted to three paired domains in the segmental plate and one unpaired domain in the tail bud (Müller *et al.,* 1996). Injection of fluoresceine–dextran into individual cells was used to demonstrate that the domains of *her1* expression defined the primordium of every other somite. The intervening regions of nonexpression were fated to become distinct somites. This pattern of expression is reminiscent to that of the pair-rule genes of *Drosophila* and supports a model of somite regulation based on the metameric pattern being generated with two distinct segments instead of one (Meinhardt, 1986).

It is possible that the periodic oscillation model in chick and the static positional information model in zebrafish represent evolutionary divergence in the strategies used for imposing segmentation within the presomitic mesoderm. More likely, the two models are not mutually exclusive. The clock and wavefront model is dependent on the synchronicity of cells within a block for oscillation to segmentation competency. This could be controlled by the alternate expression domains revealed by *her1*. The definition of boundaries between somites would be defined by the positional information and the timing of segmentation would be controlled by the periodic wave expression. A better understanding of the role of these regulatory pathways will come from the identification of genes that impose patterns of expression on these genes.

Five homologs of *E(Spl)* have been identified in mice and rats, which have been designated *Hes-1–5* (Akazawa *et al.*, 1992; Sasai *et al.*, 1992; Ishibashi *et al.*, 1993; Sakagami *et al.*, 1994; Takebayashi *et al.*, 1994). Of these, only *Hes-1, -3,* and *-5,* are expressed at the appropriate time to regulate somitogenesis or neurogenesis (de la Pompa *et al.*, 1997). A targeted mutation of *Hes-1* resulted in misregulation of neuronal differentiation markers, premature neurogenesis, and perinatal death, but phenotypically normal somites (Ishibashi *et al.*, 1995). This suggests that the two developmental systems have different requirements for *Hes-1.* The role of other members of the Hes family of factors in somitogenesis and their target genes will be important to our understanding of how segmentation is controlled.

III. The Role of Notch Signaling in Segmentation

A signaling pathway that has received a great deal of attention with regard to segmentation is that of the Notch family of receptors and their ligands. Notch has been shown to play a central role in cell fate decisions during both embryonic and adult life in *Drosophila* and *Caenorhabditis elegans* (*lin-12* and *Glp-1*) (for review, see Artavanis-Tsakonas *et al.*, 1995). In vertebrates, multiple homologs of Notch have been identified including *Xotch* in *Xenopus* (Coffman *et al.*, 1990), *Notch1, 2, 3,* and *4,* in mouse (Del Amo *et al.*, 1992; Lardelli and Lendahl, 1993; Lardelli *et al.*, 1994; Uyttendaele *et al.*, 1996), and homologous genes in rat (Weinmaster *et al.*, 1991, 1992), humans (Ellisen *et al.*, 1991; Larsson *et al.*, 1994), and zebrafish (Bierkamp and Campos-Ortega, 1993). The embryonic expression patterns of the vertebrate *Notch1, 2,* and *3* genes, suggest a role in establishing segmentation along the A/P axis. Targeted mutations in genes proposed to be involved in analogous Notch signaling pathways in *Drosophila* have provided insight into the signaling pathway(s) that controls segmentation.

A. Notch Proteins and Signal Transduction

Notch proteins function as cell surface receptors that integrate extracellular signals and direct changes in gene expression, resulting in the restriction of cell fate. The Notch proteins are similar in structure, with an extracellular domain consisting of 34 to 36 Epidermal Growth Factor (EGF) repeats, a cysteine-rich domain, and three Notch/*lin-12* repeats. The intracellular domain has six tandem ankyrin repeats, flanked by putative nuclear localization signals, followed by a PEST sequence (Wharton *et al.,* 1985; Kidd *et al.,* 1986; Yochem *et al.,* 1988; Yochem and Greenwald, 1989). Notch receptors are processed by proteolytic cleavage in the trans-Golgi network to generate two fragments, one which contains the extracellular domain, and the other, the transmembrane and intracellular domains. These two fragments are tethered at the cell surface and form the signaling-competent heterodimeric receptor (Blaumueller *et al.,* 1997; Pan and Rubin, 1997). The EGF repeats are required for Notch to bind its ligands, Serrate (Ser) and Delta (Dl). These ligands are also integral membrane proteins expressed on the cell surface. The ankyrin repeats of the intracellular domain of Notch mediate signal transduction through interactions with the cytoplasmic factor, deltex, and the zinc finger protein, Suppressor of hairless [Su(H)] (Diederich *et al.,* 1994; Fortini and Artavanis-Tsakonas, 1994; Matsuno *et al.,* 1995).

On ligand activation of Notch, there is evidence of a second cleavage of the intracellular fragment, which leads to its nuclear translocation (Fortini and Artavanis-Tsakonas, 1994; Jarriault *et al.,* 1995; Lieber *et al.,* 1993; Kopan *et al.,* 1994, 1996). It has been hypothesized that this fragment can directly regulate gene expression. In *Drosophila,* it has been found that concomitant with activation of the Notch receptor, Su(H) is released from the cytoplasm and translocates to the nucleus, where it also regulates gene expression. Targets of Su(H) include *wingless,* and members of the *enhancer of split* gene complex [E(spl)], which encodes seven related bHLH transcription factors (Kim *et al.,* 1996; Neumann and Cohen, 1996).

B. Notch Signaling in *Drosophila*

The best studied functions of Notch signaling are in specification of neuroblasts from an ectodermal monolayer and dorsal–ventral (D/V) boundary formation in the imaginal wing discs of *Drosophila.* Neuroblasts form within an initially equipotent field of cells. Expression of the proneural *achaete-scute* gene complex is restricted to a single cell in each proneural cluster (Fig. 4) through a process called lateral inhibition. As the Notch signal becomes stronger in the cells surrounding the prospective neuroblast cell, it adopts the appropriate cell fate and inhibits neighboring cells from adopting the same fate (Heitzler and Simpson, 1991; Greenwald and Rubin, 1992).

Notch signaling along the D/V boundary results in wing margin-specific gene expression and cell proliferation in the dorsal and ventral compartments (Shellenbarger and Mohler, 1978; Diaz-Benjumea and Cohen, 1995; de Celis *et al.,* 1996a). It is this type of regulation that more closely approximates the formation of A/P boundaries between individual somites in vertebrates. Generating a polarity within the cells at the wing margin requires that the Notch protein interacts

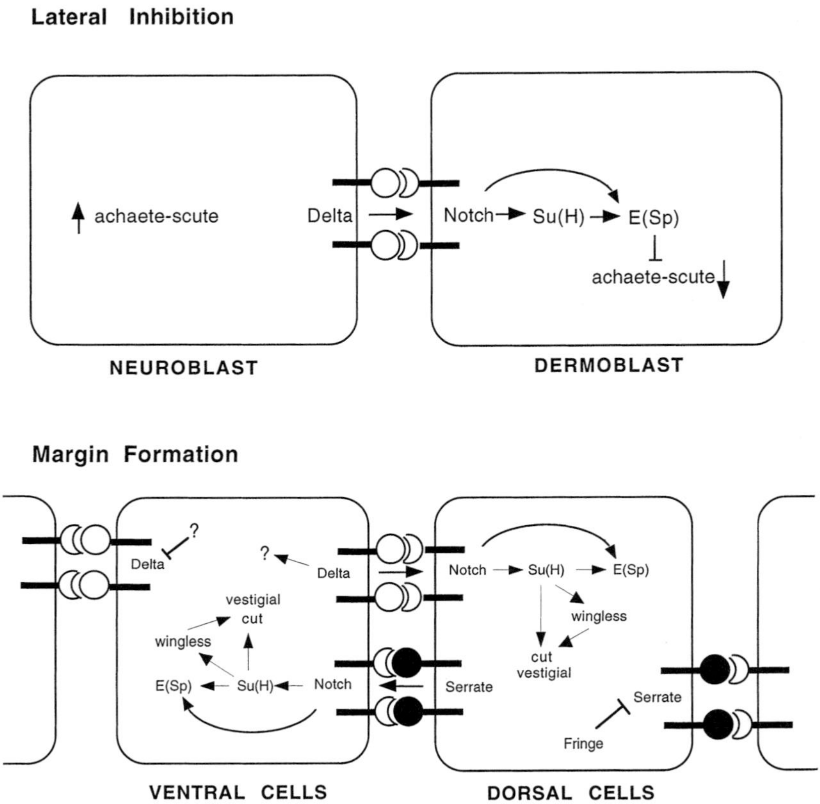

Figure 4 A comparison of the role of the Notch signaling cascade in neuroblast specification and formation of the anterior–posterior boundary of the imaginal wing disc in *Drosophila.* Arrows represent a positive regulation of transcription and flat arrows represent a negative regulation of transcription.

with both of its ligands. Initially, Dl and Ser are expressed broadly in the wing disc, but become rapidly restricted to the D/V boundary (Doherty *et al.,* 1996). Ser is expressed in dorsal cells, where it activates Notch in ventral cells, and Dl is expressed the ventral cells, activating Notch in the dorsal cells. Ser and Dl enhance each others expression, forming a positive feedback loop that presumably strengthens the D/V boundary. This feedback loop is restricted to the D/V boundary by reciprocal restriction of dorsal and ventral cells which respond to Dl and Ser (Panin *et al.,* 1997). In dorsal cells, this is accomplished by the activity of fringe (fng), a secreted protein that is expressed exclusively in the dorsal cells of the margin. Fng appears to enhance the ability of Notch to respond to Dl in dorsal cells and prevent a response to Ser (Fleming *et al.,* 1997; Panin *et al.,* 1997). The activation of Notch is interpreted intracellularly by margin-specific gene expression through Su(H) activation and expression of the E(spl) complex (de Celis *et al.*, 1996b), wingless, and cut (Kim *et al.*, 1996; Neumann and Cohen, 1996).

C. The Role of Notch Genes during Somitogenesis

The formation of somites from the anterior end of the segmental plate bears a resemblance to the establishment of wing margins in *Drosophila.* It is possible that the mechanism for forming such boundaries may involve evolutionarily conserved regulatory circuits. Consistent with this hypothesis, *Notch1* is expressed at the anterior limit of the segmental plate, just caudal to the newly formed somite (Reaume *et al.*, 1992; Williams *et al.*, 1995). It is also initially expressed throughout the new somite. Its expression within the somite diminishes and becomes restricted to the dorsal-medial aspect as the somite matures. Within cells of the paraxial mesoderm lineage, *Notch2* and *3* transcripts are detected only after the somites have formed, where they overlap with *Notch 1* expression.

Mice homozygous for a *Notch1* null allele die between 10 and 11.5 dpc due to defects unrelated to somitogenesis (Swiatek *et al.*, 1994; Conlon *et al.*, 1995). At the three to four somite stage, the *Notch* mutant littermates had continuous strips of unsegmented, but condensed, mesenchyme. By the five somite stage, the somites appeared to segment and epithelialize in a manner comparable to wild-type littermates. However, epithelialization was not complete in all somites and the intersomitic clefts were not consistently coordinated across the midline. Similar to what was observed in *paraxis* mutant mice, cell fate specification appeared to be independent of Notch1 and the segmentation defects. These results indicate that Notch1 plays a role in establishing the timing, but not the definition of, the segmental boundaries during somitogenesis. Functional redundancy among members of a gene family has been described in multiple developmental processes, including myogenesis (Rudnicki *et al.*, 1993). Firm conclusions about the role of *Notch1* in somitogenesis will require the generation of mice carrying compound null mutations with *Notch2* or *3.*

D. Delta

Homologs of *Delta* have been identified in mice (*Dll1* and *Dll3;* Bettenhausen *et al.*, 1995; Dunwoodie *et al.*, 1997), *Xenopus* (*X-Delta-1* and *X-Delta-2;* Chitnis *et al.*, 1995; Jen *et al.*, 1997), and chickens (*C-Delta;* Henrique *et al.*, 1995). In mice, Dll1 and Dll3 are expressed in a pattern consistent with their role as ligands for Notch in regulating somitogenesis. Dll1 is first detected in the primitive streak and the anterior aspect of the embryonic segmental plate mesoderm at 7 dpc (Bettenhausen *et al.*, 1995). After 9.5 dpc, expression extends caudally until all of the presomitic mesoderm expresses *Dll1.* Within the newly formed somite, *Dll1* expression becomes restricted to the posterior half. *Dll3* expression is initiated much earlier (between 5.5 and 6.0 dpc), in the cells of the epiblast prior to gastrulation (Dunwoodie *et al.,* 1997). Transcripts are found in the primitive streak during gastrulation, and subsequently become localized to the tail bud mesoderm. The highest accumulation of *Dll3* message is in the anterior aspect of the segmental plate, where it roughly overlaps with *Dll1* and *Notch1* expression. In the somites, Dll1 and Dll3 are restricted to the anterior and posterior halves, respectively. It is possible that Dll1 and Dll3 are involved in establishing segmentation by defining the anterior and posterior sides of the boundary through interaction with Notch, reminiscent of the D/V boundary of the *Drosophila* wing margin. This model would also predict that the anterior–posterior polarity of each somite could be established at the juxtaposition of Dll1 and Dll3 within each newly formed somite.

To directly examine the role of *Dll1* in somitogenesis, Hrabe de Angelis *et al.* (1997) replaced the gene with the *lacZ* gene ($Dll1^{lacZ}$). Defects in somitogenesis were observed in the mutant embryos that indicate a differential dependence on Dll1 along the rostral–caudal axis. The initiation and formation of the first somites were unaffected in the absence of *Dll1.* Beginning at 8.5 dpc (5 to 10 somites), somites in the $Dll1^{lacZ/lacZ}$ embryos were irregular in shape, similar to what was observed in *Notch* mutants. In the thoracic and lumbar somites (9.0 dpc), epithelial somites were replaced by unsegmented dermatome and sclerotome. By 10.5 dpc (30 to 35 somites), distinctive gaps were readily visible between the dermatomes, but the cells of the myotome clearly spanned multiple borders. $Dll1^{lacZ/lacZ}$ mice were inviable by 12 dpc due to severe hemorrhaging, presumably related to the role of *Dll1* in vasculature development.

The importance of members of the Delta family in somitogenesis is further supported by recent studies performed in *Xenopus* and zebrafish. The expression of a dominant–negative form of *X-Delta-2* (X-Delta-2tr) in *Xenopus* resulted in defects in segmentation, without perturbing the specification of cells to the myogenic lineage (Jen *et al.,* 1997). The overexpression of *delta D,* the zebrafish homolog, resulted in defects in the adaxial mesoderm (Dornseifer *et al.,* 1997).

Dll1 also appears to be required for the establishment of the metameric pattern of the peripheral nervous system. Neural crest cells and peripheral nerves migrate selectively through the anterior half of the sclerotome of each somite, resulting in the segmented pattern. The migration of neural crest cells through the sclerotome in the caudal half of the somite is inhibited by the presence of ephrin-B1 and ephrin-B2, which are membrane bound ligands for the Eph receptor tyrosine kinases (Wang and Anderson, 1997; Krull *et al.,* 1997). Other surface proteins proposed to play a role in this process include collagen IX (Ring *et al.,* 1996), T-cadherin (Ranscht and Bronner-Fraser, 1991), and versican (Landolt *et al.,* 1995). In $Dll1^{lacZ/lacZ}$ embryos, the DRG are unsegmented, suggesting an indiscriminate migration of neural crest cells through both halves of the somite. Similar phenotypes have been reported in *MesP2* and *Presenilin* mutants, which also do not express *Dll1.* It could be inferred from this that *Dll1* is required for the specification of the caudal half of the somite. Defining the regulatory relationship between Dll1 and the expression of Ephrin-B ligands and/or cell adhesion molecules, will be important for understanding the regulation of axon guidance in the peripheral nerves.

E. The Role of Intracellular Targets of Notch

Loss-of-function experiments in mice and *Xenopus* make a strong case for the conservation of the Notch/Delta signaling pathway in vertebrate boundary formation. Since intracellular targets of Notch activation have been characterized in *Drosophila,* this provides a fertile area for identifying genes that might regulate somitogenesis. A vertebrate homolog of Su(H) has been cloned independently as RBP-Jϰ, CBF1, and KBF2 (Matsunami *et al.,* 1989; Schweisguth and Posakony, 1992). In mice, RBP-Jϰ is expressed in the neural tube, the segmental plate, and somites (Oka *et al.,* 1995; de la Pompa *et al.,* 1997). A targeted null mutation of *RBP-Jϰ* resulted in defects in somite formation that include a decrease in size and cell density, consistent with RBP-Jϰ being activated as part of the Notch signaling pathway in the presomitic mesoderm. However, a segmentation deficiency, which is the hallmark of both the *Notch* and *Delta* null mutations, was not reported in *RBP-Jϰ*$^{-/-}$ embryos. This suggests that RBP-Jϰ may regulate the expression of a subset of the genes activated by Notch. Evidence for a RBP-Jϰ independent pathway has been described for Notch-mediated inhibition of myogenesis (Shawber *et al.,* 1996). A divergence in the role of Su(H) and RBP-Jϰ is indicated by the difference in localization of the two gene products. While Su(H) translocates from the cytoplasm to the nucleus on Notch activation, RBP-Jϰ is restricted to the nucleus (de la Pompa *et al.,* 1997). This predicts that translocation of the intracellular Notch proteolytic fragment to the nucleus is an obligate step in activation, or that an intermediate protein is required in vertebrates. The former model is supported by the observation that a RBP-Jϰ complex with intracellular Notch is required to transactivate the *Hes-1* promoter in HeLa cells (Jarriault *et al.,* 1995). It remains to be determined if this interaction is required for the proper segmentation of paraxial mesoderm.

F. The Role of *presenilin* Family Members in Segmentation

Presenilin1 has recently been identified as a modifier of Notch function during somitogenesis. Members of the presenilin family (*PS1* and *PS2*) map to chromosomal loci associated with Alzheimer's disease (Sherrington *et al.,* 1995; Levy-Lahad *et al.,* 1995; Rogaev *et al.,* 1995). Mice lacking PS1 have similarities to *Notch1* and *Dll1* mutants (Shen *et al.,* 1997; Wong *et al.,* 1997). At 9.5 dpc, the rostral somites were irregular in shape, and not always in register across the midline, while the caudal somites did not form. These embryos survived until birth and died soon thereafter, indicating that the *PS1*$^{-/-}$ embryos are not a phenocopy of *Notch* or *Dll1* mutants. The *PS1*$^{-/-}$ neonates also exhibit a caudal truncation, and fusion of the ribs and vertebrae along the A/P axis, similar to the defects described for *paraxis*$^{-/-}$ and *Mesp2*$^{-/-}$ neonates.

Clues as to how PS1 may function in the Notch signaling pathway have come from the analysis of gene markers in *PS1*$^{-/-}$ mutants. The level of *Notch1* and *Dll1* transcripts are dramatically reduced in the presomitic mesoderm and somites of the mutant (Wong *et al.,* 1997). This result appears to place PS1 upstream of both of Notch and its ligand in the signaling cascade within the paraxial mesoderm. It is not obvious how PS1 may regulate transcription based on its proposed modes of action. The PS1 protein has been detected in the Golgi and endoplasmic reticulum (ER) along with the amyloid precursor protein (APP), which is implicated in the pathogenesis of Alzheimer's disease (Sherrington *et al.,* 1995). Since APP processing is increased in patients and transgenic mice with mutations in *PS1* (Duff *et al.,* 1996; Scheuner *et al.,* 1996; Citron *et al.,* 1997), it has been hypothesized that PS1 is required for regulation of protein processing in the ER. Perhaps, a block in efficient processing and/or transport to the cell surface interrupts the positive feedback loops proposed to amplify *Dll1* and *Notch* transcriptional regulation. A decrease in *Dll1* transcription in the presomitic mesoderm and somites has been observed in the *RBP-Jκ*$^{-/-}$ embryos, supporting the link between Notch activation and *Dll1* transcription (de la Pompa *et al.,* 1997). The relationship of PS1 and Notch in somitogenesis is not clear, but it is likely that they do not lie in a simple linear regulatory pathway. It is important to note that PS1 is not required for Notch signaling in all cell types. *Notch, Dll1,* and *RBP-Jκ* null mutants are inviable between 8.5 and 10.5 dpc, consistent with a cardiac or circulatory deficiency, while the *PS1*$^{-/-}$ mutants are able to survive to birth.

What is the mechanism by which Notch regulates segmentation? Does it occur by a process of lateral inhibition or margin formation? A model of segmentation based on lateral inhibition has been forwarded by Conlon *et al.* (1995). Small differences in the expression of Notch and its ligands are amplified, resulting in the organization of mesodermal cells into a periodic arrangement of signaling and receiving cells. Overt segmentation is then triggered by an increase in cell adhesion and epithelialization along the boundary. The broad expression of *Notch* and *Dll1* in the segmental plate do not support this model. During cell fate restriction of proneural cells, transcription of *Delta* becomes restricted. A similar restriction in *Dll1* expression has not been observed in vertebrate presomitic mesoderm. However, modulation of Dll1 function could occur at the posttranscriptional level. It is clear that cell compaction alone cannot account for physical segmentation, since *paraxis*$^{-/-}$ embryos demonstrated that epithelialization can be dissociated from segmentation.

It appears that a model of regulation involving differential activation of Notch along the A/P boundary, similar to that required for establishing *Drosophila* D/V wing margin is more likely. Cells at the border could become defined by site-specific enhancement of Notch activation by ligands juxtaposed on opposite sides of the A/P boundary. Definition of the boundary in this case would be established by a modifying gene expressed on one side of the border. This model is supported by the observation that the murine homologs of *Serrate* (*Jagged1*) (Lindsell *et al.*, 1995) and *Fringe* (*lunatic Fringe*) are expressed in overlapping stripes at the anterior end of the presomitic mesoderm, which demarcate the next two intersomitic boundaries (Johnston *et al.*, 1997; Cohen *et al.*, 1997). This model fails to account for the fact that neither *Notch1* nor *Dll1* are restricted in expression along the A/P boundary, as is observed during wing margin formation. It is more likely that regulation of segmentation in vertebrates will be more complex than the simple wing margin model defined in *Drosophila*. This is underscored by the fact that most of the *Drosophila* genes in the signaling pathway have multiple vertebrate family members. The Notch and *Delta* gene families alone each have three members expressed in paraxial mesoderm. Additionally, a simple linear model would predict that null mutations in *Notch1*, *Dll1*, and *RBP-Jκ* would have a similar phenotype. Though there are common defects, each is quite distinct. Dissecting out the contribution of individual family members is the next important step.

IV. Summary

The use of gene knockout technology has led to significant advances in our understanding of the genetic regulation of somitogenesis. The genes that have been identified to date appear to regulate either epithelialization of the somite or segmentation (Fig. 5). These genes should be viewed as the first clues to the puzzle

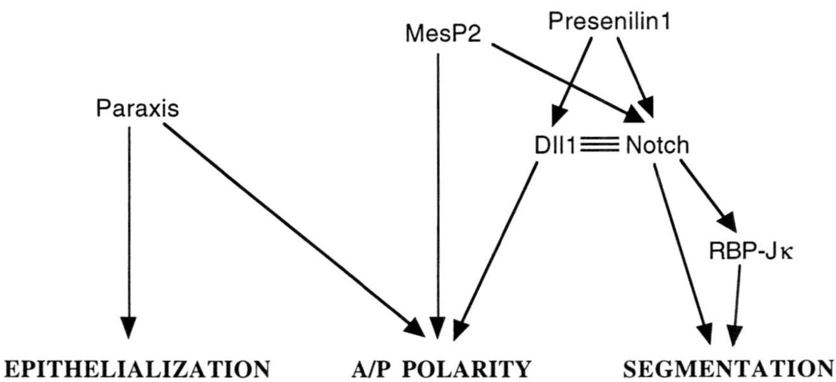

Figure 5 The network of signaling factors regulating somite formation. The interrelationship between signaling factors demonstrated to be required for epithelialization, segmentation, or establishment of the anterior–posterior compartmentalization of individual somites. Three parallel lines represent the requirement of a protein–protein interaction for the function of the gene. Arrows denote a positive but not necessarily direct regulation.

of somitogenesis and not the answers. It will be important to define the genetic interactions between these genes and their roles in morphogenetic events. As an example, the loss of the A/P polarity within the somite is a consequence of either a defect in epithelialization or segmentation. This suggests a point of convergence between the two pathways. It will be equally important to search for genes which lie downstream in these genetic pathways. It is only through the characterization of such genes that the molecular nature of somitogenesis can be fully understood.

Acknowledgments

Work in E. Olson's laboratory is supported by the National Institutes of Health, The American Heart Association, and The Robert A. Welch Foundation.

References

Akazawa, C., Sasai, Y., Nakanishi, S., and Kageyama, R. (1992). Molecular characterization of a rat negative regulator with a basic helix-loop-helix structure predominantly expressed in the developing nervous system. *J. Biol Chem.* **267,** 21879–21885.

Artavanis-Tsakonas, S., Matsuno, K., and Fortini, M. E. (1995). Notch Signaling. *Science* **268,** 225–232.

Barnes, G. L., Alexander, P. G., Hsu, C. W., Mariani, B. D., and Tuan, R. S. (1997). Cloning and characterization of chicken *Paraxis:* A regulator of paraxial mesoderm development and somite formation. *Dev. Biol.* **189,** 95–111.

Bettenhausen, B., Hrabe de Angelis, M., Simon, D., Guenet, J. L., and Gossler, A. (1995). Transient and restricted expression during mouse embryogenesis of *Dll1,* a murine gene closely related to *Drosophila Delta. Development* **121,** 2407–2418.

Biben, C., Stanley, E., Fabri, L., Kotecha, S., Rhinn, M., Lah, M., Wang, C. C., Nash, A., Hilton, D., Ang, S. L., Mohun, T., and Harvey, R. P. (1998). Murine cerberus homologue mCer-1: A candidate anterior patterning molecule. *Dev. Biol.* **194,** 135–151.

Bierkamp, C., and Campos-Ortega, J. A. (1993). A zebrafish homologue of the *Drosophila* neurogenic gene *Notch* and its pattern of transcription during early embryogenesis. *Mech. Dev.* **43,** 87–100.

Blanar, M. A., Crossley, P. H., Peters, K. G., Steingrimsson, E., Copeland, N. G., Jenkins, N. A., Martin, G. R., and Rutter, W. J. (1995). Meso1, a basic-helix-loop-helix protein involved in mammalian presomitic mesoderm development. *Proc. Natl. Acad. Sci. U.S.A.* **92,** 5870–5874.

Blaumueller, C. M., Qi, H., Zagouras, P., and Artavanis-Tsakonis, S. (1997). Intracellular cleavage of Notch leads to a heterodimeric receptor on the plasma membrane. *Cell* **90,** 281–291.

Burgess, R., Cserjesi, P., Ligon, K. L., and Olson, E. N. (1995). *Paraxis:* A basic helix-loop-helix protein expressed in paraxial mesoderm and developing somites. *Dev. Biol.* **168,** 296–306.

Burgess, R., Rawls, A., Brown, D., Bradley, A., and Olson, E. N. (1996). Requirement of the *paraxis* gene for somite formation and musculoskeletal patterning. *Nature* **384,** 570–573.

Chitnis, A., Henrique, D., Lewis, J., Ish-Horowicz, D., and Kintner, C. (1995). Primary neurogenesis in *Xenopus* embryos regulated by a homologue of the *Drosophila* neurogenic gene *Delta. Nature* **375,** 761–766.

Christ, B., Jacob, M., Jacob, H. J., Brand, B., and Wachtler, F. (1986). *In* "Somites in Developing Embryos" (R. Bellairs, D. A. Ede, and J. W. Lash, Eds.), pp. 261–275. Plenum, New York.

Christ, B., Schmidt, C., Huang, R., Wilting, J., and Brand-Saberi, B. (1998). Segmentation of the vertebrate body. *Anat. Embryol.* **197,** 1–8.

Citron, M., Westaway, D., Xia, W., Carlson, G., Diehl, T., Levesque, G., Johnson-Wood, K., Lee, M., Seubert, P., Davis, A., Kholodenko, D., Motter, R., Sherrington, R., Perry, B., Yao, H., Strome, R., Lieberburg, I., Rommens, J., Kim, S., Schenk, D., Fraser, P., St George Hyslop, P., and Selkoe, D. J. (1997). Mutant presenilins of Alzheimer's disease increase production of 42-residue amyloid beta-protein in both transfected cells and transgenic mice. *Nat. Med.* **3,** 67–72.

Coffman, C., Harris, W., and Kintner, C. R. (1990). *Xotch,* the *Xenopus* homolog of *Drosophila Notch. Science* **249,** 1438–1441.

Cohen, B., Bashirullah, A., Dagnino, L., Campbell, C., Fisher, W. W., Leow, C. C., Whiting, E., Ryan, D., Zinyk, D., Boulianne, G., Hui, C. C., Gallie, B., Phillips, R. A., Lipshitz, H. D., and Egan, S. E. (1997). Fringe boundaries coincide with Notch-dependent patterning centres in mammals and alter Notch-dependent development in *Drosophila. Nat. Genet.* **16,** 283–288.

Conlon, R. A., Reaume, A. G., and Rossant, J. (1995). *Notch1* is required for the coordinate segmentation of somites. *Development* **121,** 1533–1545.

Cooke, J., and Zeeman, E. C. (1976). A clock and wavefront model for control of the number of repeated structures during animal morphogenesis. *J. Theor. Biol.* **58,** 455–476.

Cserjesi, P., Brown, D., Ligon, K. L., Lyons, G. E., Copeland, N. G., Gilbert, D. J., Jenkins, N. A., and Olson, E. N. (1995). Scleraxis: A basic helix-loop-helix protein that prefigures skeletal formation during mouse embryogenesis. *Development* **121,** 1099–1110.

de Celis, J. F., Garcia-Bellido, A., and Bray, S. J. (1996a). Activation and function of *Notch* at the dorsal–ventral boundary of the wing imaginal disc. *Development* **122,** 359–369.

de Celis, J. F., de Celis, J., Ligoxygakis, P., Preiss, A., Delidakis, C., and Bray, S. (1996b). Functional relationships between *Notch, Su(H)* and the bHLH genes of the *E(spl)* complex: The E(spl) genes mediate only a subset of Notch activities during imaginal development. *Development* **122,** 2719–2728.

Del Amo, F. F., Smith, D. E., Swiatek, P. J., Gendron-Maguire, M., Greenspan, R. J., McMahon, A. P., and Gridley, T. (1992). Expression pattern of *Motch,* a mouse homolog of *Drosophila Notch,* suggests an important role in early postimplantation mouse development. *Development* **115,** 737–744.

de la Pompa, J. L., Wakeham, A., Correia, K. M., Samper, E., Brown, S., Aguilera, R. J., Nakano, T., Honjo, T., Mak, T. W., Rossant, J., and Conlon, R. A. (1997). Conservation of the Notch signaling pathway in mammalian neurogenesis. *Development* **124,** 1139–1148.

Denetclaw, W. F., Jr, Christ, B., and Ordahl, C. P. (1997). Location and growth of epaxial myotome precursor cells. *Development* **124,** 1601–1610.

Diaz-Benjumea, F. J., and Cohen, S. M. (1995). *Serrate* signals through *Notch* to establish a *Wingless*-dependent organizer at the dorsal/ventral compartment boundary of the *Drosophila* wing. *Development* **121,** 4215–4225.

Diederich, R. J., Matsuno, K., Hing, H., and Artavanis-Tsakonas, S. (1994). Cytosolic interaction between *deltex* and *Notch* ankyrin repeats implicates deltex in the Notch signaling pathway. *Development* **120,** 473–481.

Doherty, D., Feger, G., Younger-Shepherd, S., Jan, L. Y., and Jan, Y. N. (1996). *Delta* is a ventral to dorsal signal complementary to *Serrate,* another Notch ligand, in *Drosophila* wing formation. *Genes Dev.* **10,** 421–434.

Dornseifer, P., Takke, C., and Campos-Ortega, J. A. (1997). Overexpression of a zebrafish homologue of the *Drosophila* neurogenic gene *Delta* perturbs differentiation of primary neurons and somite development. *Mech. Dev.* **63,** 159–171.

Duff, K., Eckman, C., Zehr, C., Yu, X., Prada, C. M., Perez-tur, J., Hutton, M., Buee, L., Harigaya, Y., Yager, D., Morgan, D., Gordon, M. N., Holcomb, L., Refolo, L., Zenk, B., Hardy, J., and Younkin, S. (1996). Increased amyloid-beta42(43) in brains of mice expressing mutant presenilin 1. *Nature* **383,** 710–713.

Dunwoodie, S. L., Henrique, D., Harrison, S. M., and Beddington, R. S. (1997). Mouse *Dll3:* A novel divergent *Delta* gene which may complement the function of other Delta homologues during early pattern formation in the mouse embryo. *Development* **124,** 3065–3076.

Ellisen, L. W., Bird, J., West, D. C., Soreng, A. L., Reynolds, T. C., Smith, S. D., and Sklar, J. (1991).

TAN-1, the human homolog of the *Drosophila notch* gene, is broken by chromosomal translocations in T lymphoblastic neoplasms. *Cell* **66,** 649–661.

Elsdale, T., Pearson, M., and Whitehead, M. (1976). Abnormalities in somite segmentation following heat shock to *Xenopus* embryos. *J. Embryol. Exp. Morphol.* **35,** 625–635.

Fasel, N. J., Bernard, M., Deglon, N., Rousseaux, M., Eisenberg, R. J., and Cohen, G. H. (1991). Isolation from mouse fibroblasts of a cDNA encoding a new form of the fibroblast growth factor receptor (flg). *Biochem. Biophys. Res. Commun.* **178,** 8–15.

Fisher, A. L., Ohsako, S., and Caudy, M. (1996). The WRPW motif of the hairy-related basic helix-loop-helix repressor proteins acts as a 4-amino-acid transcription repression and protein–protein interaction domain. *Mol. Cell. Biol.* **16,** 2670–2677.

Fleming, R. J., Gu, Y., and Hukriede, N. A. (1997). *Serrate*-mediated activation of *Notch* is specifically blocked by the product of the gene *fringe* in the dorsal compartment of the *Drosophila* wing imaginal disc. *Development* **124,** 2973–2981.

Fortini, M. E., and Artavanis-Tsakonas, S. (1994). The suppressor of hairless protein participates in notch receptor signaling. *Cell* **79,** 273–282.

Fraser, S., Keynes, R., and Lumsden, A. (1990). Segmentation in the chick embryo hindbrain is defined by cell lineage restrictions. *Nature* **344,** 431–435.

Greenwald, I., and Rubin, G. M. (1992). Making a difference: The role of cell–cell interactions in establishing separate identities for equivalent cells. *Cell* **68,** 271–281.

Heitzler, P., and Simpson, P. (1991). The choice of cell fate in the epidermis of *Drosophila. Cell* **64,** 1083–1092.

Henrique, D., Adam, J., Myat, A., Chitnis, A., Lewis, J., and Ish-Horowicz, D. (1995). Expression of a *Delta* homologue in prospective neurons in the chick. *Nature* **375,** 787–790.

Hrabe de Angelis, M., McIntyre II, J., and Gossler, A. (1997). Maintenance of somite borders in mice requires the Delta homologue *Dll1. Nature* **386,** 717–721.

Ishibashi, M., Sasai, Y., Nakanishi, S., and Kageyama, R. (1993). Molecular characterization of HES-2, a mammalian helix-loop-helix factor structurally related to *Drosophila* hairy and Enhancer of split. *Eur. J. Biochem.* **215,** 645–652.

Ishibashi, M., Ang, S. L., Shiota, K., Nakanishi, S., Kageyama, R., and Guillemot, F. (1995). Targeted disruption of mammalian hairy and Enhancer of split homolog-1 (HES-1) leads to upregulation of neural helix-loop-helix factors, premature neurogenesis, and severe neural tube defects. *Genes Dev.* **9,** 3136–3148.

Jan, Y. N., and Jan, L. Y. (1993). HLH proteins, fly neurogenesis, and vertebrate myogenesis. *Cell* **75,** 827–830.

Jarriault, S., Brou, C., Logeat, F., Schroter, E. H., Kopan, R., and Isreal, A. (1995). Signaling downstream of activated mammalian Notch. *Nature* **377,** 355–358.

Jen, W.-C., Wettstein, D., Turner, D., Chitnis, A., and Kintner, C. (1997). The Notch ligand, *X-Delta-2,* mediates segmentation of the paraxial mesoderm in *Xenopus* embryos. *Development* **124,** 1169–1178.

Johnston, S. H., Rauskolb, C., Wilson, R., Prabhakaran, B., Irvine, K. D., and Vogt, T. F. (1997). A family of mammalian *Fringe* genes implicated in boundary determination and the *Notch* pathway. *Development* **124,** 2245–2254.

Keynes, R. J., and Stern, C. D. (1988). Mechanisms of vertebrate segmentation. *Development* **103,** 413–429.

Kidd, S., Kelley, M. R., and Young, M. W. (1986). Sequence of the Notch locus of *Drosophila melanogaster:* Relationship of the encoded protein to mammalian clotting and growth factors. *Mol. Cell. Biol.* **6,** 3094–3108.

Kim, J., Sebring, A., Esch, J. J., Kraus, M. E., Vorwerk, K., Magee, J., and Carroll, S. B. (1996). Integration of positional signals and regulation of wing formation and identity by *Drosophila vestigial* gene. *Nature* **382,** 133–138.

Kopan, R., Nye, J. S., and Weintraub, H. (1994). The intracellular domain of mouse Notch: A constitutively activated repressor of myogenesis directed at the basic helix-loop-helix region of MyoD. *Development* **120,** 2385–2396.

Kopan, R., Schroeter, E. H., Weintraub, H., and Nye, J. S. (1996). Signal transduction by activated *mNotch:* Importance of proteolytic processing and its regulation by the extracellular domain. *Proc. Natl. Acad. Sci. U.S.A.* **93,** 1683–1688.

Krull, C. E., Lansford, R., Gale, N. W., Collazo, A., Marcelle, C., Yancopoulos, G. D., Fraser, S. E., and Bronner-Fraser, M. (1997). Interactions of Eph-related receptors and ligands confer rostro-caudal pattern to trunk neural crest migration. *Curr. Biol.* **7,** 571–580.

Landolt, R. M., Vaughan, L., Winterhalter, K. H., and Zimmermann, D. R. (1995). Versican is selectively expressed in embryonic tissues that act as barriers to neural crest cell migration and axon outgrowth. *Development* **121,** 2303–2312.

Lardelli, M., and Lendahl, U. (1993). *Motch A* and *motch B*— two mouse Notch homologues coexpressed in a wide variety of tissues." *Exp. Cell Res.* **204,** 364–372.

Lardelli, M., Dahlstrand, J., and Lendahl, U. (1994). The novel Notch homologue mouse Notch 3 lacks specific epidermal growth factor-repeats and is expressed in proliferating neuroepithelium. *Mech. Dev.* **46,** 123–136.

Larsson, C., Lardelli, M., White, I., and Lendahl, U. (1994). The human *NOTCH1, 2,* and *3* genes are located at chromosome positions 9q34, 1p13-p11, and 19p13.2-p13.1 in regions of neoplasia-associated translocation. *Genomics* **24,** 253–258.

Levy-Lahad, E., Wasco, W., Poorkaj, P., Romano, D. M., Oshima, J., Pettingell, W. H., Yu, C. E., Jondro, P. D., Schmidt, S. D., Wang, K., Crowley, A. C., Fu, Y.-H., Guenette, S. Y., Galas, D., Nemens, E., Wijsman, E. M., Bird, T. D., Schellenberg, G. D., and Tanzai, R. E. (1995). Candidate gene for the chromosome 1 familial Alzheimer's disease locus. *Science* **269,** 973–977.

Lieber, T., Kidd, S., Alcamo, E., Corbin, V., and Young, M. W. (1993). Antineurogenic phenotypes induced by truncated Notch proteins indicate a role in signal transduction and may point to a novel function for Notch in nuclei. *Genes Dev.* **7,** 1949–1965.

Lindsell, C. E., Shawber, C. J., Boulter, J., and Weinmaster, G. (1995). Jagged: A mammalian ligand that activates Notch1. *Cell* **80,** 909–917.

Lumsden, A., and Krumlauf, R. (1996). Patterning in vertebrate neuraxis. *Science* **274,** 1109–1115.

Matsunami, N., Hamaguchi, Y., Yamamoto, Y., Kuze, K., Kangawa, K., Matsuo, H., Kawaichi, M., and Honjo, T. (1989). A protein binding to the J kappa recombination sequence of immunoglobulin genes contains a sequence related to the integrase motif. *Nature* **342,** 934–937.

Matsuno, K., Diederich, R. J., Go, M. J., Blaumueller, C. M., and Artavanis-Tsakonas, S. (1995). *Deltex* acts as a positive regulator of *Notch* signaling through interactions with the Notch ankyrin repeats. *Development* **121,** 2633–2644.

Meier, S. (1979). Development of the chick mesoblast. Formation of embyonic axis and establishment of the metameric pattern. *Dev. Biol.* **73,** 25–45.

Meinhardt, H. (1986). *In* "Somites in Developing Embryos" (R. Bellairs, D. A. Ede, and J. W. Lash, Eds.), pp. 179–189. Plenum, New York.

Müller, M., von Weizsäcker, E., and Campos-Ortega, J. A. (1996). Expression domains of a zebrafish homologue of the *Drosophila* pair-rule gene *hairy* correspond to primordia of alternating somites. *Development* **122,** 2071–2078.

Neumann, C. J., and Cohen, S. M. (1996). A hierarchy of cross-regulation involving *Notch, wingless, vestigial* and *cut* organizes the dorsal/ventral axis of the *Drosophila* wing. *Development* **122,** 3477–3485.

Nieto, M. A., Gilardi-Hebenstreit, P., Charnay, P., and Wilkinson, D. G. (1992). A receptor protein tyrosine knase implicated in the segmental patterning of the hindbrain and mesoderm. *Development* **116,** 1137–1150.

Nüsslein-Volhard, C., and Weichaus, E. (1980). Mutations affecting segment number and polarity in *Drosophila. Nature* **287,** 795–801.

Oka, C., Nakano, T., Wakeham, A., de la Pompa, J. L., Mori, C., Sakai, T., Okazaki, S., Kawaichi, M.,

Shiota, K., Mak, T. W., and Honjo, T. (1995). Disruption of the mouse *RBP-Jκ* gene results in early embryonic death. *Development* **121,** 3291–3301.

Olson, E. N., Brown, D., Burgess, R., and Cserjesi, P. (1996). A new subclass of helix-loop-helix transcription factors expressed in paraxial mesoderm and chondrogenic cell lineages. *Ann. N.Y. Acad. Sci.* **785,** 108–118.

Palmeirim, I., Henrique, D., Ish-Horowicz, D., and Pourquie, O. (1997). Avian *hairy* gene expression identifies a molecular clock linked to vertebrate segmentation and somitogenesis. *Cell* **91,** 639–648.

Pan, D., and Rubin, G. M. (1997). Kuzbanian controls proteolytic processing of Notch and mediates lateral inhibition during *Drosophila* and vertebrate neurogenesis. *Cell* **90,** 217–280.

Panin, V. M., Papayannopoulos, V., Wilson, R., and Irvine, K. D. (1997). Fringe modulates Notch-ligand interactions. *Nature* **387,** 908–912.

Pankratz, M. J., and Jäckle, H. (1993). *In* "Development of *Drosophila.*" (C. M. Bate and A. Martinez-Arias, Eds.), pp. 467–516. Cold Spring Harbor Press, Cold Spring Harbor, New York.

Paroush, Z., Finley, R. L., Jr, Kidd, T., Wainwright, S. M., Ingham, P. W., Brent, R., and Ish-Horowicz. D. (1994). Groucho is required for *Drosophila* neurogenesis, segmentation, and sex determination and interacts directly with hairy-related bHLH proteins. *Cell* **79,** 805–815.

Quertermous, E. E., Hidai, H., Blanar, M. A., and Quertermous, T. (1994). Cloning and characterization of a basic helix-loop-helix protein expressed in early mesoderm and the developing somites. *Proc. Natl. Acad. Sci. U.S.A.* **91,** 7066–7070.

Ranscht, B., and Bronner-Fraser, M. (1991). T-cadherin expression alternates with migrating neural crest cells in the trunk of the avian embryo. *Development* **111,** 15–22.

Reaume, A. G., Conlon, R. A., Zirngibl, R., Yamaguchi, T. P., and Rossant, J. (1992). Expression analysis of a Notch homologue in the mouse embryo. *Dev. Biol.* **154,** 377–387.

Ring, C., Hassell, J., and Halfter, W. (1996). Expression pattern of collagen IX and potential role in the segmentation of the peripheral nervous system. *Dev. Biol.* **180,** 41–52.

Rogaev, E. I., Sherrington, R., Rogaeva, E. A., Levesque, G., Ikeda, M., Liang, Y., Chi, H., Lin, C., Holman, K., Tsuda, T., Mar, L., Sorbl, S., Nacmias, B., Piacentini, S., Amaducci, L., Chumakov, I., Cohen, D., Lannfelt, L., Fraser, P. E., Rommens, J. M., and St. George-Hyslop, P. H. (1995). Familial Alzheimer's disease in kindreds with missense mutations in a gene on chromosome 1 related to the Alzheimer's disease type 3 gene. *Nature* **376,** 775–778.

Rudnicki, M. A., Schnegelsberg, P. N., Stead, R. H., Braun, T., Arnold, H. H., and Jaenisch, R. (1993). *MyoD* or *Myf-5* is required for the formation of skeletal muscle. *Cell* **75,** 1351–1359.

Saga, Y., Hata, N., Kobayashi, S., Magnuson, T., Seldin, M. F., and Taketo, M. M. (1996). MesP1: A novel basic helix-loop-helix protein expressed in the nascent mesodermal cells during mouse gastrulation. *Development* **122,** 2769–2778.

Saga, Y., Hata, N., Koseki, H., and Taketo, M. M. (1997). Mesp2: A novel mouse gene expressed in the presegmented mesoderm and essential for segmentation initiation. *Genes Dev.* **15,** 1827–1839.

Sakagami, T., Sakurada, K., Sakai, Y., Watanabe, T., Nakanishi, S., and Kageyama, R. (1994). Structure and chromosomal locus of the mouse gene encoding a cerebellar Purkinje cell-specific helix-loop-helix factor *Hes-3. Biochem. Biophys. Res. Commun.* **203,** 594–601.

Sasai, Y., Kageyama, R., Tagawa, Y., Shigemoto, R., and Nakanishi, S. (1992). Two mammalian helix-loop-helix factors structurally related to *Drosophila hairy* and *Enhancer of split. Genes Dev.* **6,** 2620–2634.

Scheuner, D., Eckman, C., Jensen, M., Song, X., Citron, M., Suzuki, N., Bird, T. D., Hardy, J., Hutton, M., Kukull, W., Larson, E., Levy-Lahad, E., Viitanen, M., Peskind, E., Poorkaj, P., Schellenberg, G., Tanzi, R., Wasco, W., Lannfelt, L., Selkoe, D., and Younkin, S. (1996). Secreted amyloid beta-protein similar to that in the senile plaques of Alzheimer's disease is increased *in vivo* by the presenilin 1 and 2 and APP mutations linked to familial Alzheimer's disease. *Nat. Med.* **2,** 864–870.

Schweisguth, F., and Posakony, J. W. (1992). *Suppressor of Hairless,* the *Drosophila* homolog of the mouse recombination signal-binding protein gene, controls sensory organ cell fates. *Cell* **69,** 1199–1212.

Shawber, C., Nofziger, D., Hsieh, J. J., Lindsell, C., Bogler, O., Hayward, D., and Weinmaster, G. (1996). Notch signaling inhibits muscle cell differentiation through a CBF1-independent pathway. *Development* **122,** 3765–3773.

Shellenbarger, D. L., and Mohler, J. D. (1978). Temperature-sensitive periods and autonomy of pleiotropic effects of l(1)Nts1, a conditional *notch* lethal in *Drosophila. Dev. Biol.* **62,** 432–446.

Shen, J., Bronson, R. T., Chen, D. F., Xia, W., Selkoe, D. J., and Tonegawa, S. (1997). Skeletal and CNS defects in *Presenilin-1*-deficient mice. *Cell* **89,** 629–639.

Sherrington, R., Rogaev, E. I., Liang, Y., Rogaeva, E. A., Levesque, G., Ikeda, M., Chi, H., Lin, C., Li, G., Holman, K., Tsuda, T., Mar, L., Foncin, J.-F., Bruni, A. C., Montesi, M. P., Sorbi, S., Rainero, I., Pinessi, L., Nee, L., Chumakov, I., Pollen, D., Brokkes, A., Sanseau, P., Polinsky, R. J., Wasco, W., Da Silva, H. A. R., Haines, J. L., Pericak-Vance, M. A., Tanzi, R. E., Roses, A. D., Fraser, P. E., Rommens, J. M., and St. George-Hyslop, P. H. (1995). Cloning of a gene bearing missense mutations in early-onset familial Alzheimer's disease. *Nature* **375,** 754–760.

Sosic, D., Brand-Saberi, B., Schmidt, C., Christ, B., and Olson, E. N. (1997). Regulation of *paraxis* expression and somite formation by ectoderm- and neural tube-derived signals. *Dev. Biol.* **185,** 229–243.

Swiatek, P. J., Lindsell, C. E., del Amo, F. F., Weinmaster, G., and Gridley, T. (1994). *Notch1* is essential for postimplantation development in mice. *Genes Dev.* **8,** 707–719.

Takada, S., Stark, K. L., Shea, G., Vassileva, G., McMahon, J. A., and McMahon, A. P. (1994). *Wnt-3a* regulates somites and tailbud formation in the mouse embryo. *Genes Dev.* **8,** 174–189.

Takebayashi, K., Sasai, Y., Sakai, Y., Watanabe, T., Nakanishi, S., and Kageyama, R. (1994). Structure, chromosomal locus, and promoter analysis of the gene encoding the mouse helix-loop-helix factor *HES-1.* Negative autoregulation through the multiple N box elements. *J. Biol. Chem.* **269,** 5150–5156.

Tam, P. P., and Trainor, P. A. (1994). Specification and segmentation of the paraxial mesoderm. *Anat. Embryol.* **189,** 275–305.

Uyttendaele, H., Marazzi, G., Wu, G., Yan, Q., Sassoon, D., and Kitajewski, J. (1996). *Notch4/int-3,* a mammary proto-oncogene, is an endothelial cell-specific mammalian *Notch* gene. *Development* 122, 2251–2259.

van Eeden, F. J., Granato, M., Schach, U., Brand, M., Furutani-Seiki, M., Haffter, P., Hammerschmidt, M., Heisenberg, C. P., Jiang, Y. J., Kane, D. A., Kelsh, R. N., Mullins, M. C., Odenthal, J., Warga, R. M., and Nusslein-Volhard, C. (1996). Genetic analysis of fin formation in the zebrafish, *Danio rerio. Development* **123,** 255–262.

Wang, H. U., and Anderson, D. J. (1997). Eph family transmembrane ligands can mediate repulsive guidance of trunk neural crest migration and motor axon outgrowth. *Neuron,* **18,** 383–396.

Weinmaster, G., Roberts, V. J., and Lemke, G. A. (1991). Homolog of *Drosophila Notch* expressed during mammalian development. *Development* **113,** 199–205.

Weinmaster, G., Roberts, V. J., and Lemke, G. (1992). *Notch2:* A second mammalian Notch gene. *Development* **116,** 931–941.

Wharton, K. A., Johansen, K. M., Xu, T., and Artavanis-Tsakonis, S. (1985). Nucleotide sequence from the neurogenic locus Notch implies a gene product that shares homology with proteins containing EGF-like repeats. *Cell* **43,** 567–581.

Williams, R., Lendahl, U., and Lardelli, M. (1995). Complementary and combinatorial patterns of *Notch* gene family expression during early mouse development. *Mech. Dev.* **53,** 357–368.

Wilson, V., and Beddington, R. S. P. (1996). Cell fate and morphogenetic movement in the late mouse primitive streak. *Mech. Dev.* **55,** 79–89.

Wilting, J., Kurz, H., Brand-Saberi, B., Steding, G., Yang, Y. X., Hasselhorn, H. M., Epperlein, H. H., and Christ, B. (1994). Kinetics and differentiation of somite cells forming the vertebral column: Studies on human and chick embryos. *Anat. Embryol.* **190,** 573–581.

Wong, P. C., Zheng, H., Chen, H., Becher, M. W., Sirinathsinghji, D. J., Trumbauer, M. E., Chen, H. Y., Price, D. L., Van der Ploeg, L. H., and Sisodia, S. S. (1997). *Presenilin 1* is required for *Notch1* and *Dll1* expression in the paraxial mesoderm. *Nature* **387,** 288–292.

Yochem, J., and Greenwald, I. (1989). *glp-1* and *lin-12,* genes implicated in distinct cell–cell interactions in *C. elegans,* encode similar transmembrane proteins. *Cell* **58,** 553–563.

Yochem, J., Weston, K., and Greenwald, I. (1988). The *Caenorhabditis elegans lin-12* gene encodes a transmembrane protein with overall similarity to *Drosophila Notch. Nature* **335,** 547–550.

6

Hox Genes and the Global Patterning of the Somitic Mesoderm

Ann Campbell Burke
Department of Biology
University of North Carolina
Chapel Hill, North Carolina 27599

I. Introduction

The musculoskeletal system of vertebrates arises from the embryonic mesoderm. There are four populations of mesoderm in the vertebrate embryo (Fig. 1): chordamesoderm or notochord; paraxial mesoderm or somites; lateral plate mesoderm (splanchnic and somatic); and the intermediate mesoderm. With the exception of the skull, the full axial skeleton arises from somites, as do virtually all of the striated muscles of the body. Changes in the arrangement and proportions of these basic anatomical elements account for much of the morphological variation that has appeared during the course of vertebrate evolution.

The somites are serially homologous embryonic structures. When they first form, each somite along the anterior–posterior (AP) axis is morphologically identical to every other somite, and eventually gives rise to the same cell types (muscle, bone, dermis). A great deal of past and current research focuses on the primary segmentation of the paraxial mesoderm into the epithelial somites, and the subsequent interactions that determine the ultimate differentiation state of somitic cells. Many of these studies focus on local events within the embryo, and are the subject of most of the chapters in this volume. The overall global patterning that somite

Current Topics in Developmental Biology, Vol. 47

0070-2153/00 $30.00

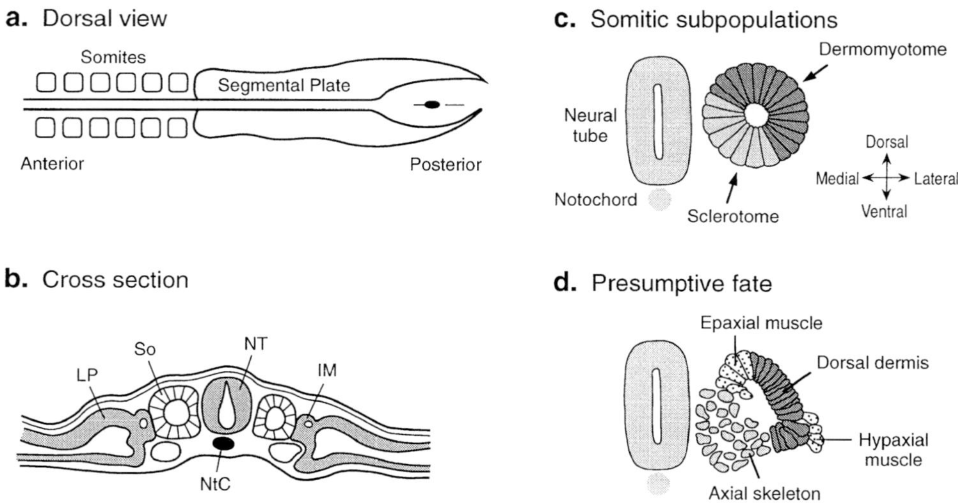

Figure 1 Mesoderm in the vertebrate embryo. (a) Dorsal view showing somites and segmental plate. (b) Cross section showing lateral plate mesoderm (LP), intermediate mesoderm (IM), and somitic mesoderm (So), as well as the neural tube (NT) and the notochord (NtC). (c) Schematic cross section of an individual epithelial somite showing the somitic subpopulations. (d) Schematic cross section of a stage VI–VII somite labeled with prospective fates. (c and d modified from Gilbert, 1997.)

cells participate in, however, is dramatically different depending on the AP position of the somite. This is true not only within individual organisms, but also between different taxa. The patterning that occurs between the stage of somite formation and final differentiation requires local interactions that convey information that ultimately results in global patterning.

The term "positional information" was coined by Wolpert in 1969 to describe the translation of local or genetic information into global morphological pattern. This term is now almost universally used to describe pattern formation (e.g., Boncinelli and Mallamaci, 1995; Slack *et al.,* 1992; Kessel and Gruss, 1991). Unfortunately, because of its generality, "positional information" tends to serve as an explanation in itself, rather than as a description of what still needs to be explained. Wolpert (1996) emphasizes that the information that determines differentiation is independent of the "positional value" of a particular cell. Therefore, correct pattern requires two levels of information.

The recognized necessity for two levels of information in tissue organization remains one of the major problems in our understanding of pattern formation. At one level, individual cells receive signals that influence their final differentiation state. This can be recognized as the short-range, local signals that tell a cell to differentiate into a myoblast rather than a chondroblast. Local signals, acting in isolation, however, do not seem to contain the amount of information necessary to form a complete organism. Local signals must perform within a context that determines overall pattern over longer distances. Information that provides larger contextual or global landmarks is needed, for example, to ensure that a muscle cell contributes to the erector spinae muscle in the back, rather than the brachioradialis muscle in the forelimb.

The somitic mesoderm provides an excellent example of the differences defined by local versus global factors since the ultimate fate of somitic cells along the axis requires both local information and global patterning. Extensive recent work on local signaling has identified many molecules and has explored those molecular interactions that influence cell behaviors including migration and differentiation. These signals are generally common to all somites regardless of axial level. Individually they are apparently independent of more global patterning mechanisms that determine the morphological differences along the AP axis.

The nature of the information that distinguishes the behavior of somites at different axial levels (such as neck versus trunk) is understood with far less detail. The aim of this chapter is to review the data that relate to the translation of local information into global patterning of the somitic mesoderm derivatives. I will first review the types of morphological variation seen among vertebrate taxa that result from evolutionary changes in global patterning in the mesoderm. Second, I will briefly summarize aspects of local, or primary patterning within individual somites that are apparently universal within vertebrates. Next, I will review the morphological fate of the somitic mesoderm determined by classic experiments on model systems that indicate which somites contribute to which adult structures. Finally, I will discuss experimental and comparative studies that have investigated molecular genetic factors that may control global patterning. These factors provide a means of understanding intrinsic proximal events in the evolution of animal form.

II. Patterns in the Mesoderm

A. Global Patterns in the Vertebrate Body Plan—Transposition

Morphological evolution among vertebrates involves many variations on a fixed, metameric body plan. The primary segmentation of vertebrate embryos, visible in the division of the paraxial mesoderm into somites, has profound effects on anatomical structure and function. An extremely basic but common form of variation is described by the term "transposition," coined by E. S. Goodrich in 1913. Transposition describes differences in the numbers of segments, or somites, contributing to different regions of the body in different animals. The axial formula of an animal is the number of somites included in each anatomical region: occipital, cervical, thoracic, lumbar, sacral, and caudal (Fig. 2). Each of these regions can be identified by its relative position along the AP axis, by the morphology of individual vertebrae, and by the relative positions of lateral structures such as the limbs.

In the 350-million-year history of the tetrapod vertebrates, the axial formula has

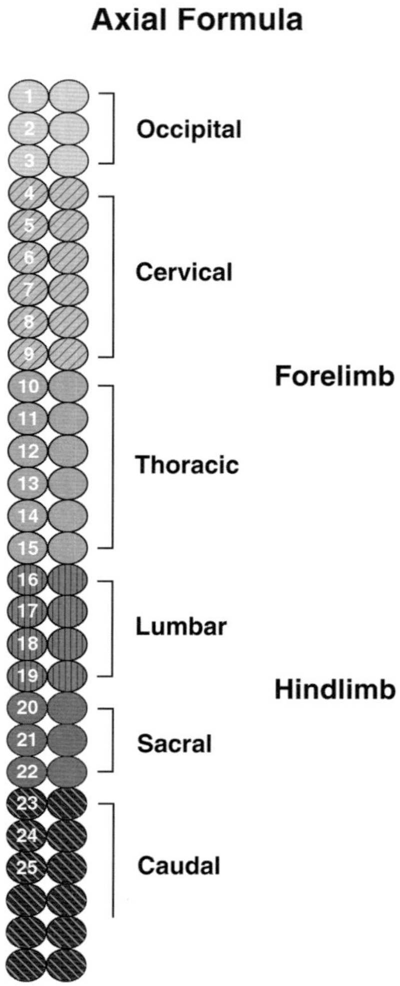

Figure 2 Schematic representation of the axial formula. The circles represent individual segments, or somites extending from anterior (top) to posterior (bottom). The axial formula is defined here as the number of segments that contribute to distinct morphological regions, which are shown in different shades.

undergone extensive evolution. The variation includes highly specialized regionalizations among mammals, like the neck of the giraffe and the prehensile tail of certain monkeys. The frogs have evolved drastically reduced axial formulas and have no more than nine presacral vertebrae. Other taxa have undergone extreme multiplication of segments. Snakes, for example, have as many as 450 vertebrae of uniform type.

Vertebral morphology varies between regions in an individual species, and also between the vertebrae of the same region of different species. Different taxonomic groups show different levels of constraint on the variation available in different regions (see review by Richardson *et al.,* 1998). In mammals, for instance, the number of cervical vertebrae is almost always 7, except in some sloths (6–9) and manatees (6). In the thoracic region of mammals, however, the number ranges from 10 to 20 (Flower, 1885). Turtles have a fixed number of 8 cervical and 10 thoracic vertebrae (Wake, 1979). Birds can have 11–25 cervical vertebrae, but generally have 5–7 thoracic (Gadow, 1933; Proctor and Lynch, 1993).

Obviously, the variation of the axial skeleton does not evolve in isolation, but as part of an integrated functional system. Aside from the anterior to posterior regionalization of the vertebrate body described above, there are also dorsal, ventral, and lateral regions. The dorsal, epaxial region is defined by the vertebrae themselves, and comprises the intrinsic muscles of the back. The hypaxial region includes the limbs and ventrolateral body wall. Certain relationships between the epaxial and hypaxial regions are generally maintained in tetrapods. For instance, the forelimb is always associated with the cervical–thoracic transition, and the hind limb with the sacral region of the axial skeleton. Because of this, placement of the limbs provides an obvious marker for transpositions between taxa.

B. Local Information and Pattern in the Somitic Mesoderm

The somitic mesoderm arises during the movements of gastrulation, and the segmental plate is a bilaterally symmetric population of mesenchyme on either side of the neural tube (thus the term paraxial). From the segmental plate (SP), epithelial somites bud off in an anterior to posterior sequence, initiating immediately behind the region of the otic placode at the early stages of neurulation. Each individual somite is subdivided into regions. One set of subdivisions is based on the eventual differentiation state of cell populations and roughly conforms to the ventromedial and dorsolateral axis of each somite. The sclerotome is ventromedial, and the dermotome and myotome, or dermomyotome, is dorsolateral. These give rise to skeleton and dermis and muscle cells, respectively (Fig. 1c,d).

The somitic subpopulations are characterized by the expression of certain genes that are now used as markers to identify the different regions (reviewed by Yamaguchi, 1997). For example, several members of the family of paired-box genes have specific expression in the somites (reviewed by Dahl *et al.,* 1997). *Pax1* and *Pax9* are expressed in sclerotome, whereas *Pax3* is expressed in dermomyotome (Deutsch *et al.,* 1988; Wallin *et al.,* 1994; Goulding *et al.,* 1994). Surgical perturbation experiments demonstrate that the dorsoventral polarity of each somite is established after segmentation as the result of interactions with the surrounding tissues. Adjacent tissues respecify young somites that are rotated 180° along the dorsal–ventral (DV) axis. Gene expression reflects their new position (Deitrich *et al.,* 1997, 1998), and they perform normally in the new context (Aoyama and Asamoto, 1988).

In the AP dimension, each somite also possesses polarity. The cranial and caudal-half sclerotome have different properties that can be visualized as differences in gene expression, cell density, and extracellular components (Stern *et al.,* 1986). Significantly, some of these properties are predetermined in the presegmental mesoderm. After experimental reversal of the rostrocaudal axis of the segmental plate, the AP polarity of the tissue is maintained, and does not respecify to its new orientation (Aoyama and Asamoto, 1988). This can be seen in the persistence of markers, the sequence of overt segmentation (reversed from the host axis), and the migratory paths taken by neural crest cells and axons of motorneurons from the spinal cord (Keynes and Stern, 1984; Bonner-Fraser and Stern, 1991). Any disruption of the primary mesodermal segmentation is reflected by these other structures, on which segmentation is apparently imposed by the somitic mesoderm.

In a process known as resegmentation (*Neugliederung;* Remak, 1855), the primary segmentation in the sclerotome is replaced by a new segmental pattern in the vertebral column. The anterior half of one sclerotome joins with the posterior half-sclerotome of the next anterior somite, to form a single vertebral centrum (Fig. 3). Thus each somite contributes to two vertebrae. The primary segmentation of the dermomyotome remains, allowing the functional necessity of an offset between the muscular and skeletal segments that enables lateral movement between the vertebra. Several experimental studies support the idea of resegmentation (e.g., Bagnall *et al.,* 1988; Goldstein and Kalcheim, 1992). There are authors who debate the reality of resegmentation, and that debate is covered in other chapters of this volume (Chap. 1, Vol. 48, by Brand-Saberi and Christ).

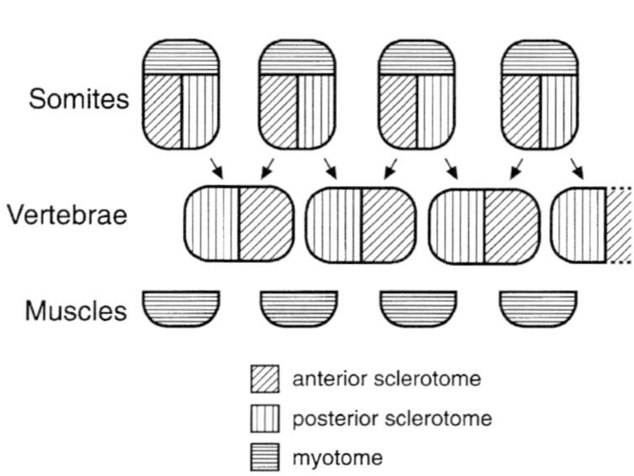

Figure 3 The theory of resegmentation predicts that the anterior-half of one sclerotome, together with the posterior-half of the next anterior sclerotome, come together to form a single vertebral centrum. The myotome retains its original segmental pattern.

C. Fate Maps from Experimental Embryology

The developmental fate and the timing of pattern determination of somitic populations in avian embryos were studied extensively by two groups in France and Germany in the 1970s and 1980s (reviewed by Gumpel-Pinot, 1984). Using a combination of techniques, including various forms of cell labeling, tissue transplants, and especially quail-chick chimeras, much progress was made toward understanding the behavior of this cell population, and fate maps have been drawn for the somitic derivatives (Fig. 4).

The stage at which cells become committed to fate and pattern is critical in understanding the control of these phenomena. The degree of correlation between the fate maps and the determination states of the somitic tissue can also be tested by experimental perturbation. As mentioned above, local signals that determine the dorsoventral identity of somites can respecify fates in young somites that have

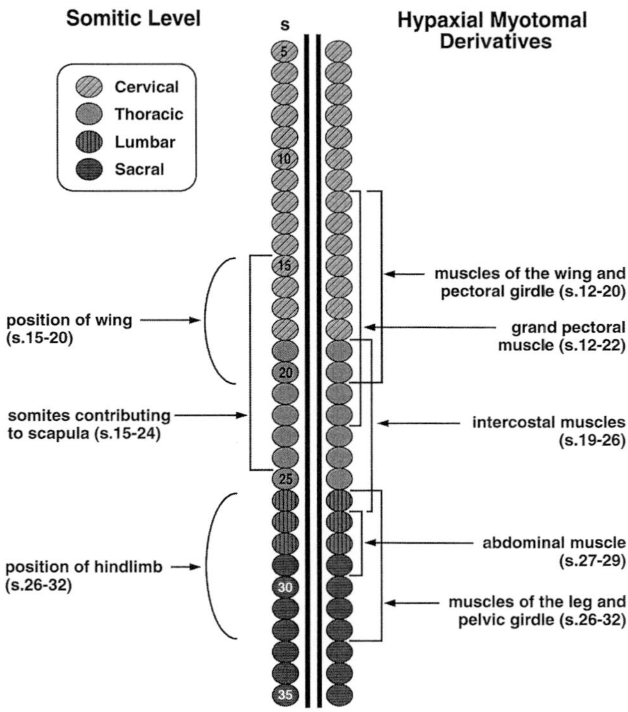

Figure 4 The axial formula of the chick is illustrated. Curved lines on the left side indicate the position of the limbs relative to somite level. Data collected from fate mapping experiments are added to show the somitic level of the myoblasts that make up various hypaxial muscle groups (shown with brackets on the right side), and the somites that contribute cells to the scapula are indicated on the left. (Modified from Gumpel-Pinot, 1984.)

been inverted 180° on the DV axis (Aoyama and Asamoto, 1988). In contrast, the AP polarity of the somitic mesoderm does not adjust after surgical reorientation, and the refractory behavior of the sclerotome is seen on both local and global levels of patterning (see below).

1. Sclerotome and Dermotome

The regionalization of sclerotomal derivatives along the AP axis was studied by Kieny *et al.* (1972) in the chick. Through a series of heterotopic grafts, the authors demonstrated that global patterning in terms of regional identity of the sclerotomal portion of the somite is determined very early, before overt segmentation. Transplantation of thoracic level segmental plate (SP) to cervical levels resulted in the presence of ribs in the neck. The corresponding half vertebrae were of thoracic type, indicating that the SP was intrinsically patterned to produce thoracic level structures. The reverse experiment, where cervical SP was transplanted to thoracic levels, resulted in cervical-half vertebra and corresponding gaps in the rib basket. Though the authors did not comment on the muscle derivatives of these transplants, a separate paper examined the derivatives of the dermotome in the same specimens (Mauger, 1972). This analysis showed that dorsal (epaxial) dermis and associated feather patterns arises from the somite, while dermis of the hypaxial regions arises from lateral plate mesoderm. In agreement with the pattern of sclerotome derivatives, the dermotome also maintains its original AP identity after transplantation, and produces feather tracts appropriate to its site of origin.

2. Myotome

In the myotome, there are two distinct myogenic lineages (Ordahl and Le Douarin, 1992). The dorsomedial lip of the dermomyotome gives rise to a lineage that will supply cells for the epaxial muscles, and the ventromedial cap of the dermomyotome will give rise to the highly migratory population of myoblasts that will form all of the hypaxial muscles. Somite transplants demonstrate that the limb muscles do arise from the somite (Chevallier, 1978; Chevallier, 1977; Chevallier and Mauger, 1977; Christ *et al.,* 1977, 1983). Furthermore, when brachial level somites are replaced by midtrunk level somites, normal limb muscles form (reviewed by Gumpel-Pinot, 1984), and the innervation is normal (Keynes *et al.,* 1987). Thus, the hypaxial muscle precursors do not contain their own global patterning information. The information for the correct axial patterning of the hypaxial muscles must reside in the lateral plate mesoderm, which forms the connective tissue in which the myotomal cells differentiate into myocytes. Murakami and Nakamura (1991) using quail-chick chimeras looked specifically at trunk muscles and concluded that this cell population shows signs of autonomous, level-specific patterning, but under the influence of the heterotopic lateral plate.

In general, surgical studies show profound differences in the timing of pattern determination in epaxial and hypaxial precursors. Global aspects of patterning are autonomously set in the somitic populations that contribute to the epaxial structures. In contrast, the hypaxial population remains plastic. The epaxial structures all develop more or less *in situ,* without migrating substantial distances and, most importantly, without mixing with any other cell population (except neural crest). The hypaxial muscles all form within a connective tissue matrix derived from the lateral plate mesoderm.

The difference between epaxial and hypaxial myotomal populations is also reflected by activity of the promoter regions of the myosin light chain gene in adult muscle. A transgenic construct that places the chloramphenicol acetyltransferase (CAT) reporter gene under control of the MCL1 promoter displays a marked anterior to posterior gradient in the expression of the CAT reporter (Donoghue *et al.,* 1991, 1992; Grieshammer *et al.,* 1992). The greater than 100-fold gradient is found only in the epaxial muscles, and is maintained in cultured isolates. The significance of this gradient is not understood, but it implies the presence of a cell-autonomous, AP identity.

If the lateral myotomal population possesses any axial identity, it is overwritten by the context of the lateral plate as the hypaxial precursors migrate from the midline. This patterning strategy makes good functional sense. The skeleton of the limb is also lateral plate derived, and the functional integrity of the limbs requires the sophisticated alignment of muscle and bone. One patterning system rather than two greatly reduces the chances of error, and greatly increases the opportunities of genetic changes leading to successful morphologies.

3. Scapula

In 1977 Chevallier demonstrated that the avian scapula is formed in part from somitic cells. This is somewhat surprising, since the rest of the appendicular skeleton. including pelvic girdle, coracoid, and all elements of the limbs proper, is formed from lateral plate tissue. Generally, an embryonic distinction was assumed to parallel the historical and functional distinction between the axial and the appendicular skeletal systems in vertebrates.

The participation of somites in the formation of the scapula requires unique behaviors from some subset of somitic cells. In chicks, Chevallier reports that cells from somites 15–24 are found in the scapular blade in an AP pattern consistent with their somitic origin (Chevallier, 1977). This stretch of somites includes the cervical to thoracic transition, and somites that are forming thoracic vertebrae and ribs are also contributing cells to the scapula (Fig. 4). The segregation of a portion of somitic cells to fulfill this role requires additional patterning information.

In the course of investigating the development and evolution of the turtle body plan, I performed extirpations of somites on early turtle embryos (Burke, 1991a). These experiments and those done by Yntema (1970) show that only extirpations of somites from the cervical series were found to cause depletions in the scapular blade of the turtle. Thus two different amniotes (chick and snapping turtle) have utilized the segmental mesoderm in an element of the appendicular skeleton, but to a different degree. These data illustrate one aspect of the evolution of developmental patterns between these two lineages.

In order to more fully understand the evolution of this developmental character, it was necessary to look at a primitive outgroup to determine the character state the chick and turtle patterns evolved from. Most salamanders have a very basic body plan that is likely to conform to the primitive body plan of tetrapods. Surgical extirpations of brachial level somites in *Ambystoma maculatum* results in larvae with totally depleted epaxial muscles in the limb region, but a normal scapula (Burke, 1991b). The experiments suggest that the primitive tetrapod scapula arises solely from lateral plate.

The *undulated* mutant in mice, now known to be one of various mutations of *Pax1,* has malformations of the axial skeleton (Balling *et al.,* 1988; Wallin *et al.,* 1994). These mice also have a scapular phenotype that may have implications for the role of the somites in the scapula of mammals (Timmons *et al.,* 1994). The ventral half of the scapular cartilage is missing, and the tissue in its place is muscle (Gruneberg, 1953). Since the rest of the appendicular skeleton is normal in these mice, the mutant phenotype suggests the participation of somitic cells in the scapula.

Placed in phylogenetic context, the data from different tetrapods suggest that a somitic contribution to the scapula represents a developmental innovation in amniotes (Fig. 5). On the basis of the primitive condition seen in salamanders, one can hypothesize that somitic cells, which primitively formed scapular muscles in the ancestor of modern amniotes, were recruited into the scapular blade. This would require that cells from the myotome change their fate from myogenic to skeletogenic. This may not be a novel or unique instance, as it has been reported

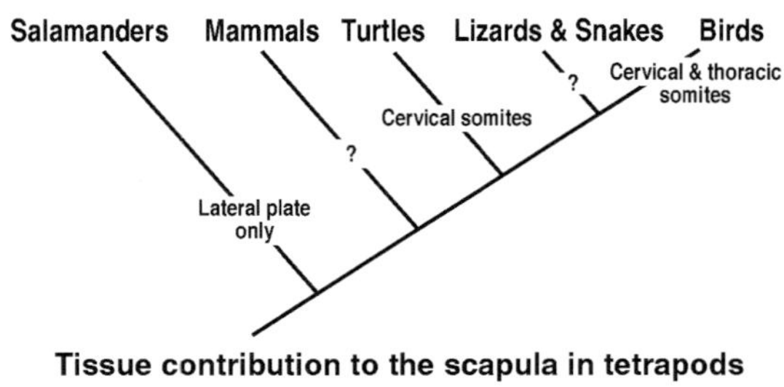

Figure 5 A cladogram illustrating the current consensus of the phylogenetic relationships between crown group tetrapods and the distribution of existing data on the developmental origin of the scapula in the indicated taxa. Salamanders, representing the amphibians, are hypothesized to have a scapula derived from lateral plate mesoderm only. Turtles and birds among the amniotes have variable somitic contributions to the scapula and the condition in the remaining taxa is unknown. See text for discussion.

that the ribs in quail–chick chimeras arise from the dermomyotome, not the sclerotome (Kato and Asymoto, 1998).

A skeletogenic fate for dermomyotomal cells has been reported in mice deficient in *Myf5,* one of the local signals in the myogenic pathway (Ott *et al.,* 1991; Tajbakhsh *et al.,* 1996). The evolution of the amniote scapula may have used this pathway to alter the fate of myotomal cells. The recruitment of the segmented mesoderm by the appendicular skeleton may have enabled the wide range of locomotor adaptations shown in this highly diverse clade (Burke, 1991b). Lineage analysis of cells within single somites will test this hypothesis (Burke, in preparation). Regardless which somitic subpopulation provides cell to the scapula, global patterning information must exist to control which somites along the axis are involved.

III. Mechanisms for Global Patterning

The establishment of the primary AP and DV axes in embryos show strong similarities as well as differences between organisms. The molecular signals so far identified in these processes tend to be conserved, but the timing and topography of their action varies with the taxa. These axes are generally set up just before and at the initiation of gastrulation and will not be discussed here (see reviews by Slack *et al.,* 1992; Boncinelli and Mallamaci, 1995; Goldstein and Freeman, 1997; Lemaire and Kessel, 1997). Once the basic beginning-middle-and-end of the embryo is established, additional signals come into play that further regionalize the body. In vertebrates one of the outcomes of this process is the axial formula.

The first essential criterion for a signaling/information system that controls global morphological differences along the axis is differential distribution of the signal to demarcate different regions. As far as we can tell with current techniques, the molecular agents that control local patterning of somites (mentioned above) are uniform along the AP axis. Therefore they do not contain global information. Evolutionary changes in these genes have an impact at the level of physiology, but can be quite independent of morphological evolution. To date the only known candidates that demonstrate significant differences in AP expression domains in the paraxial mesoderm are the *Hox* genes. (The AP gradient of MLC1-CAT mentioned above occurs in relatively mature somites, so is presumably secondary to *Hox* expression.)

A. *Hox* Genes and the Vertebrate Body Plan

Genes that affect the pattern of different body segments were first discovered in *Drosophila* (Lewis, 1978), and members of the *Hox* family of homeobox containing transcription factors have now been found in most metazoans (reviewed in Kappen and Ruddle, 1993; Duboule, 1993). Arthropods and vertebrates both have segmental body plans. Despite their independent evolution for at least 600 million years, and vast differences in the morphology of their segments (Minelli and Peruffo, 1991: Minelli, 1996), the genetic underpinnings of regional AP patterning are similar. *Hox* genes control the identity of individual segments in both phyla (Akam, 1989).

In the jawed vertebrates, the original *Hox* cluster has undergone duplication, and there are four or more clusters (see below) situated on four different chromosomes. Individual gene members of the original cluster were differentially duplicated. In tetrapods, there are at least 39 members representing 13 paralogous groups in clusters A, B, C, and D (Fig. 6). There is little doubt that the conservation of structure in the families results from some form of interdependence between the members. One of the remarkable features of the *Hox* genes is the phenomenon known as colinearity. In both arthropods and chordates, *Hox* genes show an anterior to posterior sequence of expression along the embryonic axis that is colinear with their position on the chromosomes (Lewis, 1978; Duboule and Dollé, 1989; Graham *et al.,* 1989). Colinearity is a truly remarkable instance of a spatial parallel between the genotype and phenotype.

Axial *Hox* expression in chicks and mice is generally established in two phases during embryogenesis (Deschamps and Wijgerde, 1993; Lemaire and Kessel, 1997). The first phase starts in the posterior primitive streak during its formation

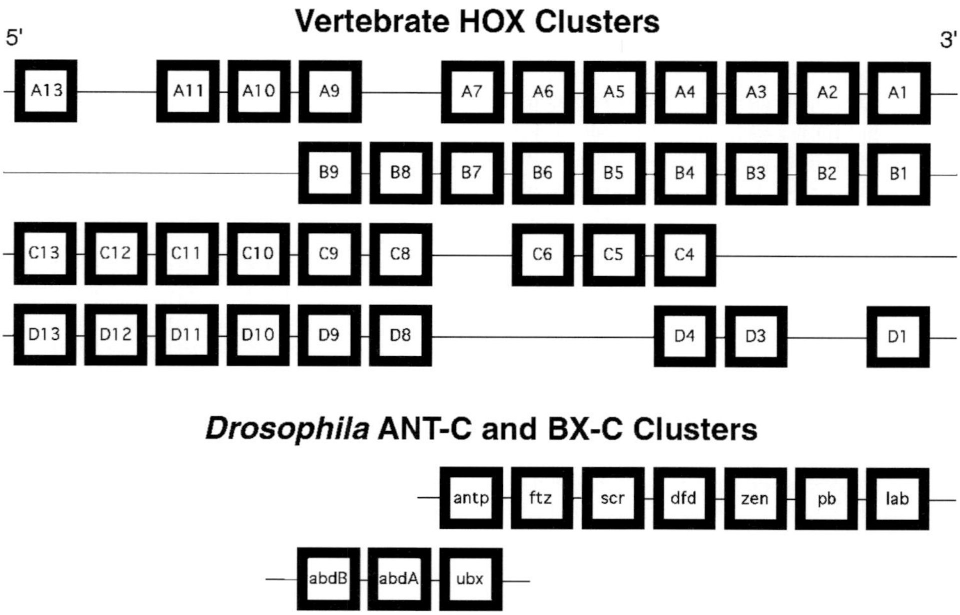

Figure 6 Schematic representation of the homeotic cluster genes (HOM-C or *Hox* genes) in *Drosophila* (bottom) and in vertebrates (top).

and extends from posterior to anterior. The second phase involves the refinement of an anterior border of expression (Gaunt, 1991; Gaunt and Strachan, 1994). *Hox* genes are expressed in a number of embryonic tissues, and generally show obvious colinear expression in the central nervous system (CNS), the paraxial mesoderm, and the gut (Krumlauf, 1994; Roberts *et al.,* 1995; Yokouchi *et al.,* 1995). The lateral plate mesoderm also expresses *Hox* genes (i.e., Cohn *et al.,* 1997), though an overall pattern has not been fully described. The expression in the paraxial mesoderm is generally offset from that of the CNS, and anterior borders of expression of individual *Hox* genes are more anterior in the nervous system than in the mesoderm.

1. The *Hox* Code

The colinear, AP expression pattern of the *Hox* genes fulfills the first requirement for agents of global information, but there are several other very compelling reasons to identify *Hox* genes as instruments of global patterning. In mice, both loss- and gain-of-function mutants that alter *Hox* gene expression in the paraxial mesoderm often, though not always, result in altered fates of the sclerotomal derivatives (reviewed in McGinnis and Krumlauf, 1992; Crawford, 1995). The changes in the morphology of individual vertebrae have been interpreted as homeotic transformations.

In 1991 Kessel and Gruss proposed that a "combination of functionally active *Hox* genes, the *Hox* code, specifies the identity of a body region, for example, a vertebral segment. Different *Hox* codes thus represent the interpretation of positional information along the anterioposterior body axis" (Kessel and Gruss, 1991, p. 101). The original hypothesis of a code was based on the results of retinoic acid perturbations in mice that caused alteration in *Hox* expression domains and morphological transformations of vertebrae (Kessel and Gruss, 1991). Quantitative analysis of shape change in vertebrae is consistent with a cumulative, combinatorial *Hox* code. Using Fourier analysis, Johnson and O'Higgins (1996) found a strong correlation between degree of shape change and the number of *Hox* genes active in two adjacent cervical vertebrae in the mouse.

Transgenic manipulations of *Hox* genes in mice also support the notion of a *Hox* code. The morphologies generated in these experiments, however, are often difficult to interpret or predict. Models of homeotic mutations from *Drosophila* are based on the idea of "posterior prevalence" wherein a gene with a more posterior expression boundary dominates the more anterior genes (Duboule and Morata, 1994). Such a model predicts that loss-of-function mutations will result in anterior transformations of segments, whereas gain-of-function will generate posterior transformations. Though many of the mouse *Hox* mutants display homeotic transformations of vertebral types, they are inconsistent and so far unpredictable. As an example, a null mutant of *Hox a-4* changes the *Hox* code of the third cervical vertebra to that of the second (Kostic and Capecchi, 1994; Ramirez-Solis *et al.,*

1993). The morphology of this mutant conforms to the *Hox* code: the third cervical is transformed to that of the second, an anterior transformation as predicted by the loss-of-function mutation. However, as reported in the same paper, the knockout of *Hox a-6* results in a posterior transformation as the seventh cervical is transformed into the first thoracic.

There are now many reports of double and triple *Hox* mutants in the literature. There are examples of nonallelic, noncomplementation; cumulative effects; synergies or redundancy between paralogue members (see Horan *et al.,* 1995). The task of reviewing all of the mutants is a large one, and, as yet, there are no models that can explain all of the results (Maconochie *et al.,* 1996, and see discussion of Crawford, 1995, below).

B. Comparative Data and Evolutionary Implications

The morphological results of the genetic experiments mentioned above are strong evidence of a causal role for *Hox* genes in the patterning of individual vertebrae. These data leave open, however, the question of how *Hox* genes are acting at the level of global pattern (i.e., the axial formula), and thus what proximal role they play in the evolution of vertebrate body plans. The experimental association of altered *Hox* expression boundaries and vertebral transformations in mice could simply reflect a role for *Hox* genes in primary segmentation. Specific vertebral morphology would then be determined entirely by downstream genes, and evolutionary transposition would occur independent of the *Hox* code. This patterning system would be analogous to that seen in insects. Comparing flies to butterflies, Carroll and co-workers (reviewed by Carroll *et al.,* 1995) have shown that the normal expression boundary of *Ubx* is fixed at the third thoracic segment (T3) in both taxa. Genes that are downstream of *Ubx* cause wings to form on T3 in lepidoptera (butterflies), whereas haltares form in diptera (e.g., flies). The evolutionary changes governing segment morphology in these taxa do not involve changes in the regulation of *Hox* genes, but rather changes in genes acting downstream that act to model the segment.

If the segmental patterning system in vertebrates functioned in the way described above for insects, we would find that the *Hox* expression borders were fixed to somite number, regardless of changes in the axial formula. Testing the evolutionary role of *Hox* genes in vertebrate diversification requires studies that compare the development of body regions in different vertebrates.

Chicks and mice provide the necessary variation for a comparative test of the evolutionary role of *Hox* genes in vertebrate transposition. Since the fifth somite in all amniotes contributes partly to the occipital region of the skull, and partly to the atlas, or first cervical vertebrae (Goodrich, 1930), chicks and mice start their vertebral series with the same somite. The axial formula of chicks includes 14 cervical vertebra and 7 thoracic, whereas the formula in mice is 7 cervical and 13 thoracic. Therefore, though they begin at the same point, the morphological and functional boundary between neck and trunk is transposed in these two taxa by seven segments. It is important to point out here that vertebrate morphologists assume that the axial regions (i.e., cervical versus thoracic) are homologous in these two organisms. That is to say, at some point in their history, mammals and birds have a common ancestor with an axial formula that included both neck and trunk. The differences in the number of segments in these regions between birds and mammals have evolved since their phylogenetic lineages diverged. Thus the regions are homologous, even if individual vertebrae are not.

Comparative analyses that map *Hox* gene expression borders to morphological regions in the chick and mouse embryo demonstrate that gene expression patterns are transposed consistently with morphology (Gaunt, 1994; Burke *et al.,* 1995). In other words, specific *Hox* genes are always expressed at specific morphological boundaries along the AP axis regardless of changes in the number of segments that contribute to each morphological region in different species of vertebrates (Fig. 7). *Hox c-6,* for instance, is expressed in the first thoracic vertebrae of mice (somite 12), chicks (somite 19), and geese (somite 21, Fig. 8). In fishes and amphibians, there is no distinct morphological transition between cervical and trunk vertebrae. However, *Hox c-6* expression is aligned with the segment that provides the most posterior spinal nerve that innervates the forelimb of *Xenopus* and the pectoral fin of the zebrafish. In amniotes, the last segment to contribute a spinal nerve to the brachial plexus is the first thoracic, indicating that the axial level of *Hox c-6* expression is morphologically consistent between all these vertebrates (Burke *et al.,* 1995) (Fig. 8).

The phylogenetic consistency of this relationship between gene expression and morphology confirms a causal role for *Hox* genes in AP regionalization, and gives a molecular level to the homology of body regions in tetrapods. These data also demonstrate that the evolutionary changes that cause transposition in vertebrates occur upstream of the *Hox* genes. A similar situation is seen in the crustaceans. In these organisms *Hox* expression is decoupled from segment number. In contrast to the insects, *Ubx* expression is correlated with regional morphological changes between segments with feeding and locomotor appendages in different taxa (Averof and Patel, 1997).

The structure and expression pattern of the *Hox* family in the zebrafish *Danio rerio* has been extensively studied (Prince *et al.,* 1998a,b). These authors report additional members of the gene family in zebrafish that have not been found in amniotes. The expression boundaries of the *Hox* genes in the trunk are far more compressed along the AP axis than is seen in chicks and mice. The morphology of the axial skeleton of the zebrafish does not show the same degree of regionalization characteristic of many amniotes. However, changes do exist along the axis, and the tail is a very complex structure. What is perhaps most interesting is the fact that the *Hox* gene boundaries of paralogues 1–10 all occur within approximately the first 14 somites (Fig. 9). *Hox c-11* expression in the paraxial mesoderm falls at the level of somite 17, which is approximately the position of the vent or cloaca, marking the transition from trunk to tail (Prince *et al.,* 1998b). The first 14 somites

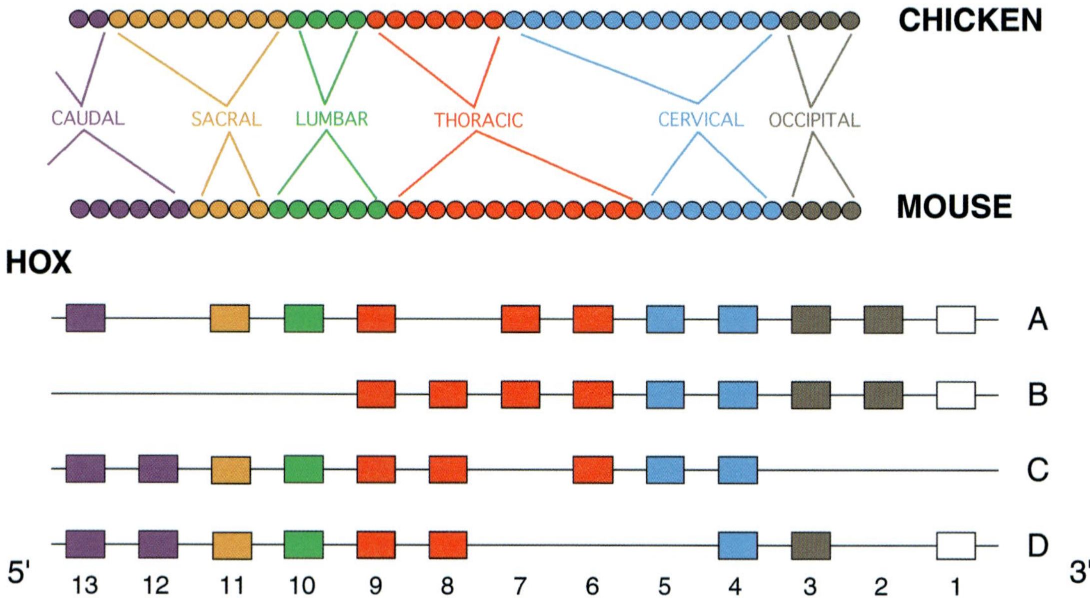

Figure 7 The axial formula of the chick and the mouse are compared to demonstrate the transposition between homologous anatomical regions in these two taxa. Especially noteworthy is the position of the cervical–thoracic transition. Below, the schematized *Hox* clusters are shaded to correspond to the morphological region shown above, in which specific groups of *Hox* paralogues have their anterior borders of expression. See text for discussion.

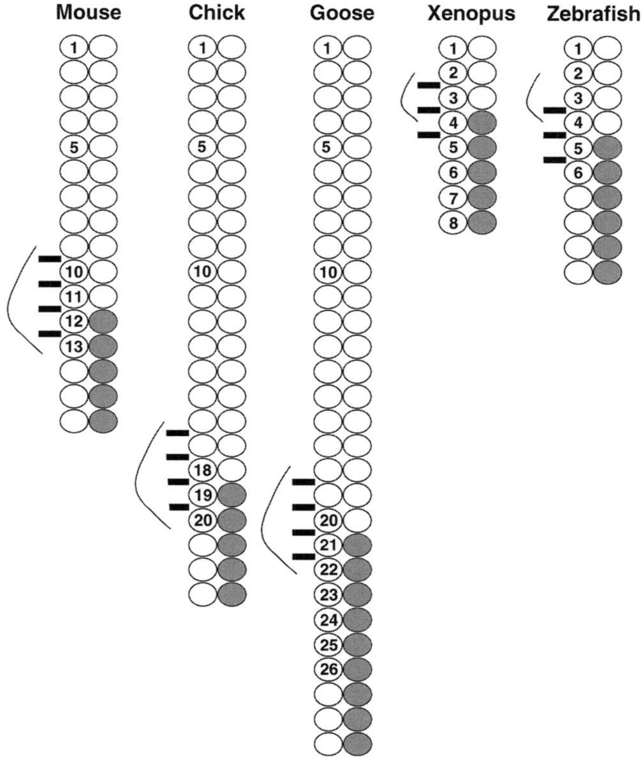

Figure 8 Axial formulas for a range of different vertebrates showing the somite level of the anterior border of *Hoxc-6* expression (shaded segments). The horizontal bars represent the spinal nerves that will form the brachial plexus and innervate the forelimb/pectoral fin.

are the only somites in the zebrafish that are put in place during the epiboly stages of gastrulation. The posterior 30–40 segments arise from the tail bud (Müller *et al.,* 1996). The possibility that these *Hox* boundaries are significant to AP patterning in the fish needs additional work. In chick and mouse, anterior expression boundaries of paralogue 10 members are associated with the lumbosacral transition, paralogue 11 members with the sacral region, and paralogue 12 and 13 are found in the caudal region (Figs. 7 and 9).

To fully understand the role of *Hox* genes in the evolution of vertebrate axial anatomy, some notion of the expression patterns in an ancestral form is necessary. This can be estimated by out-group comparison to the appropriate taxa. The zebrafish is not an adequate representative of a primitive vertebrate axial formula, because this species is itself a highly derived teleost. *Hox* genes in protochordates have also been the subject of study. *Amphioxous* is known to posses a single cluster of *Hox* genes that are not expressed in the paraxial mesoderm (Garcia-Fernández and Holland, 1994). This indicates that duplications of the cluster must

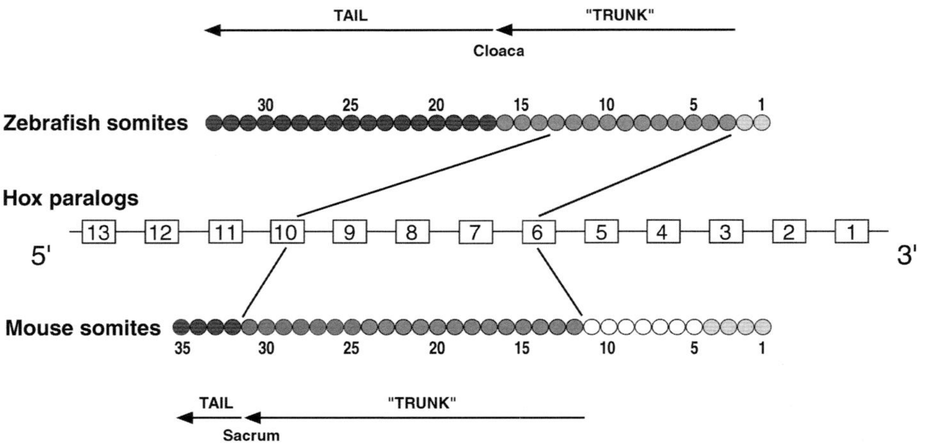

Figure 9 The position of the anterior expression borders of *Hox* paralogue groups 6 through 10, relative to the transition between "trunk" and tail in the axial formula of zebrafish (top) and mouse (bottom). Zebrafish data were redrawn from Prince *et al.* (1998b).

have occurred in the lineage leading to the agnathans (modern lamprey and hagfish) and jawed vertebrates because these taxa have multiple clusters. The precise timing of the event is unclear (Sharman and Holland, 1998).

1. Relevance of the Paraxial *Hox* Code to Lateral Structures

As mentioned above, the axial structures derived from the somites do not function in isolation, but must be patterned in such a way as to work in harmony with lateral structures, particularly the paired appendages. This brings into question how axial (somite) patterning is coordinated with lateral structures in the global context. One answer discussed above lies in the fact that the lateral plate dictates lateral myotomal patterning. However, this does not account for the fact that the limb buds are always positioned adjacent to specific axial regions. Many species of fish seem to have flexibility in the positioning of their fins relative to the main body axis (Parenti and Song, 1996). However, this flexibility is apparently unavailable to tetrapods, which have a consistent placement of the forelimb at the cervicothoracic transition (site of particular *Hox* expression) and the hind limb at the lumbosacral transition (site of a different set of *Hox* expression). Both of these regions can be transposed independently in different taxa but are never uncoupled from their respective limb (except in the case of total limb loss). The three-way correlation between *Hox* expression boundaries, axial transitions, and limb placement requires some explanation.

Experimental evidence from several laboratories has shown that initial limb bud outgrowth can be elicited anywhere in the flank by the ectopic expression of fibroblast growth factors (FGFs) (Cohn *et al.,* 1995; Ohuchi *et al.,* 1998; Crossley *et al.,* 1996; Vogel *et al.,* 1996). Cohn *et al.* (1997) further showed that the axial *Hox* expression does not change as a result of ectopic FGF induced limbs, but that the expression of several *Hox* genes in the lateral plate of the flank is affected. The changes result in the re-creation of the normal forelimb-level pattern of *Hox* gene expression at the ectopic site in the lateral plate. As a result, the authors conclude that the *Hox* genes are independently regulated in somitic and lateral plate mesoderm and that this independence may have been an instrumental step in the evolution of paired appendages. This argument has been expanded by linking the history of lateral plate *Hox* expression to expression in the gut (Coates and Cohn, 1998). Both endoderm and splanchnic mesoderm contribute to the gut, which is also regionalized by *Hox* genes (Roberts *et al.,* 1995; Yokouchi *et al.,* 1995). Coates and Cohn (1998) suggest that regionalized *Hox* expression originally present in the gut was co-opted by the somatic lateral plate during the evolution of paired fins. The pre-existing *Hox* boundaries could be used as orientation for fin placement. This would explain the independent regulation of the lateral plate and somitic *Hox* code.

Without some form of information exchange, it is not clear how a lateral plate *Hox* code could be entirely independent of the axial code when the derivatives of the two tissues maintain a constant positional relationship throughout tetrapod evolution. Communication between *Hox* expression in the neural tube, the paraxial mesoderm, and the lateral mesoderm would account for concerted development. Some communication between *Hox*-expressing tissues has been reported. Somitic mesoderm transplanted to hindbrain levels can influence *Hox* expression in the rhombomeres (Itasaki *et al.,* 1996), and thoracic level somites transplanted to the level of the sacrum affects *Hox* expression in the neural tube (C. Lance-Jones, 1998, personal communication).

C. *Hox* Function

Obviously, both upstream regulators and downstream targets of *Hox* genes are at play in the generation of morphological variation in evolution. Developmentally these form two questions, on the one hand, what turns the *Hox* genes on and establishes their pattern of expression, and, on the other, what is it that *Hox* genes themselves actually do. In *Drosophila,* many parts to both these questions are understood. For instance, *Pair-rule* and *Gap* genes initiate *Hox* expression, and members of the *Polycomb* and *trithorax* groups maintain and tailor their expression. Known downstream targets include "realizator" genes (Garcia-Bellido, 1975), whose products are involved in the process of differentiation, as well as other regulatory genes (Graba *et al.,* 1997). In vertebrates however, very little is known about *Hox* regulation in either direction. It is striking that a gene family that has monopolized so much thought and effort by such a wide range of biologists remains so elusive.

1. Possible Upstream Factors

It is generally accepted that the events leading to general AP patterning occur during the process of gastrulation. It is important to point out that as cells undergo the characteristic shape changes and migrations that constitute gastrulation, they are subject to local signals, both chemical and physical. Apparently these local signals initiate genetic responses that will result in global pattern. Several factors are known to influence *Hox* expression *in vivo.* Manipulation of the mammalian homologues of the *Drosophila Polycomb* and *trithorax* genes can alter *Hox* gene expression boundaries and generate "homeotic" vertebral transformations (see van der Lugt *et al.,* 1996; Schumacher *et al.,* 1996; Yu *et al.,* 1995). The knockout of *Cdx1,* a homologue of the *Drosophila* gene *caudal,* causes alterations of at least five different *Hox* expression boundaries and anterior transformations in the mouse (Subramanian *et al.,* 1995). *Xcad-2* in *Xenopus* also effects *Hox* expression and may act in concert with retinoic acid (Epstein *et al.,* 1997).

Retinoic acid (RA) is notorious as a potential morphogen, and many studies have demonstrated a mechanistic relationships between RA, its receptors, and *Hox* genes (reviewed by Conlon, 1995). This relationship between RA and *Hox* genes may be unique to the chordate lineage, because it has not been found among the arthropods or echinoderms. *Hox* expression is sensitive to RA in tunicates and *Amphioxous* (reviewed by Shimeld, 1996). Among the vertebrates, RA binding sites have been found in *Hox* regulatory regions, and concentration gradients of RA have been found in the right time and place to potentially direct the early expression of *Hox* genes. For example, RA is found in Hensen's node in the chick and the dorsal lip of the blastopore in *Xenopus* (Hogan *et al.,* 1992; Chen and Solursh, 1994). Current models propose that the anterior, (3′) *Hox* paralogue members are directly regulated by RA. The initial activation could then result in the temporal and spatial colinear expression of the *Hox* family along the AP axis (Durston *et al.,* 1989; Conlon, 1995).

The proximal genetic changes that regulate *Hox* expression could be either cis- or trans-acting factors. These regulatory elements are one of the targets for evolutionary changes that result in variations in the axial formulas between taxa. There is currently excellent evidence of differences in cis-regulatory elements for *Hox c-8* in chicks and mice (Belting *et al.,* 1998).

2. Downstream Targets and Models of *Hox* Function

The collective experimental and comparative data mentioned above are compelling evidence for the existence of a *Hox* code. Once a *Hox* code is established by upstream factors, however, it requires interpretation, and *Hox* genes must effect local morphology via downstream targets. At least 19 candidate targets for the *Drosophila Hox* genes have been identified including other transcription factors, signaling molecules, and structural genes (Graba *et al.,* 1997). Estimates predict as many as 85–170 target genes for *Ubx* (Mastick *et al.,* 1995). In comparison to our knowledge of *Drosophila,* downstream targets for vertebrate *Hox* genes are extremely scarce. In the face of this absence, there are only a few theories about how the genes actually control morphology. This is the question that cuts to the heart of the issue of global patterning in somites.

The detailed mechanistic understanding of homeotic transformations in *Drosophila* has led to the implicit assumption that a genetic switch which triggers a different, segment-specific developmental program causes homeotic morphologies in vertebrates. One example of a programmatic interpretation of *Hox* gene function comes from interpretations of the *Hox a-2* knockout in mice. This phenotype, which involves morphology in the first and second branchial arches, has been interpreted as the re-creation of an ancestral condition, resulting in the production of an atavistic palatoquadratate ("pterygoquadrate" of Rijli *et al.*, 1993; Mark *et al.*, 1995). Smith and Schneider (1998) have pointed out that detailed morphological analysis of this phenotype is more accurately interpreted as the result of generic properties of *Hox* genes, rather than an altered genetic "groundplan" reflecting an ancestral condition. Generic properties include genetic and epigenetic effects on cell migration, proliferation, and differentiation. Mutant phenotypes therefore result from these generic responses rather than specific programs.

In several papers Duboule (1994, 1995) has also argued for a less programmatic interpretation of *Hox* function. He maintains that the so-called homeotic transformations of vertebrae in *Hox* knockouts may result simply from heterochronic shifts in rates of cell proliferation. The difference between one vertebra and the next may result from "realizator" genes that act on rates and patterns of cell division in the somitic sclerotome, for instance. Using similar reasoning, Crawford (1995) has argued that the vertebral transformations seen in many *Hox* knockouts may be the result of intercalary regeneration rather than true homeotic transformations. The regeneration occurs as the embryo attempts to recover from the positional discontinuities that result from the absent expression domain of one of the *Hox* genes in the "code."

These models define a proximal role for individual *Hox* genes that is contextual and generic and does not require segment-specific targets. The identified target genes in both *Drosophila* and vertebrates, referred to above, are consistent with the notion that the proximal role of *Hox* genes is to effect generic properties of cell behavior.

IV. Conclusions

Our understanding of global patterning remains superficial. However, the *Hox* gene family now provides a solution to the conceptual difficulty of uniting two hierarchical levels of patterning information in the somitic mesoderm. Both local and global information can be accounted for by the phenomenon of temporal and spatial colinearity. The action and interaction of *Hox* genes represents the accumulation of local signals that result in global patterning.

Colinearity is the single most important aspect of *Hox* gene behavior because it endows the *Hox* genes with the capability of global patterning (Duboule, 1994). On the other side of the coin, a role in global patterning saddles the *Hox* cluster with a considerable evolutionary constraint. There is approximately 1200 million years in the collective evolutionary life span of the *Hox* clusters (just counting the lineage distance between chordates and arthropods). The maintenance of cluster structure over this time span can only be due to a profound functional burden (Akam, 1989). Primary body organization seems adequate grounds for such a constraint.

Within our current understanding of pattern formation there is still a great deal to be worked out. The role of the *Hox* genes in establishing the body plan is often attributed to the ever-elusive positional information. For instance, "the body plan of the future organism . . . obviously requires positional information. In abstract terms, positional information within the embryo has first to be created, then transmitted to daughter or neighboring cells and finally translated into regional identity. *Hox* genes are believed to be implicated in this last process of translating positional information into regional identity along the rostrocaudal axis of the trunk" (Boncinelli and Mallamaci, 1995).

Understanding how *Hox* genes translate positional information into regional identity requires a better picture of the proximal roles *Hox* genes play in normal development. The combined use of classic morphological perturbation experiments with manipulations at the molecular genetic level enable us to directly test proposed molecular function. Although the role of a single *Hox* gene is immaterial outside of the context of colinearity, molecular methods for loss- and gain-of-function of individual genes can provide clues to downstream targets. The use of comparative studies that specifically address the role of *Hox* genes in animals with distinct morphological variations on a common body plan can also test proposed functional roles, as well as demonstrate proximal incidents of genetic changes that result in morphological evolution.

Acknowledgments

I thank Timothy Dansdill for critical reading of the manuscript, and Susan Whitfield for execution of the illustrations. The thought and work that went into this chapter are dedicated to the memory of Pere Alberch, whose own work made a place for it, and whose passing is a major loss. The author's work is currently supported by NIH R29-HD35932-01, and a Basil O'Connor Starter Scholar Research Award from the March of Dimes.

References

Akam, M. (1989). Hox and HOM: Homologous gene clusters in insects and vertebrates. *Cell* **57,** 347–349.

Aoyama, H., and Asamoto, K. (1988). Determination of somite cells: Independence of cell differentiation and morphogenesis. *Development* **104,** 15–28.

Averof, M., and Patel, N. (1997). Crustacean appendage evolution associated with changes in Hox gene expression. *Nature* **388,** 682–686.

Bagnall, K. M., Higgins, S. J., and Sanders, E. J. (1988). The contribution made by a single somite to the vertebral column: Experimental evidence in support of resegmentation using the chick–quail chimaera model. *Development* **103,** 69–85.

Balling, R., Deutsch, U., and Gruss, P. (1988). *Undulated,* a mutation affecting the development of the mouse skeleton, has a point mutation in the paired-box of *Pax 1. Cell* **55,** 531–535.

Belting, H., Shashikant, C., and Ruddle, F. (1998). Modification of expression and cis-regulation of Hoxc8 in the evolution of diverged axial morphology. *Proc. Nat. Acad. Sci. U.S.A.* **95,** 2355–2360.

Boncinelli, E., and Mallamaci, A. (1995). Homeobox genes in vertebrate gastrulation. *Curr. Opin. Genet. Dev.* **5,** 619–627.

Bronner-Fraser, M., and Stern, C. (1991). Effects of mesodermal tissues on avian neural crest cell migration. *Dev. Biol.* **143,** 213–217.

Burke, A. C. (1991a). The development and evolution of the turtle body plan: Inferring intrinsic aspects of the evolutionary process from experimental embryology. *Am. Zool.* **31,** 616–627.

Burke, A. C. (1991b). "Proximal Elements in the Vertebrate Limb: Evolutionary and Developmental Origin of the Pectoral Girdle." Plenum, New York.

Burke, A. C., Nelson, C. E., Morgan, B. A., and Tabin, C. (1995). Hox genes and the evolution of vertebrate axial morphology. *Development* **121,** 333–346.

Carroll, S., Weatherbee, S., and Langeland, J. (1995). Homeotic genes and the regulation and evolution of insect wing number. *Nature* **375,** 58–61.

Chen, Y., and Solursh, M. (1994). A concentration gradient of retinoids in the early *Xenopus laevis* embryo. *Dev. Biol.* **161,** 70–76.

Chevallier, A. (1977). Origine des ceintures scapulaires et pelviennes chez l'embryon d'oiseau. *J. Embryol. Exp. Morphol.* **42,** 275–292.

Chevallier, A. (1978). Etude de la migration des cellules somitiques dans le mesoderme somatopleural de l'ebauch de l'aile. *Wilheim Roux's Arch.* **184,** 57–73.

Chevallier, A. K., and Mauger, M. A. (1977). Limb–somite relationship: Origin of the limb musculature. *J. Embryol. Exp. Morphol.* **41,** 245–258.

Christ, B., Jacob, H. J., and Jacob, M. (1977). Experimental analysis of the origin of the wing musculature in avian embryos. *Anat. Embryol.* **150,** 171–186.

Christ, B., Jacob, M., and Jacob, H. J. (1983). On the origin and development of the ventrolateral abdominal muscles in the avian embryo. *Anat. Embryol.* **166,** 87–101.

Coates, M., and Cohn, M. (1998). Fins, limbs, and tails: Outgrowths and axial patterning in vertebrate evolution. *BioEssays* **20,** 371–381.

Cohn, M. J., Izpisúa-Belmonte, J. C., Abud, H., Heath, J. K., and Tickle, C. (1995). Fibroblast growth factors induce additional limb development from the flank of chick embryos. *Cell* **80,** 739–746.

Cohn, M. J., Patel, K., Krumlauf, R., Wilkinson, D. G., Clarke, J. D. W., and Ticke, C. (1997). Hox-9 genes and vertebrate limb specification. *Nature* **387,** 97–101.

Conlon, R. A. (1995). Retinoic acid and pattern formation in vertebrates. *Trends Genet.* **11,** 314–319.

Crawford, M. (1995). Transformations in null mutants of Hox genes: Do they represent intercalary regenerates? *BioEssays* **17,** 1065–1073.

Crossley, P. H., Minowada, G., MacArthur, C. A., and Martin, G. R. (1996). Roles for FGF8 in the induction, initiation and maintenance of chick limb development. *Cell* **84,** 127–136.

Dahl, E., Koseki, H., and Balling, R. (1997). Pax genes and organogenesis. *BioEssays* **19,** 755–765.

Deschamps, J., and Wijgerde, M. (1993). Two phases in the establishment of HOX expression domains. *Dev. Biol.* **156,** 473–480.

Deutsch, U. *et al.* (1988). Pax 1, A member of a paired box homologous murine gene family, is expressed in segmented structures during development. *Cell* **53,** 617–625.

Dietrich, S., Schubert, F. R., and Lumsden, A. (1997). Control of dorsoventral pattern in the chick paraxial mesoderm. *Development* **124,** 3895–3908.

Dietrich, S., Schubert, F. R., Healy, C., Sharpe, P. T., and Lumsden, A. (1998). Specification of the hypaxial musculature. *Development* **125,** 2235–2249.

Donoghue, M. J., Merlie, J. P., Rosenthal, N., and Sanes, J. R. (1991). A rostrocaudal gradiant of transgene expression in adult skeletal muscle. *Proc. Nat. Acad. Sci. U.S.A.* **88,** 5847–5851.

Donoghue, M. J., Morris-Valero, R., Johnson, Y. R., Meril, J. P., and Sanes, J. R. (1992). Mammalian muscle cells bear a cell-autonomous, heritable memory of their rostrocaudal position. *Cell* **69,** 67–77.

Duboule, D., Ed. (1993). "Guidebook to the Homeobox Genes." Oxford Univ. Press, Oxford.

Duboule, D. (1994). Temporal colinearity and the phylotypic progression: A basis for the stability of a vertebrate Bauplan and the evolution of morphologies through heterochrony. *Dev. Suppl.* 135–142.

Duboule, D. (1995). Vertebrate Hox genes and proliferation: An alternative pathway to homeosis? *Curr. Opin. Genet. Dev.* **5,** 525–528.

Duboule, D., and Dollé, P. (1989). The structural and functional organization of the murine HOX gene family resembles that of *Drosophila* homeotic genes. *EMBO J.* **8,** 1497–1505.

Duboule, D., and Morata, G. (1994). Colinearity and functional hierarchy among genes of the homeotic complexes. *Trends Genet.* **10,** 358–364.

Durston, A. J., Timmermans, J. P. M., Hage, W. J., Hendriks, H. F. J., de Vries, N. J., Heideveld, M., and Nieuwkoop, P. D. (1989). Retinoic acid causes an anteroposterior transformation in the developing central nervous system. *Nature* **340,** 140–144.

Epstein, M., Pillemer, G., Yelin, R., Yisraeli, and Fainsod, A. (1997). Patterning of the embryo along the anterior–posterior axis: The role of the caudal genes. *Development* **124,** 3805–3814.

Flower, W. (1885). "An Introduction to the Osteology of the Mammalia." A. Asher & Company, Amsterdam.

Gadow, H. F. (1933). "The Evolution of the Vertebral Column." Cambridge Univ. Press, Cambridge.

Garcia-Bellido, A. (1975). Genetic control of wing disc development in *Drosophila. Ciba Found. Symp.* **29,** 161–178.

Garcia-Fernández, J., and Holland, P. W. (1994). Archetypal organization of the amphioxus Hox gene cluster. *Nature* **120,** 407–413.

Gaunt, S. (1991). Expression pattern of mouse *hox* genes: Clues to an understanding of developmental and evolutionary strategies. *BioEssays* **13,** 505–512.

Gaunt, S. J. (1994). Conservation of the Hox code during morphological evolution. *Int. J. Dev. Biol.* **38,** 549–552.

Gaunt, S. J. and Strachan, L. (1994). Forward spreading in the establishment of a vertebrate Hox expression boundary: The expression domian separates into anterior and posterior zones, and the spread occurs across implanted glass barriers. *Dev. Dyn.* **199,** 229–240.

Gilbert, S. F. (1997). Developmental Biology. 5th ed., Sunderland, MA: Sinauer Assoc.

Goldstein, R. S., and Kalcheim, C. (1992). Determination of epithelial half-somites in skeletal morphogenesis. *Development* **116,** 441–445.

Goldstein, B., and Freeman, G. (1997). Axis specification in animal development. *BioEssays* **19,** 105–116.

Goodrich, E. S. (1913). Metameric segmentation and homology. *Q. J. Micrsc. Sci.* **59,** 227.

Goodrich, E. S. (1930). "Studies on the Structure and Development of Vertebrates." Macmillan, London.

Goulding, M., Lumsden, A., and Paquette, A. J. (1994). Regulation of Pax-3 expression in the dermamyotome and its role in muscle development. *Development* **120,** 957–971.

Graba, Y., Aragnol, D., and Pradel, J. (1997). *Drosophila* Hox complex downstream targets and the function of homeotic genes. *BioEssays* **19,** 379–388.

Graham, A., Papalopulu, N., and Krumlauf, R. (1989). The murine and *Drosophila* homeobox gene complexes have common features of organisation and expression. *Cell* **57,** 367–378.

Grieshammer, U., Sassoon, D., and Rosenthal, N. (1992). A transgene target for positional regulators marks early rostrocaudal specification of myogenic lineages. *Cell* **69,** 79–93.

Gruneberg, H. (1953). Genetical studies on the skeleton of the mouse XII. *J. Genet.* **52,** 441–455.

Gumpel-Pinot, M. (1984). Muscle and skeleton of limbs and body wall. *In* "Chimeras in Developmental Biology," (N. M. Douarin and A MacLaren, Eds.), pp. 281–310. Academic Press, London.

Hogan, B. L. Thaller, C., and Eichele, G. (1992). Evidence that Hensen's node is a site of retinoic acid synthesis. *Nature* **359,** 237–241.

Horan, G., Kovacs, E., (1995). Mutations in paralogous Hox genes result in overlapping homeotic transformations of the axial skeleton: Evidence for unique and redundant function. *Dev. Biol.* **169,** 359–372.

Itasaki, N., Sharpe, J., and Morrison, A. (1996). Reprogramming Hox expression in the vertebrate hindbrain: Influence of paraxial mesoderm and rhombomere transposition. *Neuron* **16,** 487–500.

Johnson, D., and O'Higgins, P. (1996). Is there a link between changes in the vertebral "hox code" and the shape of vertebrae? A quantitative study of shape change in the cervical vertebral column of mice. *J. Theor. Biol.* **183,** 89–93.

Kappen, C., and Ruddle, F. H. (1993). Evolution of a regulatory gene family: HOM/HOX genes. *Curr. Biol.* **3,** 931–938.

Kato, N., and Aoyama, H. (1998). Dermomyotomal origin of the ribs as revealed by extirpation and transplantation experiments in chick and quail embryos. *Development* **125,** 3437–3443.

Kessel, M., and Gruss, P. (1991). Homeotic transformations of murine vertebrae and concomitant alteration of Hox codes induced by retinoic acid. *Cell* **76,** 89–104.

Keynes, R. J., and Stern, C. D. (1984). Segmentation in the vertebrate nervous system. *Nature* **310,** 786–789.

Keynes, R. J., Stirling, R. V., Stern, C. D., and Summerbell, D. (1987). The specificity of motor innervation of the chick wing does not depend upon the segmental origin of muscles. *Development* **99,** 565–575.

Kieny, M., Mauger, A., and Sengel, P. (1972). Early regionalization of the somatic mesoderm as studied by the development of the axial skeleton of the chick embryo. *Dev. Biol.* **28,** 142–161.

Kostic, D., and Capecchi, M. R. (1994). Targeted disruption of the murine hoxa-4 and Hoxa-6 genes result in homeotic transformations of the vertebral column. *Mech. Dev.* **46,** 231–247.

Krumlauf, R. (1994). Hox genes in vertebrate development. *Cell* **78,** 191–201.

Lemaire, L., and Kessel, M. (1997). Gastrulation and homeobox genes in chick embryos. *Mech. Dev.* **67,** 3–16.

Lewis, E. B. (1978). A gene complex controling segmentation in *Drosophila. Nature* **276,** 565–570.

McGinnis, W., and Krumlauf, R. (1992). Homeobox genes and axial patterning. *Cell* **68,** 283–302.

Maconochie, M., Nonchev, S., Morrison, A., and Krumlauf, R. (1996). Paralogous Hox genes: Function and regulation. *Annu. Rev. Genet.* **30,** 529–556.

Mark, M., Rijli, F. M., *et al.* (1995). Alteration of Hox gene expression in the branchial region of the head causes homeotic transformations, hindbrain segmentation defects, and atavistic changes. *Semin. Dev. Biol.* **6,** 275–284.

Mastick, G. S., McKay, R. *et al.* (1995). Identification of target genes regulated by homeotic proteins in *Drosophila melanogaster* through genetic selection of Ultrabithorax protein binding sites in yeast. *Genetics* **139,** 349–363.

Mauger, A. (1972). Role du mesoderme somitique dans le developpement du plumage dorsal chez l'embryon de poulet. *J. Embryol. Exp. Morphol.* **28,** 313–341.

Minelli, A. (1996). Segments, body regions, and the control of development through time. *New Perspectives on the History of Life* **20,** 55–61.

Minelli, A., and Peruffo, B. (1991). Developmental pathways, homology and homonomy in metameric animals. *J. Evol. Biol.* **3,** 429–445.

Müller, M., Weizacker, E., and Campos-Ortega, J. (1996). Expression domains of a zebrafish homologue of the *Drosophila* pair-rule gene *hairy* correspond to primordia of alternating somites. *Development* **122,** 2071–2178.

Ohuchi, H., Takeuchi, J., *et al.* (1998). "Correlation of wing–leg identity in ectopic FGF-induced chimeric limbs with the differential expression chick Tbx5 and Tbx4. *Development* **125,** 51–60.

Ordahl, C. P., and Le Douarin, N. M. (1992). Two myogenic lineages within the developing somite. *Development* **114,** 339–353.

Ott, M., Bober, E., Lyons, G., Arnold, H., and Buckingham, M. (1991). Early expression of the myogenic regulatory gene, *myf-5,* in precursor cells of skeletal muscle in the mouse embryo. *Development* **111,** 1097–1107.

Parenti, L., and Song, J. (1996). Phylogenetic significance of the pectoral-pelvic fin association in acanthomorph fishes: A reassessment using comparative neuroanatomy. *In* "Interrelationships of Fishes," (M. Stiassny, L. Parenti, and G. Johnson, Eds.), pp. 427–444. Academic Press, San Diego.

Prince, V. E., Moens, C. B., Kimmel, C. B., and Ho, R. K. (1998a). Zebrafish Hox genes: Expression in the hindbrain region of wild-type and mutants of the segmentation gene, *valentino. Development* **125,** 393–406.

Prince, V. E., Joly, L., *et al.* (1998b). Zebrafish Hox genes: Genomic organization and modified colinear expression patterns in the trunk. *Development* **125,** 407–420.

Procter, N. S., and Lynch, P. J. (1993). Manual of Ornithology: Avian structure and function. New Haven: Yale Univ. Press.

Ramirez-Solis, R., Zheng, H., Whiting, J., Krumlauf, R., and Bradley, A. (1993). Hoxb-4 (Hox-2.6) mutant mice show homeotic transformation of a cervical vertebra and defects in the closure of the sternal rudiments. *Cell* **73,** 279–294.

Remak, R. (1855). "Untersuchungen über die Entwicklung der Wirbeltiere." Reimer, Berlin.

Richardson, M., Allen, A., Wright, G., Raynaud, A., and Hanken, J. (1998). Somite number and vertebrate evolution. *Development* **125,** 151–160.

Rijli, F. M., Mark, M., Lakkaraju, S., Dierich, A., Dolle, P., and Chambon, P. (1993). A homeotic transformation is generated in the rostral branchial region of the head by disruption of Hoxa-2, which acts as a selector gene. *Cell* **75,** 1333–1349.

Roberts, D., Johnson, R., Burke, A., Nelson, C., Morgan, B., and Tabin, C. (1995). Sonic hedgehog is an endodermal signal inducing Bmp-4 and Hox genes during induction and regionalization of the chick hindgut. *Development* **121,** 3163–3174.

Schumacher, A., Faust, C., and Magnuson, T. (1996). Positional cloning of a global regulator of anterior–posterior patterning in mice. *Nature* **383,** 250–253.

Sharman, A. C., and Holland, P. W. (1998). Estimation of Hox gene cluster number in lampreys. *Int. J. Dev. Biol.* **42,** 617–620.

Shimeld, S. (1996). Retinoic acid, HOX genes and the anterior axis in chordates. *BioEssays* **18,** 613–616.

Slack, J., Isaacs, H., Johnson, G., Lettice, L., Tannahill, D., and Thompson, J. (1992). Specification of the body plan during *Xenopus* gastrulation: Dorsoventral and anteroposterior patterning of the mesoderm. *Dev. Suppl.* 143–149.

Smith, K. K., and Schneider, R. A. (1998). Have gene knockouts caused evolutionary reversals in the mammalian first arch? *BioEssays* **20,** 245–255.

Stern, C. D., Sisodíya, S. M., and Keynes, R. J. (1986). Interactions between neurites and somitic cells: Inhibition and stimulation of nerve growth in the chick. *J. Embryol. Exp. Morphol.* **91,** 209–226.

Subramanian, V., Meyer, B., and Gruss, P. (1995). Disruption of the murine homeobox gene *Cdx1* affects axial skeletal identities by altering the mesodermal expression domains of Hox genes. *Cell* **83,** 641–653.

Tajbakhsh, S., Rocancourt, S. D., and Buckingham, M. (1996). Muscle progenitor cells failing to respond to positional cues adopt non-myogenic fates in myf-5 null mice. *Nature* **384,** 266–270.

Timmons, P. M., Wallin, J., Rigby, P. W. J., and Balling, R. (1994). Expression and function of Pax 1 during development of the pectoral girdle. *Development* **120,** 2773–2785.

van der Lugt, N., Alkema, M., Berns, A., and Deschamps, J. (1996). The Polycomb-group homolog Bmi-1 is a regulator of murine gene expression. *Mech. Dev.* **58,** 153–164.

Vogel, A., Rodriguez, C., and Izpisua-Belmonte, J. C. (1996). Involvement of FGF-8 in initiation, outgrowth, and patterning of the vertebrate limb. *Development* **122,** 1737–1750.

Wake, D. (1979). The endoskeloton: The comparative anatomy of the vertebral column and ribs. *In* "Hyman's Comparative Vertebrate Anatomy." 3rd ed., pp. 192–237. The Univ. Chicago Press, Chicago, Illinois.

Wallin, J., Wilting, J., Koseki, H., Fritsch, R., Christ, B., and Balling, R. (1994). The role of Pax-1 in axial skeleton development. *Development* **120,** 1109–1121.

Wolpert, L. (1969). Positional information and the spatial pattern of cellular differentiation. *Theor. Biol.* **25,** 1–47.

Wolpert, L. (1996). One hundred years of positional information. *Trends Genet.* **12,** 359–364.

Yamaguchi, T. P. (1997). New insights into segmentation and patterning during vertebrate somitogenesis. *Curr. Opin. Genet. Dev.* **7,** 513–518.

Yntema, C. L. (1970). Extirpation experiments on the embryonic rudiments of the carapace of *Chelydra serpentine. J. Morphol.* **132,** 235–244.

Yokouchi, Y., Nakazato, S., Yamamoto, M., Goto, Y., Kameda, T., Iba, H., and Kuroiwa, A. (1995). Misexpression of Hoxa-13 induces cartilage homeotic transformation and changes cell adhesiveness in chick limb buds. *Genes Dev.* **9,** 2509–2522.

Yu, B., Hess, J., Horning, S., Brown, G., and Korsmeyer, S. (1995). Altered Hox expression and segmental identity in MII-mutant mice. *Nature* **378,** 505–508.

7

The Origin and Morphogenesis of Amphibian Somites

Ray Keller
Department of Biology
University of Virginia
Charlottesville, Virginia 22903

Current Topics in Developmental Biology, Vol. 47

0070-2153/00 $30.00

The origin and development of the amphibian somitic mesoderm is summarized and reviewed with the goal of identifying issues most profitably pursued in these organisms. The location of the prospective somitic mesoderm as well as the cell movements bringing this tissue into its definitive position varies among amphibians. These variations have implications for the tissue interactions patterning the embryo, the design of the gastrulation movements, the role of the somitic mesoderm in early patterning and morphogenic processes, and the nature of the developmental pathway leading to somites. The presegmentation morphogenesis, the process of segmentation, and the subsequent, postsegmentation morphogenesis of the somitic mesoderm also varies considerably among amphibians. Although segmentation in amphibians shares what may be highly conserved and general patterning mechanisms with other vertebrates, the somitic developmental pathway as a whole is not conservative and has been capable of accommodating the use of a number of quite different morphogenic processes, all leading to very similar ends. The major challenges in studying amphibian somitogenesis are to develop molecular markers for major components of the somite, to determine the derivatives of the somite with better cell tracing experiments, and learning to work with the small dermatomal and sclerotomal cell populations found in most species. A potential advantage is that the diversity of somitogenesis among the amphibians makes this group ideal for studying the evolution of developmental processes. In addition, many amphibians allow direct observation of somitogenesis with great resolution and permit biomechanical analysis of tissues participating in morphogenesis, thus making it possible to analyze cellular mechanisms of morphogenesis in ways not possible in most other systems.

I. Introduction

The dominant theme emerging from investigations of somite development in amphibians is that the morphogenic processes involved are diverse. The origin and subsequent morphogenesis of this category of mesoderm appears to be highly variable, especially in regard to formation of the myotomes. A second theme is that relatively little is known about the development of the dermatome and sclerotome, in part because these regions of the somite are relatively inconspicuous and because development of vertabrae from sclerotome is delayed until near the end of the larval stage in the commonly studied species of amphibian. Third, relatively little is known about the patterning of the somite into myotome, sclerotome, and dermatome, both in terms of tissue interactions and the signaling molecules involved. Again this may be due in part to poor morphological description of the development of the latter two components of the somite. Fourth, the amphibian system also offers great opportunities. Amphibian somite patterning and morphogenesis have been studied, for the most part, with methods available 15 to 20 years ago. The subject could be revisited very profitably with modern techniques and with the reagents and concepts uncovered in recent work on other systems. Such studies would verify and extend what is known of the cellular basis of somite morphogenesis and uncover the molecular basis of somite patterning and morphogenesis in amphibians. In this regard, the amphibian has the advantage that the major events in normal and experimentally manipulated somite formation can be visualized and recorded directly with videomicroscopy of explanted tissues (Wilson *et al.,* 1989; Shih and Keller, 1992a,b). Also, the variations in somitogenesis among amphibians may offer a special opportunity for analysis of how regulatory, patterning, and morphogenic, force-producing processes evolve and are integrated into developmental pathways. The amphibians offer unmatched opportunities to explore these questions: the material is available, the methods are available, and the questions are compelling.

Xenopus laevis is the best understood system and thus will be used as a reference point from which other species will be examined. However, it should be understood that *X. laevis* is not representative of the amphibians in general and is the prevailing model system among amphibians largely because of its ease of care and breeding in the lab. The following discussion will also center on the cell and tissue morphogenesis of the somitic mesoderm.

II. Origins: Fate Maps and Movements of the Presomitic Mesoderm Are Variable among Amphibians

It is important to understand the origins and movements of the prospective somitic mesoderm for several reasons. First, if the somitic mesoderm is to be manipulated experimentally, it is important to know where it originates, what tissues it interacts with, and how it moves into its definitive position. The geometry involved in formation and movement of the somitic mesoderm is complex and often misunderstood, which has led to substantial misjudgments in design and interpretation of experiments. Second, the origin and movements of somitic mesoderm varies among amphibians. As pointed out above, the availability of this variation among experimentally accessible animals may be one of the advantages that the amphibian system can bring to the study of somitogenesis.

Fate maps of the amphibians studied in detail thus far show that the prospective somitic mesoderm, along with the rest of the prospective mesoderm and some of the prospective endoderm of the archenteron roof, lies in the involuting marginal zone (IMZ). The IMZ is an annular region that encircles the embryo and lies between a large mass of prospective endoderm at the vegetal pole and a large region of prospective ectoderm at the animal pole (Fig. 1). During gastrulation, the IMZ rolls over the blastoporal lip, or involutes, and thus brings the prospective endoderm and mesoderm beneath the prospective ectoderm. However, the position of the IMZ on the embryo and the shape and location of the prospective somitic mesoderm within the IMZ, are variable among amphibians (Fig. 1, left column). In addition, the timing and types of movements and cell behaviors that bring the somitic mesoderm into existence as a separate tissue layer vary among amphibians (Fig. 2).

A. *Xenopus:* Somitic Mesoderm Originates Primarily from the Deep Mesenchymal Layer of the Involuting Marginal Zone

To understand the variations in origins of the somitic mesoderm, it is necessary to know that in all amphibians studied in detail, the IMZ consists of a layer of superficial epithelial cells and one or more layers of deep, mesenchymal cells. Prospective mesoderm, including somitic mesoderm, and part of the prospective endoderm are distributed in one or both of these layers in patterns that vary with the species (Fig.1, left column).

In *Xenopus,* with some exceptions, all or most of the IMZ is covered by a thin monolayer of epithelial cells, the suprablastoporal prospective endoderm (SPE), fated to involute and become the endodermal lining of the roof of the archenteron (Fig. 1, right). The floor of the archenteron is formed from the subblastoporal prospective endoderm (SBE), which is composed of the epithelial cells lying vegetal to the blastopore. The SBE does not move to form the archenteron floor; rather the IMZ rolls over it and extends posteriorly across it, thus covering it over during gastrulation (Keller, 1975, 1976; Keller *et al.,* 1992). Recently, using biotinylation of proteins on the cell surface as a marking method, it has been found that some

---

Figure 1 Fate maps of selected amphibian species are shown in the surface view at the onset of gastrulation, as seen from the left side, dorsal sides to the right (left column), and for midsagittal sections of stereotyped midneurulae, dorsal sides to the right (right column). Of special interest is the relationship of the prospective somitic mesoderm (S, red) and suprablastoporal endoderm (SPE, yellow) lying in the involuting marginal zone (IMZ) of the early gastrula. Note that most of the prospective somitic mesoderm lies in the deep layer of the IMZ of *Xenopus,* with some individuals having no prospective mesoderm in the superficial epithelial layer of the IMZ (as shown), and others having scattered prospective mesodermal cells in the superficial layer (not shown; cf. Fig. 2). In contrast, large amounts of prospective mesoderm is present in the superficial layer of the IMZ of other species of anurans and the urodeles studied thus far. During gastrulation, the IMZ involutes to form the archenteron roof (AR) or the gastrocoel roof (GR) (right column). In the case of *Xenopus,* all or most of the roof is lined with endoderm (white arrow), covering the deep notochordal and somitic mesoderm. In the case of the other anurans studied, the notochord (white arrow) and part of the somites (black arrow) still line the roof of the gastrocoel after gastrulation; these mesodermal tissues are removed from the lining of the gastrocoel and join the underlying deep layer during the neurula stage by mechanisms that vary between species of anurans (see Fig. 2). In the case of the urodeles, most or all of the somitic mesoderm that lies between the prospective notochord and the subblastoporal endoderm (SBE) is somehow removed from the superficial layer in the course of gastrulation and early neurulation, and lies beneath the endoderm (white arrow); thus only the notochord (black arrow) is exposed in the superficial layer of the gastrocoel roof of the neurula. In the posterior region, however, prospective mesoderm, possibly somitic in

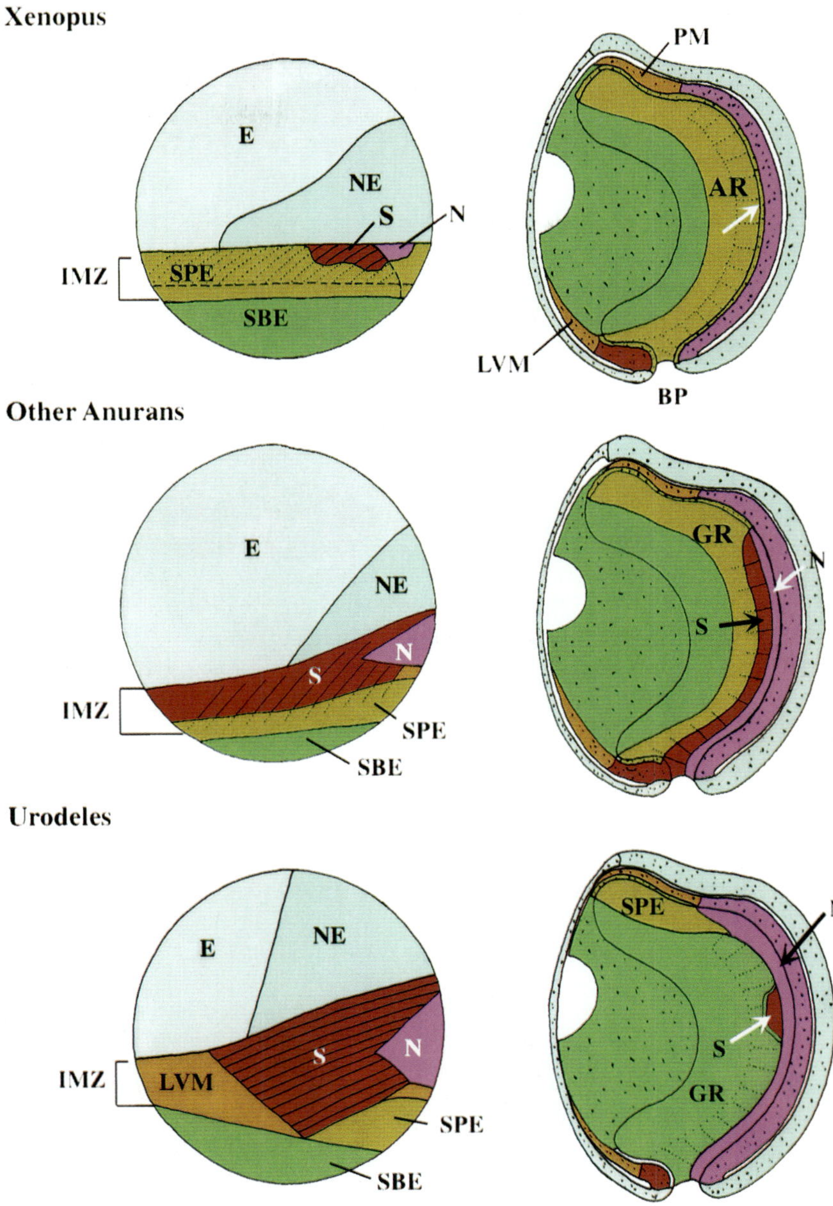

fate, is in the process of moving to the deep region. Also indicated are the blastopore (BP), the epidermis (E), the lateral/ventral mesoderm (LVM, orange), the notochord (N, purple), the neural ectoderm (NE, blue), the prechordal mesoderm (PM, orange), the subblastoporal endoderm (SBE, green). Note that the diagrams do not reflect the typical sizes of the eggs, and the stylized neurulae ignore differences of anatomical detail not relevant to somitic mesoderm formation. The fate maps are patterned after Keller (1991) and Vogt (1929).

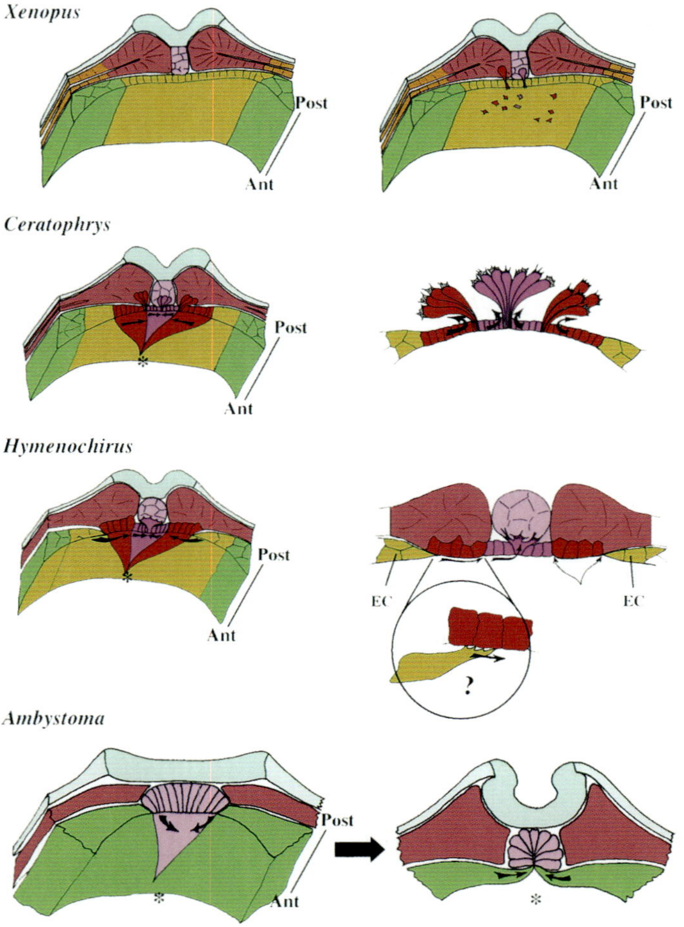

Figure 2 Diagrams show the variations among amphibians in regard to how prospective somitic mesoderm (red) and prospective notochordal mesoderm (lavender) are removed from the lining of the roof of the gastrocoel, leaving an endodermal (yellow, green) lining of the definitive archenteron. In *Xenopus,* there appear to be two variations. In some, no mesodermal cells are found in the superficial endodermal roof of the archenteron (left); in others, scattered prospective somitic and notochordal cells are found in the roof and are removed, probably by ingression of individual cells (right). In *Ceratophrys ornata,* a central zone of prospective notochord (lavender, left) is bounded on both sides by lateral zones of prospective somitic mesoderm (red, left). These are removed to the deep layer by local ingression, occurring progressively from anterior-to-posterior. As the cells in each zone ingress, others move medially to take their places (arrows, left), and subsequently, these also ingress; eventually, the endodermal epithelium on both sides meets and fuses at the midline to form the archenteron roof (asterisk). The ingressing cells take on the bottle shape and have signs of protrusive activity at their deep ends, features often associated with ingressing cell populations (right). In *Hymenochirus,* a similar trizonal organization of prospective mesodermal cells is found on the roof (left), but in this anuran, cells ingressing into the notochord spread broadly into the notochord (right) rather than forming the typical bottle cell shape. In addition, the ventral-medial cells of the somitic mesoderm appear to be

but not all *Xenopus* embryos have prospective somitic and notochordal cells, up to 30 or so of each, in the superficial epithelial layer (Fig. 2, right) (see Minsuk and Keller, 1997). The significance of this variable pattern of mesoderm location in *Xenopus* will be discussed below.

During gastrulation of *Xenopus*, the prospective endodermal epithelium of the IMZ and the underlying deep prospective mesoderm, including the somitic and notochordal mesoderm, involute over the blastoporal lip and undergo convergent extension (narrowing and elongation) together, thus elongating the body axis. After involution, the endodermal epithelium of the IMZ lines the roof of the archenteron, and the somitic mesoderm, as well as the notochord, both lie deep to this epithelium in their definitive positions as the middle layer of the embryo (Fig. 1, right; Fig. 2, left). In those embryos having prospective somitic or notochordal cells in the superficial epithelium, these cells move out of the epithelial layer and into the deep mesenchymal layer by a stage-specific ingression beginning at stage 15 (Minsuk and Keller, 1997) (Fig. 2, right). Since segmentation begins at stage 17, the late neurula, the epithelial component of the prospective somitic mesoderm is de-epithelializing and joining the deep somitic mesoderm shortly before its segmentation.

B. Other Anurans Studied Thus Far, Including *Bombina, Bufo, Ceratophrys,* and *Hymenochirus,* Have Greater Superficial Epithelial Contributions to the Somitic Mesoderm

1. Fate Maps of Other Anurans

In other anuran species studied thus far, the prospective somitic mesoderm, as well as the prospective notochord, occupy large areas in the superficial epithelial layer of the IMZ (see Vogt, 1929; Pasteels, 1942; Purcell and Keller, 1993; Minsuk and Keller, 1996). Although the geometry of their distribution varies somewhat between species, the essential features of the origins of the prospective somitic mesoderm in non-*Xenopus* species can be illustrated with a generic anuran fate map, patterned after that for *Bombina* by Vogt (1929) (other anurans, Fig. 1, left column). The prospective posterior notochord and substantial part of the prospective

---

part of the epithelium facing the gastrocoel cavity (double arrow, right), and they are mysteriously covered over in the second half of the neurula stage by the "endodermal crests" (ec) lying laterally, without the somite cells changing shape or showing other activity. One possible mechanism for this "relamination" is that the somitic cells may become nonepithelial *in situ,* without moving, thus providing an adhesive surface to support migration of the endodermal epithelium medially (right, inset). In contrast to the anurans, precocious ingression of much of the prospective somitic mesoderm appears to take place in *Ambystoma,* a urodele, leaving only notochord lining the gastrocoel roof (left). The notochord is removed by an ingression (asterisk, right). Also indicated are prospective neural ectoderm (darker blue), epidermis (lighter blue); suprablastoporal endoderm (yellow), subblastoporal endoderm (green).

somitic mesoderm, probably the medial edges of most of the somites, lie exposed in the superficial epithelial layer of the animalmost part of the IMZ. The prospective endoderm of the archenteron roof (SPE) occupies a variable amount of the superficial layer at the lower edge of the IMZ, where it covers deep prospective mesodermal tissues. These mesodermal tissues include, in order from the dorsal to the ventral sectors of the IMZ, the prospective head, heart, and lateral–ventral mesoderm (other anurans, Fig. 1, left column). As in *Xenopus,* the IMZ of these species undergoes involution during gastrulation, but after involution the superficial components of the prospective notochordal and somitic mesoderm still line the roof of the gastrocoel (other anurans, Fig. 1, right column). (Because the cavity is not yet lined completely by endoderm, it is properly referred to as a gastrocoel instead of an archenteron, the latter being defined as a cavity lined with endoderm.) These parts of the prospective notochord and somitic mesoderm are removed from the gastrocoel roof during neurulation by several mechanisms, which are discussed in the next section.

2. Urodeles

The urodeles studied thus far display yet another variation in the origin and morphogenesis of somitic mesoderm. The late blastula tends to be relatively thin-walled, and the prospective areas of the IMZ, including the prospective somitic mesoderm, occupy a relatively large area on the surface of the embryo (Fig. 1) (Vogt, 1929; Delarue *et al.,* 1992, 1996). Although it has not been quantitated, it appears that a relatively large proportion, perhaps half or more, of the prospective somitic mesoderm lies in the superficial layer of the IMZ of most urodeles. In a double labeling study of the contributions of both deep and superficial layers in *Pleurodeles,* about 50% of the anterior myotomes come from the superficial layer but decreases to less than 20% for the posterior myotome (Delarue *et al.,* 1992). For comparison, about 30% of the notochord forms from the superifical layer. Unlike the anurans, which have the suprablastoporal prospective endoderm covering the vegetal edge of the mesoderm around much or all of the dorsoventral extent of the IMZ (SPE, Fig. 1), the only such endoderm in the urodeles is the prospective pharyngeal endoderm covering the prospective prechordal mesoderm at the dorsal side (SPE, Fig. 1, left) (Vogt, 1929). Interestingly, the suprablastoporal endoderm covering the prechordal plate is an invariant feature of the superficial IMZ of all amphibians studied thus far (Fig. 1, left column). All other sectors of the IMZ may or may not have SPE in the superficial layer. Like the anurans discussed above, the urodeles gastrulate by involuting the IMZ, but involution, by itself, should leave the prospective somitic mesoderm in the superficial layer of the gastrocoel roof, as in the anurans (Fig. 1, right column). However, in the urodele much or most of the somitic mesoderm is removed from the superficial IMZ in the course of its involution. How and exactly when this occurs remains a mystery (Fig. 1, right column), one that will be discussed further in the next section.

Whether the urodelean amphibians differ in regard to the fate map of the somitic mesoderm is not known in any detail but the published fate maps based on several different species are generally similar (cf. Vogt, 1929, Pasteels, 1942, and Delarue *et al.,* 1992).

3. The Mechanisms of Removing Prospective Mesoderm from the Epithelial Layer Vary among Amphibians

In *Ceratophrys ornata,* the Argentine horned frog, the roof of the gastrocoel in the trunk region consists of three types of zones: (1) a central prospective notochordal zone (Figs. 2 and 3); (2) bilateral zones of prospective somitic mesoderm on both sides of the prospective notochord (Figs. 2 and 3); and (3) a posterior, circumblastoporal zone (Fig. 3). In the midneurula, both the prospective notochordal and somitic cells begin ingressing in the anterior region, adopting the classic bottle shape, leaving the epithelial layer, and crawling inside to join their deep components (Fig. 2). Initially the cells of the gastrocoel roof appear uniform (Fig. 3A), but as this ingression occurs, the central notochordal zone and bilateral somitic zones become morphologically defined (Fig. 3B). Ingression of the cells involves constriction of their apices and expansion of their basal ends, thus forming "bottle-shaped" cells, which then detach from the superficial epithelial layer and crawl deep (Fig. 2, right). These zones narrow and disappear as ingression proceeds (Fig. 3C), thus bringing the bilateral areas of prospective endoderm on each side together at the dorsal midline where they fuse, forming a definitive archenteron roof (asterisk, Fig. 2). This stage-specific ingression proceeds from anterior-to-posterior. The third, circumblastoporal zone (see c, in Fig. 3C) forms around the blastopore and is the site of ingression of cells into the posterior somitic mesoderm (Purcell and Keller, 1993).

In *Hymenochirus boettgeri,* another anuran and a close relative of *Xenopus,* the prospective somitic and notochordal cells are also represented in the superficial layer of the gastrocoel roof after its involution, and there they form morphological zones similar to the ones seen in *Ceratophrys* (Fig. 2, left; Fig. 3D,E). However, the behaviors removing the cells from the surface layer into the deep region are different. The bilateral zones of prospective somitic mesoderm consist of cells that simultaneously appear to be part of the deep somitic tissue and also a part of a contiguous epithelial cell layer lining the roof of the gastrocoel; that is, they appear to be contiguous laterally with the prospective endodermal epithelium and medially with the notochordal epithelial component (Fig. 2; Fig. 3E). Somehow during neurulation, the somitic cells are covered over by the lateral endodermal cells in a movement called "relamination" (Fig. 2, right). The somitic cells do not change shape, become bottle shaped, or ingress; instead the lateral "endodermal crests" somehow migrate or are pulled over the somitic cells. How this occurs presents some problems, however, since the outer surface of the epithelial component of the somitic population should be nonadhesive if it is epithelial in nature.

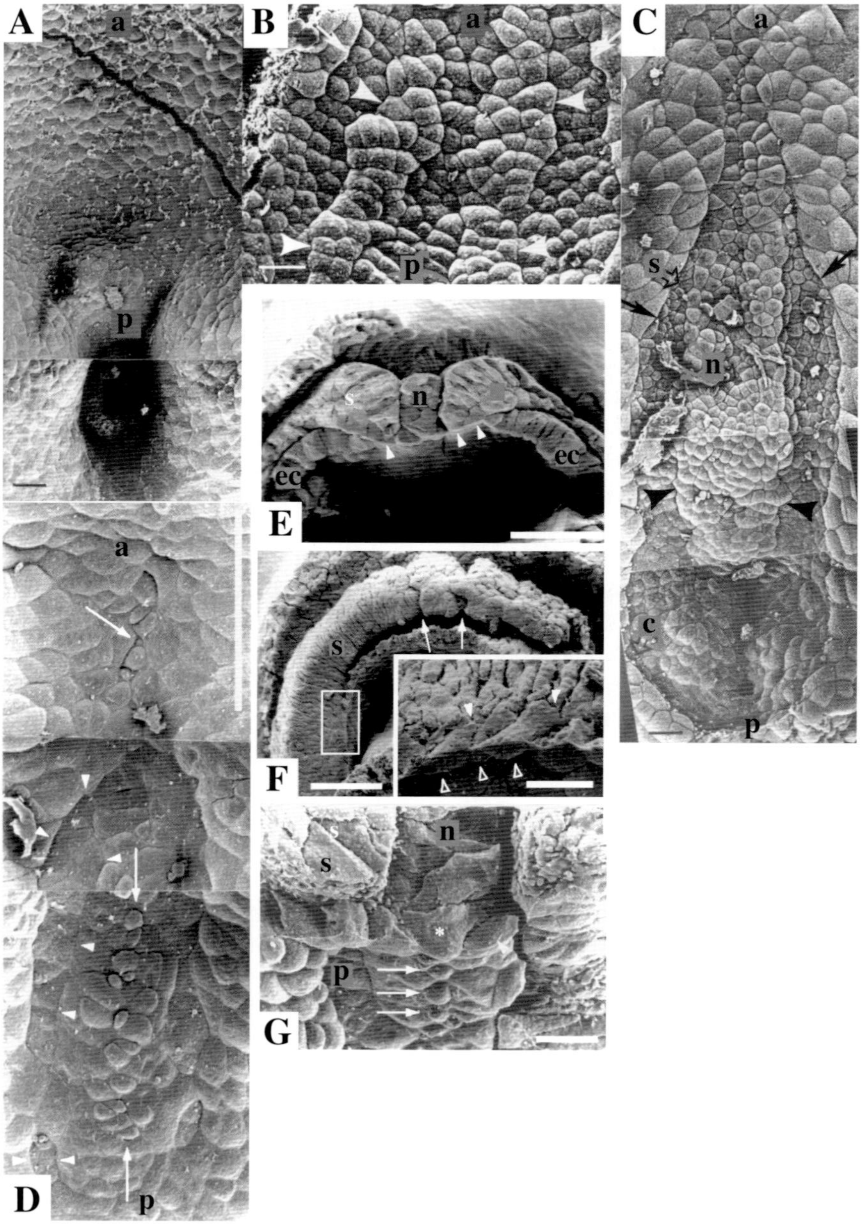

Figure 3 Scanning electron micrographs show the uniform appearance of the gastrocoel roof of *Ceratophrys ornata* in the early neurula (A), and the development of the medial zone of ingressing notochord (within in pointers, B) and lateral zones of ingressing somitic mesoderm (between pointers and arrows, B). As ingression occurs, the zones of notochord (n) and somitic mesoderm (s) narrow anteriorly (C). A third, circumblastoporal zone (c) of ingression of cells, mostly into posterior somites is

One possibility is that the somitic cells de-epithelialize *in situ,* making the outer surfaces suitable as a substrate for the endodermal epithelial cells on either side to migrate across (Fig. 2, right). Of special interest is the fact that the process of relamination occurs only several somite-lengths ahead of segmentation (Fig. 3F), which is less than 2 hr in this species. Note also that the ventral, medial somitic cells already participating, or destined to participate in relamination are the only cells of the somite that have not adopted an elongate morphology (Fig. 3E). The remaining cells have begun the elongation characteristic of somite formation in many species. Thus somite morphogenesis is well underway before all the final complement of cells have been added to the ventral aspect of the somitic mesoderm.

Like the somitic cells, the notochordal cells of *Hymenochirus* display a variant form of ingression. They also ingress but do so without becoming bottle shaped. Instead, they appear to spread on the ventral surface of the deep component of the notochord as they leave the surperficial epithelial layer (Fig. 2, right; Fig. 3G). As they ingress, and as the bilateral zones of prospective somitic mesoderm are covered over during relamination, the lateral endodermal crests are drawn together in the midline where they fuse to form the definitive archenteron roof. Again, these processes begin anteriorly in a stage specific manner (stage 15, a midneurula) and proceed posteriorly (Minsuk and Keller, 1996). Despite the variations in the mechanism used to remove somitic as well as notochordal mesoderm from the superficial layer of the gastrocoel, specific cell behaviors in all species show the anterior–posterior progression common to later aspects of somite formation.

The urodeles, as represented by *Triturus* (Vogt 1929) appear to be precocious in removing most of the prospective somitic mesoderm from the superficial IMZ during involution, such that at the end of gastrulation, only prospective notochord remains on the roof of the gastrocoel along most of the length of the body axis (Fig. 2, right). Vogt's (1929) vital dye marks showed that the somitic mesoderm shears beneath the lateral endodermal crests (the edges of the vegetal or sub-blastoporal endoderm) sometime during gastrulation or early neurulation, but he could not determine how this happened. Subsequent vital dye mapping of the

---

also shown. Scanning electron micrographs (D–G) show various aspects of somitic mesoderm formation in *Hymenochirus.* A ventral view of the gastrocoel roof of a stage 18 (late) neurula (D, anterior at the top) shows bilateral regions of somitic mesoderm exposed on the roof of the gastrocoel (white pointers) and a central region of notochord ingression, indicated by a line of constricted, smaller cell apices (white arrows). A transverse fracture at stage 15 (E) shows the zones of somitic mesoderm (s) exposed on the roof of the gastrocoel (pointers), the lateral endodermal crests (ec), and the notochord (n). Note that the dorsal cells of the somitic mesoderm are elongated and tilted medially whereas the ventral cells participating in relamination are cuboidal. A parasagittal fracture at stage 19 (F) shows that the somitic mesoderm (s) is beginning to segment anteriorly (last formed somite is between arrows) while relamination of its surface component is occurring just several somite-lengths posteriorly (box, enlarged in inset). Scale bars = 25 μm in the inset and 100 μm elsewhere. Also indicated are the blastopore (bp) and the anterior (a) and posterior (p) ends. A–C from Purcell and Keller, 1993, and D–G from Minsuk and Keller, 1996.

Mexican axolotl embryo showed that most of the prospective somitic mesoderm probably ingresses out of the superficial epithelial layer into the deep region at the corners of the blastopore during gastrulation (Lundmark, 1986), such that after blastopore closure, all or most of the somitic mesoderm has been removed from the roof of the gastrocoel, leaving only the notochord exposed on its surface (Fig. 1, right; Fig. 2). Delarue and associates (1996) found that in *Pleurodeles waltl* prospective somitic and notochordal cells in the superficial layer of the IMZ ingressed into the deep region *before* involution. In this study, labeling both deep and superficial layers showed that the involuted trunk mesoderm of *Pleurodeles* is reduced to one layer of cells and fully half or more of the cells at the midline are derived from the superficial layer. In the ventral region of the IMZ of the urodele gastrula, large amounts of ventral–lateral mesoderm (Fig. 1, right), and probably considerable posterior somitic mesoderm as well, is thought to leave the surface by ingression of individual bottle cells during gastrulation (Holtfreter, 1943). In all species, however, nothing is known about the details of this ingression, other than the fact that it must occur precociously relative to the ingression of somitic mesoderm in anurans. In those sectors of the IMZ having no prospective endoderm in the superficial layer, the entire archenteron roof is formed by the vegetal, subblastoporal endoderm. The subblastoporal vegetal endoderm forms the "endodermal crests," which move from both sides to the dorsal midline where they meet and fuse to form a continuous archenteron roof at a later stage of development.

Like the somitic mesoderm, the superficial component of the prospective notochord of urodeles shows yet another variation on the process of moving to the deep region (Fig. 2, right). First, a large part, half or more, of the notochord is exposed on the surface of urodeles, more of the total than in any of the anurans studied thus far. Second, some authors believe that it is removed by a process that resembles an invagination, like neurulation, rather than an ingression of individual cells (Lofberg, 1974; Brun and Garson, 1984). In this view, the notochord is initially rather broad in the late-midneurula; its epithelial cells elongate and become bottle shaped, but instead of ingressing, they retain their connections to one another, and form an infolding (Fig. 2, right). However, the sides of the invagination are apposed to one another, and therefore they seal off to form a solid rod, instead of a tube (Fig. 2, right) (Lofberg, 1974; see also, Brun and Garson, 1984 and Ruffini, 1925). As the rod closes, the endodermal crests on both sides meet, seal, and form the definitive roof of the archenteron (asterisk, Fig. 2, right).

The major feature that differs between the urodele and anuran patterns of somitic mesoderm behavior is that all the prospective somitic mesoderm between the lateral edges of the notochord and the vegetal endoderm ingresses precociously at the gastrula stage, rather than in the neurula as in anurans. Thus only the medial notochordal zone is left on the surface of the gastrocoel roof of the urodele neurula, whereas in the anuran there are three zones of prospective mesoderm on the roof of the gastrocoel, a central notochordal one and two lateral, somitic ones. How the urodele and anuran embryos differ in regard to the third, circumblastoporal region of ingression near the blastopore is not known. This region has been described only superficially in the anurans (Purcell and Keller, 1993; Minsuk and Keller, 1996), and more work needs to be done. Whether the urodele has an equivalent region is not known.

C. Significance of Variations in Somitic Mesoderm Formation

1. Do the Late Additions of Superficial Cells to the Somitic Mesoderm Have a Specific Fate?

Are the late epithelium-derived additions to the somitic mesoderm targeted to particular components of the so'mites? Since the superficial cells are added to the ventral or ventromedial region of the somite, one might expect them to form sclerotomal derivatives. However, cell tracing experiments have shown that many of them have a myotome fate in both *Hymenochirus* and *Xenopus* (Minsuk and Keller, 1996, 1997). Failure to find them in the sclerotome may be due to the small size and late development of this population of cells (described below), or it may be that these cells are targeted to other fates despite the fact that they enter the somite where sclerotome should form.

2. The Somitic Differentiation Pathway Is Not a Rigid, Conservative One

The somitic differentiation pathway is not a rigid one that precludes the exercise of a variety of morphogenic cell behaviors in its course. The diversity of morphogenic processes that add cells to the somitic mesoderm in the species examined thus far is impressive. More variants will probably be seen as more species are investigated, particularly those that have very different types of somites, such as the legless caecelians (Gymnophiona) (Wake, 1970; Wake and Wake, 1985). In addition, the diverse mechanisms of adding cells to the deep somitic mesoderm occur just prior to segmentation. In *Xenopus,* ingression may or may not occur, and if it does, only an hour and a quarter separates stage 15, when the mesoderm begins ingression, and stage 17 when segmentation begins (Nieuwkoop and Faber, 1967). In *Hymenochirus,* prospective somitic cells are undergoing relamination to join the segmental plate about an hour and several somite lengths before segmentation. As will be described below, there is evidence that molecular and morphological events preparatory to segmentation are already underway at the levels where cells are still being added to the somitic mesoderm (see Section VIII.A).

3. Do Variations of Fate Maps Reflect Differences in Tissue Induction or in Response?

The variable representation of the prospective somitic mesoderm, as well as other mesoderm, in the epithelial layer of the IMZ may reflect different efficiencies of inductive interactions in different species. The classic experiments of Nieuwkoop showed that signals emanating from the vegetal endoderm induce both mesodermal and endodermal tissues of the IMZ (see Nieuwkoop, 1969a,b; Sudarwati and Nieuwkoop, 1971). The fact that the variation in the amount of prospective endoderm in the superficial layer is at the upper or animal edge of the IMZ suggests that endoderm of the IMZ is induced by a signal of variable effectiveness emanting from the vegetal endoderm and passing animally, in planar fashion, through the epithelial layer. However, the prospective mesoderm animal to the superficial endoderm must be accounted for as well. One explanation is that the entire superficial layer is first induced to form mesoderm and then variable amounts of it are converted to endoderm, working upward from the vegetal edge of the IMZ. A second explanation is that the same signal induces both, but at different concentrations. In any case, the fate maps suggest that induction of prospective endoderm in the IMZ is most effective in *Xenopus,* where nearly all the IMZ is prospective endoderm, and least effective in the urodeles, where endoderm, the pharyngeal endoderm, is formed only in the dorsal sector of the blastula. However, there is no direct evidence that this is the geometry and signaling regime that accounts for suprablastoporal endoderm induction. Also, it is not known whether the effectiveness of the putative endodermalizing signal from the vegetal endoderm is stronger or if the responding tissue is more sensitive in those animals having large amounts of endoderm in the IMZ. In addition to the problem of explaining the variation in amount of endoderm in the superficial IMZ, the differences in mechanism of removal of the prospective mesoderm from the superficial layer must be accounted for. Interspecific grafting may illuminate this issue. Would a *Xenopus* animal cap grafted to the vegetal endoderm of *Hymenochirus* show the *Hymenochirus* or *Xenopus* pattern of endomesodermal induction and morphogenesis?

Despite the fact that the key variable among amphibians is whether the superficial IMZ makes somitic mesoderm or endoderm, not much is known of the molecular aspects of endoderm versus mesodermal induction and patterning. Recently a number of factors have been implicated in the development of endoderm in *Xenopus,* including Xsox17 (Zhang and King, 1996; Hudson *et al.,* 1997; Horb and Thomsen, 1997), and chordin and noggin, several components of the Spemann organizer that are also involved mesodermal and neural development (Jones *et al.,* 1993; Sasai *et al.,* 1996). For an insightful discussion of endodermal versus mesodermal induction see Hudson and others (1997).

4. Are the Initial Differences in Involuting Marginal Zones Dynamic (Morphogenic) Determinations or Tissue-Type Determinations?

Whether differences in amount of prospective mesoderm in the superficial layer represent differences in cell type determination or morphogenic determination is not known. The fate maps depicted in Fig. 1 only indicate what cells normally become, not the state or nature of their commitment. The fate of the superficial cells could be determined directly or indirectly, and in the latter case the initial determination may be of a morphogenic nature.

On the one hand, endodermalizing signals acting on the superficial layer of the IMZ could induce it to become endodermal tissue. Part of this cell-type determination could be a linked morphogenic behavior; the cells would remain epithelial and not ingress or otherwise leave the superficial epithelial layer. Those cells not receiving or not responding to the endodermalizing signal, usually the ones farther animally in the IMZ, would remain mesodermal in cell-type determination. This cell-type determination would also entail a linked morphogenic behavior: to de-epithelialize and move into the deep region by one of the methods described above. In this scenario, cell type determination is primary and early, and entails a linked morphogenic behavior that follows.

Alternatively, the primary event could be the determination of a type of cell behavior, which would, in turn, determine cell fate. For example, there is evidence that there are strong mesoderm inducing signals acting within the deep mesenchymal cell environment of the IMZ all through gastrulation (Domingo and Keller, 1995). Thus the "endodermalizing" pathway could be one in which the primary event is stabilization of the epithelial organization of the superficial cells, which would then remain as an intact, superficial sheet, out of reach of the deep mesodermalizing signals throughout gastrulation. Eventually they would default to an endodermal phenotype or be induced to follow this path by additional signals. In contrast, the primary event in the "mesodermalizing" pathway would be destabilization of the epithelial organization, allowing the superficial cells to enter the deep layer; there they would fall under the influence of mesodermalizing signals and secondarily form mesoderm. In fact, superficial cells of the IMZ of *Xenopus,* most of which are normally fated to form superficial endoderm, form mesodermal tissues when grafted into the deep region of the IMZ (Shih and Keller, 1992a).

These alternatives may be reflected by the types molecules expressed in the deep and superficial epithelial regions of the IMZ, and whether they correlate with tissue type differentiation or cell behavior. For example, Xsox17 is an endoderm-specific marker in *Xenopus* (Hudson *et al.,* 1997). Is it expressed in the *entire* superficial layer of *Xenopus,* or is it absent in those few cells destined to ingress to form mesoderm? Do the cells destined to ingress express mesodermal markers typical of axial and paraxial mesoderm prior to ingression? Answers to these questions may reveal the differences in regulatory steps that lead to superficial endoderm, superficial prospective somitic mesoderm, and deep prospective mesoderm in the IMZ. The same applies to mesoderm of other prospective fates. These questions may be difficult to answer in *X. laevis,* which has only a few prospective mesodermal cells in the superficial IMZ, making them difficult to identify in RNA *in situ* hybridizations; species having large amounts of mesoderm in the superficial layer may offer advantages in this regard. Unfortunately, not enough layer-specific and tissue-specific markers are available for *X. laevis* and fewer still for other species. Advances in these areas await cloning and characterization of genes controlling endodermal and mesodermal cell fates or serving as markers of these tissues in a broader selection of amphibian species.

III. A Life before Segmentation: The Geometry, Behavior, and Function of the Prospective Somitic Mesoderm during Gastrulation of *Xenopus*

The broader point of this section is that the segmental plate of *Xenopus,* as it begins segmentation in the late neurula, is not an amorphous tissue that has yet to display an axial character or participate in an organized morphogenic process but one that has already shown highly organized cell motility and produced a pattern of forces essential for early morphogenesis. In addition, the evidence shows that the prospective somitic mesoderm has had a major and thus-far underestimated role in both morphogenesis and embryonic patterning prior to segmentation. To develop these ideas, in the first section below the geometry of how the prospective somitic mesoderm lies in the IMZ of the gastrula will be described in some detail. Understanding this geometry is a preface to the next section, in which the mapping of the body axes map on to the prospective somitic tissue is used to understand the highly oriented cell behavior expressed in this region, and how it functions biomechanically to bring the somitic mesoderm into its definitive position. Finally, in the last section, the role of the somitic mesoderm in organizing neural and dorsal mesodermal development is re-evaluated. This section will focus on *Xenopus,* which is the best understood species in this regard.

A. The Geometry of the Prospective Somitic Mesoderm

The mapping of the future body axes on to the prospective somitic tissue in the early gastrula is often misunderstood, especially in relation the to dorsoventral aspect of the gastrula. Although this subject seems complex when first confronted, it is elegantly simple when understood. Both fate mapping and videomicroscopy of the developing somites in explants show that the somitic mesoderm and notochordal mesoderm lie in the IMZ with their future anterior–posterior and lateral–medial axes oriented as shown in Fig. 4A,B (see Keller *et al.,* 1992). Note that the prospective somitic mesoderm lies all around the lateral and ventral aspect of the blastopore. The common perception that future dorsal tissues lie in the dorsal sector of the gastrula and future ventral ones lie in the ventral sector of the gastrula is false. Note also that the dorsal–ventral dimension of the IMZ of the gastrula does not correspond to the dorsal–ventral (medial–lateral) body axis in later develop-

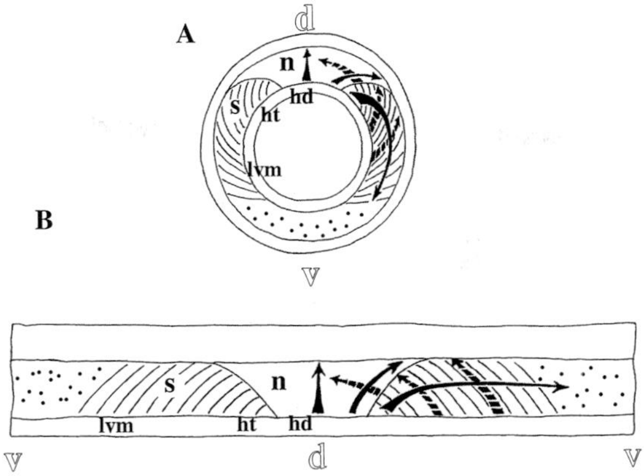

Figure 4 The future polarities of prospective mesodermal tissues and the polarities of the gastrula stage embryo are compared on the fate map of the early *Xenopus* gastrula (A) and of a "giant" explant of the IMZ laid out in planar fashion (B). Both are depicted as they would appear without the overlying suprablastoporal prospective endoderm, thus exposing the deep mesoderm. The dorsal side (d) and ventral side (v) of the gastrula are indicated with open letters. The future anterior-to-posterior (p) axes of the somitic (s) and notochordal (n) mesoderm are indicated with solid arrows pointing posteriorly. The future lateral-to-medial (m) axes of the same tissues are indicated by broken arrows pointing medially. The approximate orientation of the prospective anterior somites is indicated by parallel lines; the prospective posterior ones are shaded. The prospective head (hd), heart (ht), and lateroventral mesoderm (lvm) forming the leading edge of the mesodermal mantle are also shown. Adapted from Keller *et al.,* 1992.

ment but instead corresponds to the future anterior–posterior axis of the somitic mesoderm.

The future mediolateral or dorsoventral axis of the *Xenopus* body plan lies transverse to the anterior–posterior axes in the notochordal and somitic mesoderm. Thus in the prospective anterior mesoderm where the first six or seven future somites bound the notochord, the future medial edge of the prospective somitic mesoderm abuts the lateral boundary of the prospective notochord and the future lateral edge abuts the prospective lateral–ventral mesoderm, which lies at the vegetal edge of the IMZ near the vegetal endoderm (Fig. 4A,B). This axial organization pertains until segmentation, when the cells of the somitic mesoderm are reorganized, as described below (see Section IV).

The remaining, more posterior prospective somitic mesoderm likewise has its future lateral edge lying vegetally, near the vegetal endoderm, and its future medial edge animally in the IMZ. However, its medial edge does not abut the future notochord at the early gastrula stage but does so only after involution. This mesoderm involutes to form a compact, thickened circumblastoporal mass during closure of the blastopore; as it does so, its animal edge is the last involuted and therefore the part that lies next to the blastoporal lip. Then the notochord pushes the blastopore ventrally, as described below, and the last involuted animal edge of the somitic mesoderm shears around both sides of the blastopore and comes to bound the lateral edge of the notochord (see Keller *et al.,* 1989, 1992). The vegetal edge of this mesoderm involutes first, and therefore it lies laterally after involution, forming the lateral edge of the segmental plate, next to the lateral plate.

Note that the ventral and lateral mesodermal tissues of prospective heart, lateral plate, and ventral body wall mesoderm all lie vegetal to the somites and at the vegetal edge of the IMZ of the early gastrula. These tissues, along with the prospective head mesoderm, comprise the early involuting, migratory mesoderm that spreads out at the leading edge of the mesodermal mantle (see Figure 5 of Keller, 1991). Thus ultimately being "ventral" in the body plan has nothing to do with whether the cells lie dorsally or ventrally within the IMZ of the gastrula, but whether they lie vegetally within the IMZ, involute early, and migrate, or lie animally in the IMZ, involute later, and converge and extend. Thus the "dorsal" tissues of notochord and somite are located high in the IMZ, involute late, and participate in convergent extension. The "ventral" tissues of head, heart, and lateral and ventral mesoderm are located vegetally in the IMZ, involute early, and participate in migration on the blastocoel roof.

There are some indications that this geometry of the prospective somitic mesoderm seen in *Xenopus* is not universal among the amphibians. The fate mapping results of Vogt (1929) and Delarue and others (1992) on urodeles places more of the somitic mesoderm in the dorsolateral and lateral IMZ with the posterior somitic mesoderm lying animal to the anterior somitic mesoderm. This geometry implies a different design of gastrulation movements and also suggests that the signals patterning the IMZ may be somewhat different in geometry than those patterning *Xenopus.* However, more detailed mapping is necessary to know for sure, and further work should be done on both *Xenopus* and other species, both anuran, urodelean, and gymnophion amphibians, to clarify this issue.

B. Organized Cell Behavior within the Prospective Somitic Mesoderm: Cell Intercalation

As pointed out above, the mediolateral axes of the future notochord and anterior somitic mesoderm can be envisioned as arcs spanning the dorsal side of the IMZ with their lateral ends anchored near the vegetal endoderm. These arcs also define the lines along which convergence occurs during convergent extension (Keller *et al.,* 1992). Convergent extension is driven by a pattern of expression of several cell behaviors, collectively called mediolateral intercalation behavior (MIB). MIB has been studied by videomicroscopy of giant, open-faced explants of the entire IMZ, or a large part of it, in culture and in embryos (see Shih and Keller, 1992b,c; Keller *et al.,* 1992; Domingo and Keller, 1995; Lane and Keller, 1997). MIB con-

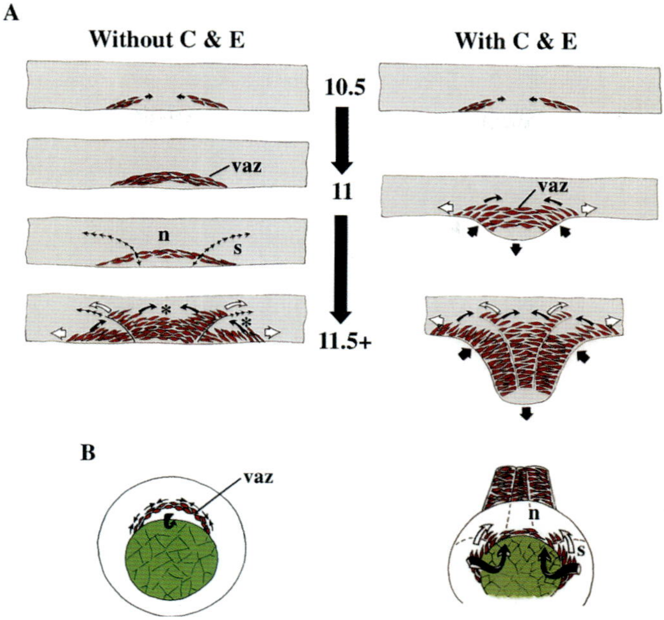

Figure 5 Diagrams show the origin and progress of the deep mesodermal cell behavior driving convergent extension (C & E) movements, as seen by videorecording of large open-faced explants of the IMZ of *Xenopus* early gastrulae. Convergent extension occurs as deep mesodermal cells show bipolar, mediolaterally oriented protrusive activity, elongation parallel to the mediolateral axis, and intercalation between one another along this axis, to form a narrower and longer array of cells. This complex of behaviors is called mediolateral intercalation behavior (MIB). The progress of MIB is indicated by red fusiform shapes on gray outlines of the explants or embryos. Diagrams (A) show the progression of MIB in explants that have been mechanically prevented from converging and extending (left column), or allowed to converge and extend (right column); in the last case, the distortion of the explant is indicated by the change in shape of the gray outlines of the explants and black arrows outside the explants. For orientation, the prospective axes in these explants are like those illustrated in Fig. 4B. Diagrams of the vegetal view of the embryo, dorsal above, show how convergence of the vegetal alignment zone (vaz) in the early midgastrula brings about involution (B) and how progressive expression of MIB drives continued involution of the IMZ during subsequent gastrulation (C) of the whole embryo. The white arrows indicate progression of MIB in the anterior–posterior direction and the black, curved arrows indicated progression of MIB in the lateral–medial direction (compare with axes shown in Fig. 4B). The notochord (n) and somitic mesoderm (s) are shown. See the text for a description. Based on Shih and Keller, 1992b, Domingo and Keller, 1995, and Lane and Keller, 1997.

sists of bipolar, medially and laterally directed protrusive activity, mediolateral cell elongation, and mediolateral cell intercalation to form a narrower, longer array. Expression of MIB in two types of explants is illustrated diagramatically: (1) those that have not been allowed to converge and extend (Fig. 5, left column), a format which makes it easier to understand the pattern of expression in the context of the future embryonic axes (cf. Fig. 4); and (2) those that have been allowed to converge and extend (Fig. 5A, right column), a format which makes it easier to understand the mechanical function of MIB in gastrulation. At the early midgastrula stage (10.5), MIB begins bilaterally, in what will become the anterior somitic mesoderm on both sides, and progresses medially, toward the dorsal midline (arrows, Fig. 5, stage 10.5), where the two arcs meet, forming the vegetal alignment zone (VAZ) (Fig. 5, stage 10.5–11). As expression of MIB in the VAZ brings about its convergence and extension, the anterior end of the unconstrained explants bulge anteriorly (Fig. 5A, stage 11, right). The notochordal–somitic mesodermal boundary forms within the VAZ at stage 11, and from this origin, it progresses posteriorly, describing the shape shown in the fate map of the notochord in explants not allowed to converge and extend (small arrows, Fig. 5A, left column, stage 11–11.5). From this time onward, MIB progresses posteriorly along the lateral edges of both the somitic and the notochordal territories (white arrows, Fig. 5A, stage 11.5). From this lateral origin in the somitic mesoderm, MIB progresses medially toward the notochord, and from its lateral origin at the boundary of the notochord, MIB progresses medially toward the dorsal midline (black arrows inside the explants, Fig. 5A, stage 11–11.5).

In vivo, the VAZ forms as an arc across the dorsal lip at the early midgastrula (stage 10.5), and therefore its initial convergence generates a hoop stress that is thought to cause the dramatic involution of the dorsal IMZ at this stage (Fig. 5B) (see Lane and Keller, 1997). During the remainder of gastrulation, the posterior progression of MIB at the lip of the blastopore generates a posteriorly directed wave of convergence and thus hoop stress, arcing across the dorsal IMZ. This posteriorly progressing hoop stress drives involution of the IMZ (Fig. 5C) and is capable of doing so without the other components of the gastrulation machinery (see Keller and Jansa, 1992; Keller *et al.,* 1992).

These relationships between MIB, convergent extension, and the eventual formation of somites, can be seen in an open-faced explant partially retarded in its convergent extension and allowed to develop to the point of somite segmentation (Fig. 6). The elongated cells that have expressed MIB are found in the lateral regions of the first five somites (s, pointers, Fig. 6) and across the anterior of the notochord and along the lateral margins of the notochord (n, Fig. 6). Note that the axis of cell elongation and the axis of cell intercalation are parallel to the mediolateral axes of the somites, and these axes describe arcs spanning the dorsal mesoderm (dashed line, Fig. 6). Note also that segmentation is not dependent on extension; the "somites" are very short and very wide, as one might expect if extension was retarded. This independence of segmentation and extension was also seen in the tail somitic mesoderm of *Rana* (Elsdale and Davidson, 1983). In the embryo, the expression of MIB is ongoing in the postinvolution notochordal and somitic mesoderm, with the result that the starting state of somitogenesis in *Xenopus* is a population of mediolaterally elongated cells lateral to the notochord.

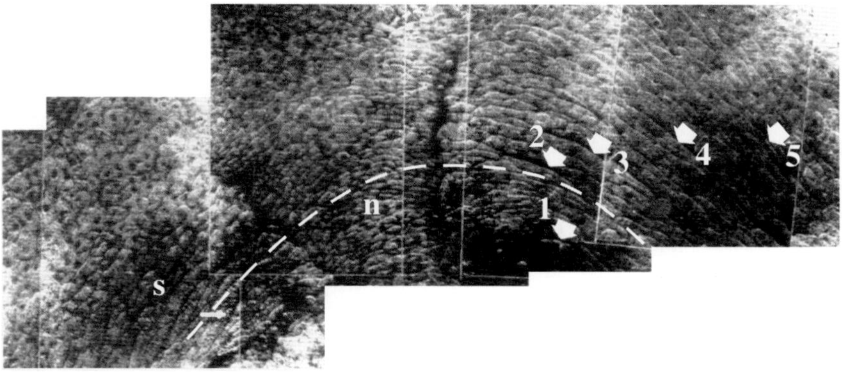

Figure 6 A videomicrograph of an open-faced explant shows the notochord (n) and somitic mesoderm (s), segmented to form the first five somites (arrows numbered 1–5). The explant has been retarded in its extension. Normally the notochord would be less than two cells across at this stage, and much longer, and the somitic mesoderm would be about 10 cells across and much longer. Note the array of tissues and the notochordal–somitic mesodermal boundary approximates the situation diagrammed in Fig. 5, stage 11.5, without C & E. From Shih and Keller (1992c).

C. The Role of Prospective Somitic Mesoderm in Organization of Morphogenesis and Differentiation

The somitic mesoderm appears to have a much stronger role in organizing and executing early morphogenesis and tissue differentiation than previously realized. Appreciation of its role begins with revision of the concept of how the Spemann organizer works. Since the organizer experiment of Spemann and Mangold (1924) (see Spemann, 1938), the organization of the gastrula tissues has been envisioned in terms of signals that originate dorsally in the organizer and pass laterally and ventrally through the IMZ, recruiting cells that would otherwise not have done so to participate in formation of the dorsal axial and paraxial tissues. In this scenario, the organizer provides a positive signal that acts on neutral tissue that will default to "ventral" mesoderm if not acted on. This notion was expressed in the "three signal model" of mesoderm organization, the third signal being the "horizontal" one in the IMZ (see Smith *et al.,* 1985).

However, recent work shows that both neural and mesodermal tissues are patterned, at least in part, by interaction of inhibitory signals emanating from the organizer that antagonize "ventral" signals that left unchecked will assure formation of "ventral" tissues of epidermis and lateral or ventral mesoderm. The ventralizing factors include BMP-4 and Wnt-8 and the dorsalizing, inhibitor factors include chordin and noggin, which bind and inhibit BMPs, and Frzb-1, a secreted, truncated Wnt receptor that binds and inhibits Wnt-8 (see Piccolo *et al.,* 1996; Zimmerman *et al.,* 1996; Leyns *et al.,* 1997; Wang *et al.,* 1997; Hoppler and Moon, 1998). Pertaining specifically to the somitic mesoderm, Hoppler and Moon

(1998) present evidence that its boundary with the notochord is determined by a cooperative mechanism involving Frzb-1 and Wnt-8 and BMP-4.

The revised concept of how the organizer works may still be somewhat off the mark, but it accounts for the "ventrally derived" positive signals that act dorsally, and encourages a more objective evaluation of ventral-to-dorsal interactions, specifically the role of the somitic mesoderm in organizing mesoderm. Because the fate map of the prospective notochordal mesoderm lies dead center in the Spemann organizer and the prospective somitic mesoderm lies largely outside it, and because of the prevalence of the dorsal, positive signal hypothesis, the evidence supporting a strong patterning role for the somitic mesoderm, both before and after its segregation from the notochordal mesoderm, has been overlooked. The evidence shows that after the establishment of the notochordal–somitic mesodermal boundary the somitic mesoderm may have a more dominant and independent role than the notochord in organizing and executing the movements driving involution and embryonic axis extension, as well as in neural induction. The evidence also shows that the embryo suffers early and dramatically from absence of somitic mesoderm, whereas it suffers later and much less from absence of notochord. Moreover, the organization of cell movements within the dorsal tissues is from lateral-to-medial rather than the reverse, a geometry that is difficult to explain if they are organized solely by signals emanating from the dorsal midline.

In normal patterning, convergent extension of the notochord is dependent on presence of the somitic mesoderm whereas the somitic mesoderm can converge and extend without the notochord. When involuted notochord and somitic mesoderm are cut apart, each with its own piece of overlying neural plate attached, the somitic mesoderm can continue to converge and extend while the notochordal mesoderm ceases extension (Wilson *et al.,* 1989; Wilson, 1990). Thus somitic mesoderm appears to be able to independently converge and extend whereas the notochord is dependent on the somitic mesoderm to maintain this type of morphogenesis, at least until the early neurula stages. When the notochord of *Xenopus* becomes independent of the somitic mesoderm in this regard, if ever, is not known. Notochord is often seen extended beyond somitic mesoderm in cultured explants, but somitic mesoderm is always nearby, and no instance of notochord convergent extension by itself can be recalled (R. Keller, unpublished observations). Extended tissue formed by animal caps induced with growth factors or by lithium-induced dorsalization often appear to consist mostly of notochord (Kao and Elinson, 1988), suggesting that this tissue can extend on its own. But the induction of notochord by these means undoubtedly bypasses normal control mechanisms that regulate these processes, including requirements normally met by contact with the somitic mesoderm. The evidence for this is that *in vivo* and in cultured explants MIB and notochord and somite differentiation is progressive, occurring from lateral to medial and from anterior to posterior (Shih and Keller, 1992c; Domingo and Keller, 1995; Lane and Keller, 1997), whereas when induced by growth factors or lithium treatment, convergent extension and associated cell behaviors appear to occur everywhere at once (see Kao and Elinson, 1988).

Embryos develop surprisingly well with only the somitic mesodermal component of the dorsal mesoderm. When the notochord is removed from urodeles, the extension during neurulation is nearly normal; it is only in the tailbud stages, a period in which notochordal vacuolation and swelling is the primary mechanism of extension (at least in *Xenopus;* see Adams *et al.,* 1991), that these notochordectomized animals show reduced extension (Kitchen, 1949). Embryos lacking a notochord due to UV irradiation of the fertilized egg are shorter than normal embryos but nevertheless do gastrulate and converge and extend (Clarke *et al.,* 1991). The difference between the extension of these animals and normal ones could be accounted for by the difference in extension rates between notochord, which is faster, and somitic mesoderm, which is slower (Keller *et al.,* 1989; see discussion below and Fig. 8). In other studies using the same treatment to ventralize, embryos lacking notochords appeared to have nearly normal axial extension (see Figure 10 of Malacinski and Youn, 1982). The reasons for the differences in extension observed in these two studies is not apparent. However, both studies show that substantially normal gastrulation, including involution and convergent extension, and neural induction occur when the somitic mesoderm is the major derivative of Spemann's organizer. It is only when the somitic mesoderm disappears in the graded series of UV-ventralized embryos that convergent extension fails and the normal anisotropic involution of the IMZ fails. The result is a symmetrical, shallow involution and a truncated symmetrical archenteron (Scharf and Gerhart, 1980).

Organizer activity, including induction of neural tissue, is often attributed to the prospective notochordal component of the organizer, but the evidence shows that embryos with prospective somitic mesoderm alone can induce and form neural tissue, including a neural tube in the amphibian (Bautzman, 1926; Cohen, 1938; Malacinski and Youn, 1981, 1982; Clarke *et al.,* 1991). In explants, somitic mesoderm can induce neural tissue (Jones and Woodland, 1989). Null mutations of the *HNF-3β* gene in the mouse results in absence of the node and the notochord, but nevertheless substantial organizer activity, anterior–posterior patterning, and neural tissue persists (Ang and Rossant, 1994). When the embryonic shield, nominally the teleost equivalent of the node/dorsal lip/organizer, was removed microsurgically from zebrafish embryos at the germ ring stage, substantial axial organization and neural induction occurred without notochordal mesoderm (Shih and Fraser, 1996). Analysis of floating head (flh) mutant fish embryos shows that the notochord precursor cells form somitic tissue (muscle) instead of notochord, but these fish have a neural tube (Talbot *et al.,* 1995; Melby *et al.,* 1996; Halpern *et al.,* 1995). In the UV ventralization series in *Xenopus,* loss of neural tissue is correlated with loss of somitic tissue (see Malacinski *et al.,* 1974; Scharf and Gerhart, 1980; Kao and Elinson, 1988).

To summarize, the cell behaviors driving extension of the axis do not begin at the dorsal midline where the traditional focus on the Spemann organizer would suggest they should, but bilaterally, in the prospective anterior somitic mesoderm. Second, the prospective somitic mesoderm shows highly organized spatial and temporal patterns of motility very early in development, somewhat before the notochord, and these behaviors are independent of the notochord's presence. Third, the reverse it not true, at least during early axial elongation; the notochord appears to be dependent on the somitic tissue for its morphogenesis. Fourth, it is clear that convergent extension of only the somitic mesoderm, in absence of the notochord, can bring about the normal anisotropic involution of the IMZ and axial elongation. In contrast, lithium-treated, dorsalized embryos in which most of the posterior mesoderm is notochord, rarely involute properly but generally form "proboscis" embryos (Kao and Elinson, 1988). Finally, neural tissue is induced in embryos developing without notochordal mesoderm.

IV. Segmentation and Somite Formation in *Xenopus laevis:* Cell Elongation and Rotation

Somite formation has been studied in *Xenopus* with histological sections (Hamilton, 1969; Kielbowna, 1981), scanning electron microscopy (SEM) (Youn *et al.,* 1980; Youn and Malacinski, 1981a), and with videomicroscopy of cultured explants (Wilson *et al.,* 1989; Wilson, 1990; Shih and Keller, 1992c). The latter studies describe the progress of the cell behaviors leading up to and including segmentation, and these will be discussed first, followed by the SEM and histological studies. Most of the following discussion focuses on the myotome, the largest and by far the best understood part of the somite.

A. Postgastrula Progression of Somitic Cell Behaviors: Videomicroscopy of Explants

Near the end of gastrulation, the entire notochord and prospective anterior somitic mesoderm, about 12 somites worth, have converged, involuted, extended, and have come to lie beneath the neural plate on the dorsal side of the embryo (Fig. 7A). The prospective posterior somitic mesoderm has converged, thickened and involuted but has not extended. Instead it forms a thick ring around the lateral-ventral aspect of the closed blastopore (see Keller and Danilchik, 1988) (Fig. 7A). The somitic and notochordal mesoderm, along with the overlying neural plate, can be isolated by cutting around the periphery of the archenteron as shown (Fig. 7A–C). If the suprablastoporal endodermal epithelium is peeled off, the underlying notochord and somitic mesoderm are exposed for recording of cell behaviors with

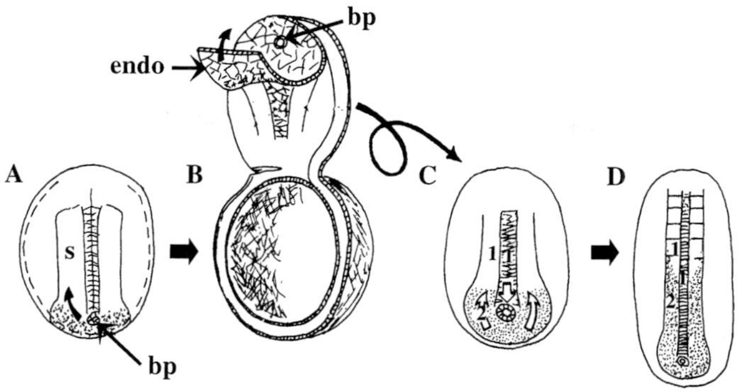

Figure 7 The dorsal view of the late gastrula/early neurula (stage 12.5/13), depicted as if the neural plate were transparent (A), shows the configuration of the notochordal (n) and the somitic mesoderm (s) that have converged and extended (unshaded), and the prospective somitic mesoderm lying ventrally and laterally around the blastopore (shaded) that will join the segmental plate during neurulation. This mesoderm can be prepared for videomicroscopy by cutting along the edge of the neural plate into the periphery of the archenteron (dashed lines, A), folding forward the neural plate/dorsal mesoderm, and peeling off the underlying endodermal roof of the archenteron, exposing the mesoderm (B). When explanted into culture (C), the anterior somitic mesoderm continues convergent extension, but not as fast as the notochord, with the result that the notochord shears posteriorly with respect to the somitic mesoderm (see number markers in C, D), pushes the blastopore (bp) posteriorly and the prospective somitic mesoderm lying ventral to the blastopore moves around both sides of the blastopore to join the posterior segmental plate (open, curved arrows, C, D). Based on Wilson *et al.,* 1989, and Keller *et al.,* 1989.

videomicroscopy (Fig. 7B,C) (see Wilson *et al.,* 1989; Wilson, 1990). The notochord has completed its involution and has undergone convergent extension to become about four cells wide; it abuts the blastopore at its posterior end. In contrast, the anterior somitic mesoderm has converged and extended, but instead of abutting the blastopore, its posterior region is continuous with the posterior somitic mesoderm, which has just involuted and forms a thick collar of cells around the blastopore (Fig. 7C). After involution, both the notochord and the somitic mesoderm continue to converge and extend by mediolateral cell intercalation. However, the somitic mesoderm extends less rapidly than the notochord because more of its convergence is channeled into thickening rather than extension. As a result, the faster-extending notochord shears posteriorly with respect to the slower-extending somitic mesoderm and pushes the blastopore posteriorly. As it does so, the prospective posterior somitic cells stream around the blastopore and into the posterior segmental plate (open arrows, Fig. 7C,D) (see Wilson *et al.,* 1989). These same movements are seen *in vivo;* dye marks placed ventral to the blastopore form "comets" around both sides of the blastopore to join the somitic mesoderm (Keller, 1975, 1976). Note that the failure of the "medial edges" of the future posterior somites to bound the notochord in the early gastrula fate map (Fig. 1A; shaded regions of Fig. 4), is now resolved as the posterior somitic mesoderm slides anteriorly along the sides of the notochord. Note also that this movement is similar in some respects to the regression of Hensen's node posteriorly between the somitic tissue during bird (Schoenwolf *et al.,* 1992) and mammalian gastrulation (Smith *et al.,* 1994).

The anterior–posterior progression of cell behaviors continues during segmentation and somite formation (Fig. 8). First, the anterior presomitic mesoderm continues convergent extension by the mediolateral cell intercalation that was begun in the gastrula stage, both extending and thickening as it converges (Fig. 8, C & E zone). Lying posterior to the zone of mediolateral cell intercalation, prospective posterior somitic mesoderm located around the blastopore undergoes radial intercalation as it thins and streams around both sides of the end of the notochord to

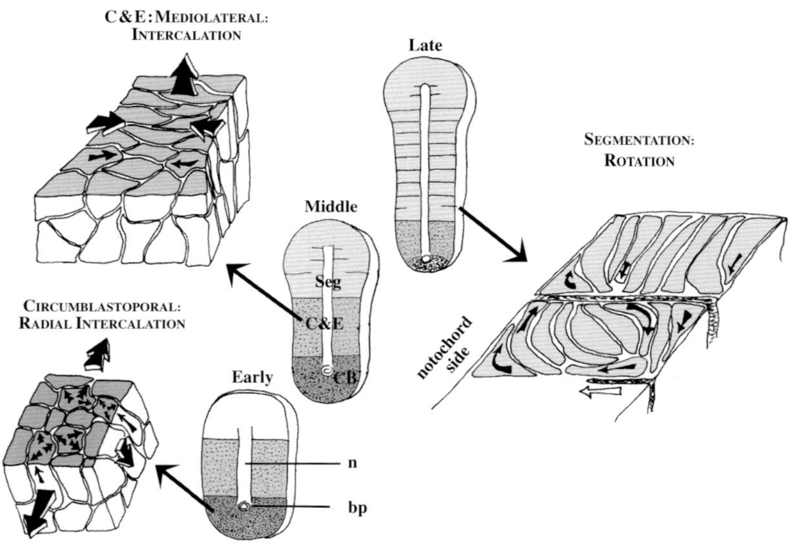

Figure 8 Diagrams summarize the cell behaviors progressing from anterior to posterior in the prospective somitic mesoderm during neurulation and beyond in *Xenopus* (center column). The early neurula/late gastrula somitic mesoderm displays a region of convergent extension and thickening (C & E, medium gray), characterized by mediolateral elongation of the cells and mediolateral cell intercalation (top left). Posterior to this region is a circumblastoporal (CB, dark gray) region in which radial intercalation is the predominant cell behavior; cells in this zone intercalate along the radius of the embryo to form a thinner array of greater area (bottom left). As the embryo nears the end of neurulation, the region of C & E is replaced anteriorly by segmentation (light gray), in which somites are formed (right); intersomitic furrows form from lateral to medial (white arrow), and the mediolaterally elongated cells simultaneously rotate (black arrows) to lie with their long axes parallel to the anterior–posterior embryonic axis (right). Modified from Wilson *et al.,* 1989 and Wilson, 1990.

join the segmental plate (Fig. 8, CB zone). As the entire explant extends, cells are fed out of the CB zone into the C & E zone, as both of the zones move posteriorly. As the C & E zone moves posteriorly, the cells left behind anteriorly begin segmentation at stage 17, a mid-to-late neurula stage. The intersomitic furrows form laterally first, and move medially between the previously elongated cells, until they reach the notochord (Fig. 8). As the intersomitic furrow appears, the cells of the newly forming somite begin a rotation; the cells adjacent to the notochord move anteriorly, the ends next to the previously formed somite move laterally, and the cells bounding the newly formed intersomitic furrow move medially, toward the notochord, generating a rotation movement (Fig. 8). The result of these movements is that cells that were originally oriented mediolaterally rotate to span the full length of the newly formed somite, achieving the state described by Hamilton (1969). Most of these cells appear to form myotome (see discussion below).

Light micrographs illustrating some of these cell behaviors are shown in Fig. 9. These micrographs show the resolution of cell shape and position that is possible with time-lapse recording of explants of *Xenopus* dorsal mesoderm under epi-illumination. Details of cell contact behavior and protrusive activity can be visualized by labeling the cells with fluorescent dextrans or GFP (green fluorescent protein) and recording behavior with low light fluorescent microscopy and imaging processing (Shih and Keller, 1992b,c; Domingo and Keller, 1995).

Based on histological measurements and wound marking experiments, Elsdale and Davidson (1983) showed that the process of forming somites from the tail mesoderm paralleled the events seen by Wilson on the anterior trunk somites. They divided the somitic mesoderm in the tail into a posterior "packing zone" containing about 20 pairs of somites. Anterior to the packing zone is a "zone of extension" in which the somitic mesoderm, consisting of about six prospective somites, extends eight-fold. Anterior to the zone of extension is a "prepatterned" zone, consisting of three somites that are not yet segmented but about to do so. The anterior boundary of the zone of extension is the site of the heat shock sensitivity seen in other studies (Elsdale *et al.,* 1976). The packing zone appears to correspond to the posterior circumblastoporal zone and the zone of extension and the prepatterned zone to the convergent extension zone of Wilson and colleagues (1989). Thus the same organizing schemes seem to be followed through both anterior trunk and tail somitogenesis.

B. Scanning Electron Microscopy and Histology: A Three-Dimensional View

Histological analysis (Hamilton, 1969; also see Schroeder, 1970) and scanning electron microscopy (SEM) (Malacinski and Youn, 1981; Youn and Malacinski, 1981a,b) have been used to describe *Xenopus* somite formation in *Xenopus* in three dimensions. During the neurula period, when the somitic mesoderm is

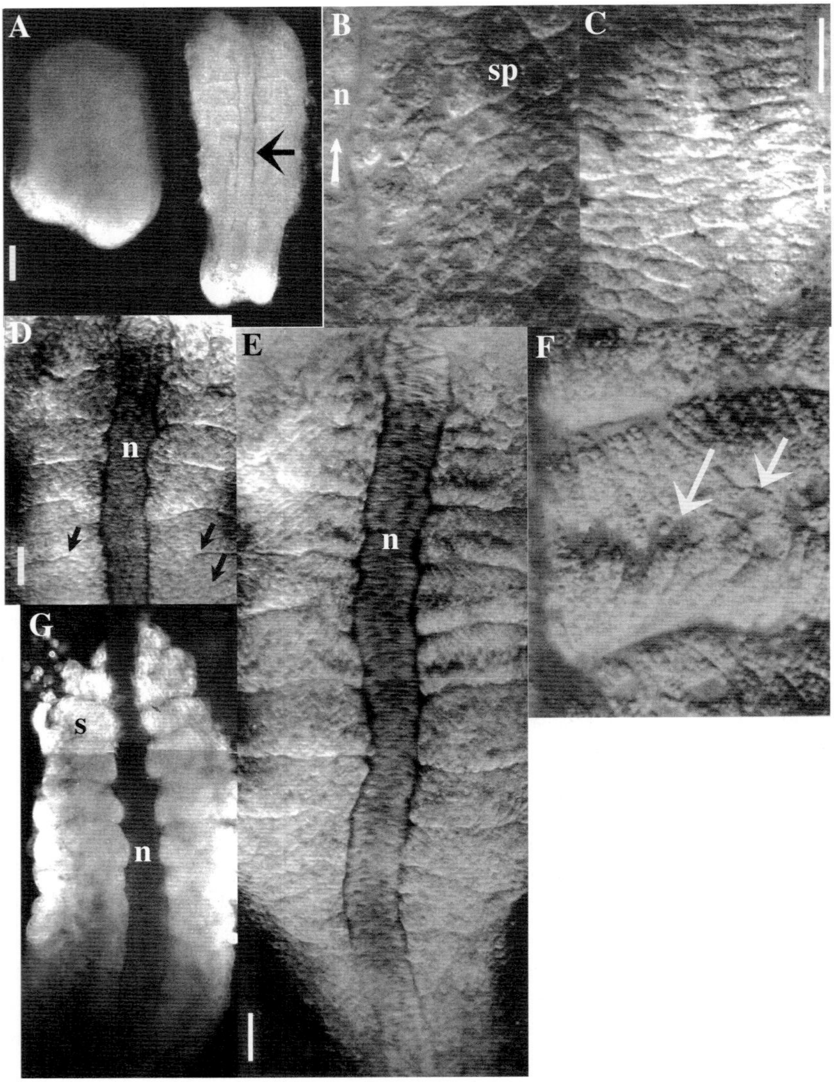

Figure 9 Light micrographs of living explants of *Xenopus* embryos made as indicated in Fig. 7. They show a late gastrula (stage 12.5) explant shortly after it was made (A, left) and after it has converged and extended by stage 23 (early tailbud) (A, right). Compare these explants to the diagrams "Early" and "Late," respectively, in Fig. 8, center column. The arrow indicates the notochord. Prospective somitic cells in the posterior somitic mesoderm of stage 13 are not elongated (B) whereas those in the central region of stage 16 (midneurula) are elongated and aligned mediolaterally (C). The arrows point anteriorly. A low magnification video micrograph shows the anterior somitic mesoderm lying on both sides of the notochord (n) at stage 19. Anteriorly, segmentation has occurred, and posteriorly, segmentation is underway, with the intersomitic furrows beginning laterally and proceeding medially toward the notochord (arrows, D). By stage 23, segmentation is advanced; cells in the older, anterior

undergoing the mediolateral intercalation described above, it is also undergoing changes in cell shape and arrangement, many of them in parallel with similar changes in the overlying neural plate. The nonnotochordal mesoderm, including the prospective somitic cells, form two layers of cuboidal cells, separated by an-interface, at the midneurula stage. In the somitic region, the interface between the two layers is referred to as a premyocoel, although it never expands to form a cavity (arrow, Fig. 10A). As the neurula stages proceed, the somitic cells elongate around the interface, their long axes lying horizontally in the case of the medial ones lying next to the notochord, and vertically in the case of those next to the gut endoderm and the folding neural plate (double-headed arrows, Fig. 10B). Somehow there is an asymmetry between the behavior of the cells above and below the interface between the two layers, and as a result, a thin layer of cuboidal cells appears to be pulled medially over the lateral surface of the somitic mesoderm (arrows, Fig. 10B,C); these cells are thought to become dermatome (see Schroeder, 1970; Youn and Malacinski, 1981a; Keller, 1976). The elongated cells appear to be prospective myotome on the basis of their movements in subsequent stages (described below). The elongating myotome cells appear to form buttresses that push on the ventral side of the folding neural plate (dashed arrow, Fig. 10C) (Schroeder, 1970), but there is no direct evidence for such forces. A small, loosely organized population of polymorphic, generally rotund cells is seen later, by late tailbud (stage 30), at the ventral-medial edge of the somite; these are thought to be the sclerotome. It is not known where these cells are located at the early and mid tailbud stages, such as those illustrated here, when all the cells in this region of the somite are elongated and resemble prospective myotome cells. As will be discussed further below, not much is known of the formation and morphogenesis of the sclerotome of most amphibians, including *Xenopus*.

Segmentation follows on the heels of these cell shape changes, beginning anteriorly at stage 17 and progressing posteriorly. In a thorough and elegant SEM study, using multiple planes of fracture through the somitic mesoderm and segmental plate, Youn and Malacinski (1981a) described the three dimensional aspects of segmentation (Fig. 11). Note that the segmental plate is tall, having thickened by the cell processes depicted in Fig. 10. It has three surfaces that abut tissues medial to it: one surface abuts the neural tube, one the notochord, and one the archenteron roof (Fig. 11A). Note also that the somites are progressively more chevronlike toward the posterior, and a dorsal medial projection extends posteriorly (Fig. 11C,D), making a much more complex shape than was apparent in the

---

somites have completed rotation and are now oriented with their long axes anteroposteriorly instead of mediolaterally, and the newly formed ones have begun rotation (E); anterior is at the top. The rotated cells, which will form myotome, span the anterior–posterior aspect of the somites (F); arrows indicate nuclei. The muscle monoclonal antibody marker 12-101 is expressed in newly formed somites (s) and anterior part of the segmental plate (sp), and it declines gradually in the posterior segmental plate (G). Bar in A = 200 μm; bars in E and F = 100 μm. Revised from Wilson, 1990.

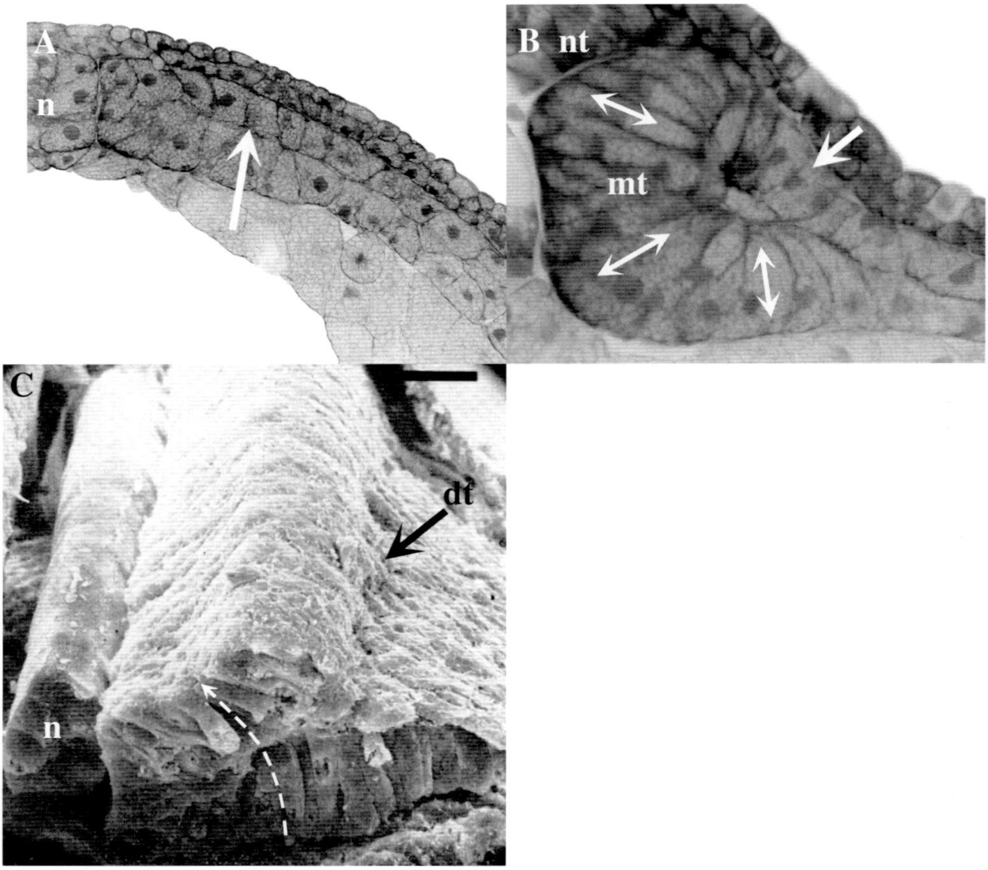

Figure 10 Confocal micrographs of fluorescently labeled *Xenopus* embryos show the cell columnarization from the early neurula stage (A) to the late neurula stage (B) in transverse view. The interface between the two layers of somitic mesoderm, the so-called premyocoel, is indicated with an arrow (A). The future dermatome is indicated with an arrow (B). Double-ended arrows show the elongation of the future myotome cells (B). An SEM (C) shows the morphology of the somitic mesoderm just prior to segmentation, including the thin sheet of cuboidal or squamous cells over the lateral surface thought to form dermatome (arrow, C). Comparable views are shown for *Rana* (Fig. 15C) and the urodele (Fig. 16C). The dashed arrow shows the direction of elongation of the "buttresses" of future myotomal cells. A and B, courtesy of Lance Davidson; C is from Youn and Malacinski, 1981a.

videomicrographic analysis above. Cell shapes and orientations were analyzed in three fracture planes with SEM (Fig. 12).

In this analysis, Youn and Malacinski (1981a) agree with Wilson and associates (1989) that the intersegmental furrows originate laterally and proceed medially toward the notochord. The fissures appear as the elongated myotome cells begin to rotate such that their long axes come to lie transverse to their original orientation

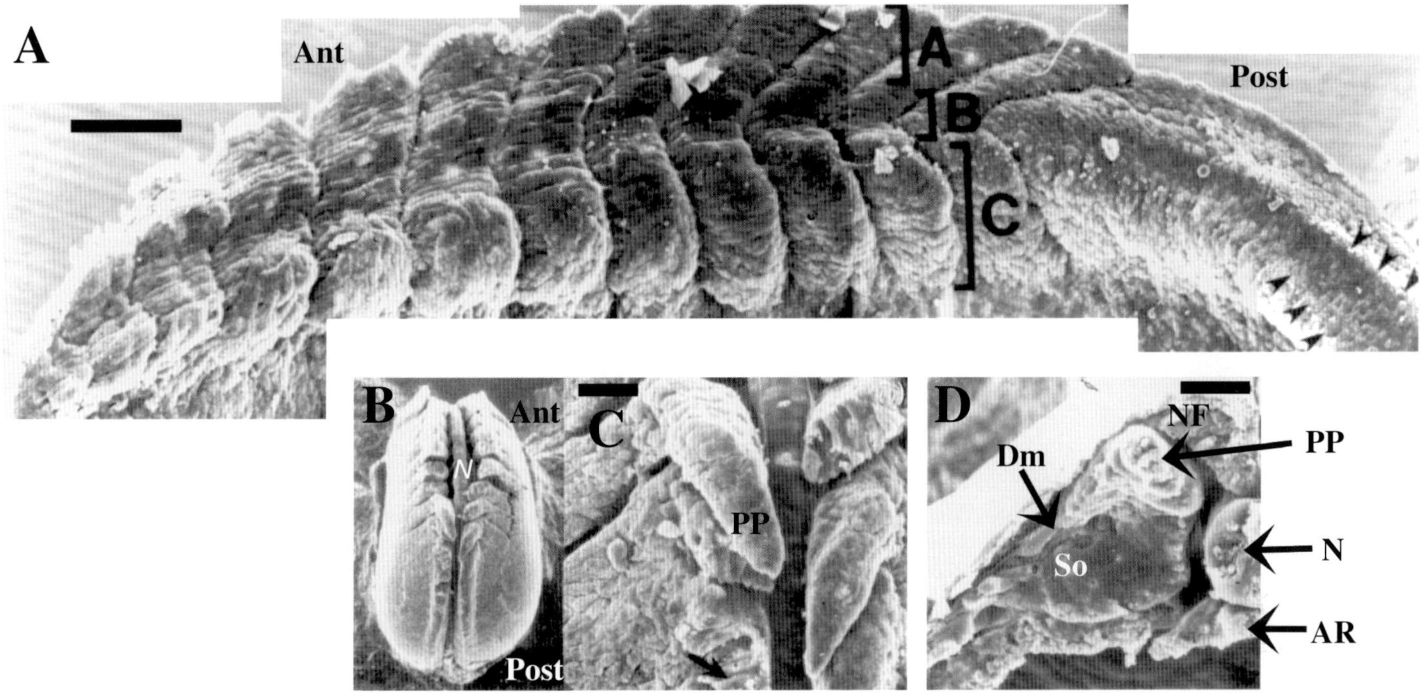

Figure 11 An SEM shows a medial view of the right somite files and segmental plate of an early tailbud *Xenopus* embryo, anterior to the left (A). The somitic mesoderm is divided into a region bounding the neural tube (A in A), the notochord (B in A), and the endodermal roof of the archenteron (C in A). These three regions are separated by arrowheads in the segmental plate at the posterior (right) of the figure. A low magnification dorsal view (B) shows the broad V-shaped groove formed by the rising banks of somites/somitic mesoderm on both sides of the notochord (N). A higher magnification dorsal view shows posterior projection (PP) of the dorsal medial region of the newly formed somite (C). A transverse view of the posterior surface of a somite (D) shows the endodermal roof of the archenteron (AR), the dermatome (Dm), the notochord (N), neural fold (NF), the posterior projection (PP), and the somite (So). Bar = 100 μm in A; bars = 50 μm in C and D. From Youn and Malacinski, 1981a.

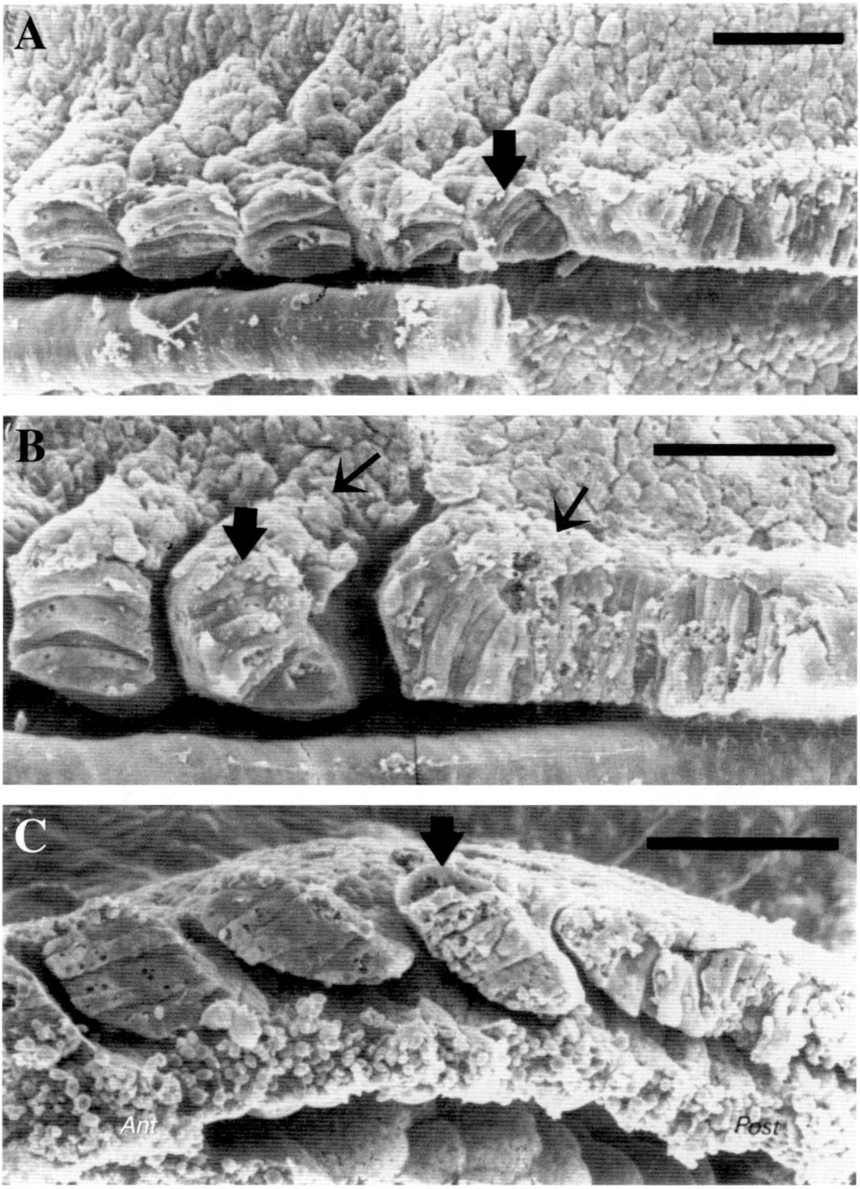

Figure 12 SEMs show frontal fractures through somites and segmenting mesoderm of *Xenopus* early tailbud (stage 22–24) at dorsal (A), middle (B), and ventral (C) fracture planes defined by the same letter designations in Fig. 11A). Note the rotation of the future myotome in newly forming somites (broad arrows) and the sheet of dermatome over the lateral surface of the somites/segmental plate (thin arrows). Anterior is at the right; bars = 100 μm. From Youn and Malacinski, 1981a.

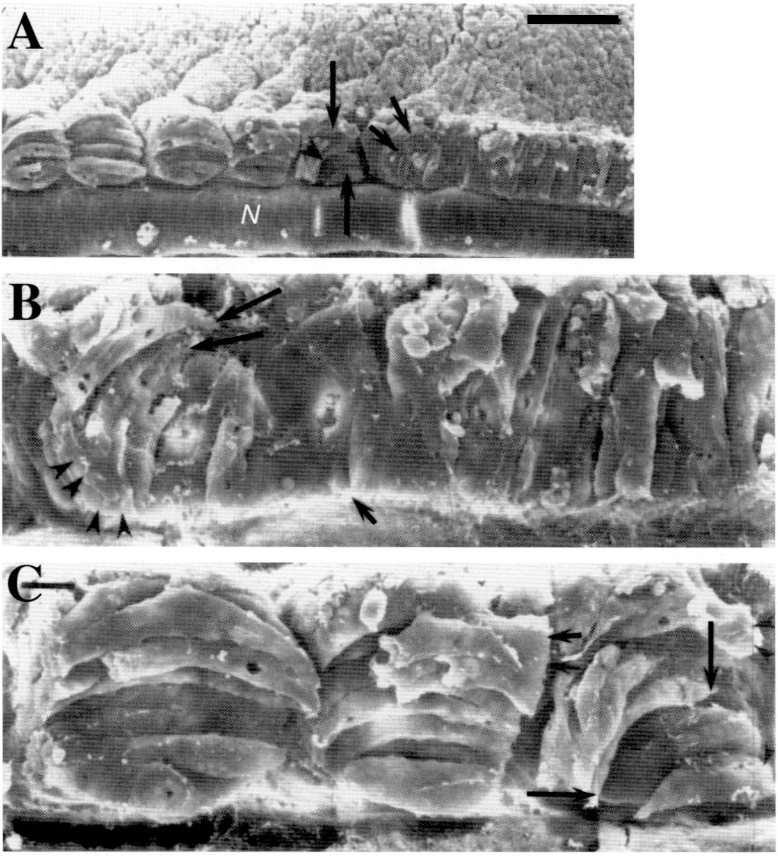

Figure 13 SEMs show details of changes in myotomal cell orientation and rotation during segmentation. A frontal fracture shows the mediolateral orientation of cells in the segmental plate (A,B, right sides) and anterior–posterior orientation in the somites (A,B, left sides). Enlargements of the region of ongoing segmentation (B,C) show that cells rotate as segmentation occurs. Note that the protrusions of the cells indicated by the arrows in B and C are the type thought to function in cell rotation. Note also that cells in the anterior of a forming somite have partially completed rotation at a time when those in the posterior region have not begun, nor has the segmental boundary formed (B). Notochord (N). From Youn and Malacinski, 1981a.

and parallel to the anterior–posterior axis of the embryo (Fig. 12A,B). The fact that the rotation begins with fissure formation means that the anterior region of the segmental plate comprising the anterior edge of the next somite is already participating in a process intimately related to segmentation before its posterior region, which will begin these movements later with the formation of the next intersegmental furrow (Fig. 13). Thus both video and SEM analysis agree that the cell rotation movement may actually define the intersomitic furrow or is intimately associated with its formation.

There is a dorsoventral temporal sequence to the rotation process. Of the three zones defined above, the middle zone, bounding the notochord, rotates first, followed by the cells dorsally and ventrally (Youn and Malacinski, 1981a). These authors originally saw no evidence of the presomitic "somitomeres," circular arrangements of cells preceding and predicting the positions of future somites seen in nearly all vertebrates reported on by Meier and Jacobson (reviewed in Jacobson, 1988), but later, these were seen (see Figure 3 of Malacinski *et al.,* 1989). Somitomeres will be discussed in more detail below.

C. Myotome and Muscle Differentiation

The elongated myotomal cells of *Xenopus* span the full length of the somite after their rotation, and they eventually differentiate myofilaments, sarcomeres, and a sarcotubular system as mononucleate cells (see Kielbowna, 1966, 1980; Muntz, 1975; Blackshaw and Warner, 1976). Although mononucleate, these cells are polyploid (Kielbowna, 1966). Multinucleate myotubes do appear later in *X. laevis,* at stage 45. At one time this was thought to occur by fusion of the uninucleate myotube cells with satellite cells that intercalate between them at later stages (Kielbowna, 1966; 1980; also see Muntz, 1975). Youn and Malacinski (1981a) suggest that the protrusions seen on the myotomal cells at the earlier tailbud stages may be involved in cell fusion. However, more recently Boudjelida and Muntz (1987) strongly suggest that such fusion does not occur, or at least is not a major event in multinucleation. Their electron microscopic data do not support the fusion hypothesis, and satellite cells were observed to invade the myotome only after multinucleation had begun. Moreover, DNA synthesis and mitotic division of the myocyte nuclei were not observed. Instead, they observed decreasing size and presence of constrictions of the nuclei during multinucleation, facts that they suggest support the idea of multinucleation by amitotic division. They note that this conclusion is consistent with the finding that myoblasts of *Xenopus* contain up to octaploid quantities of DNA whereas the adult myonuclei have diploid quantities of DNA (Kielbowna, 1966; Kielbowna and Koscielski, 1979). Thus it appears that the myotubes develop as uninucleate, polyploid cells, which then become multinucleate muscle fibers primarily by amitotic division. In contrast, there appears to be no evidence for this mechanism in amniotes in which cell fusion accounts for multinucleation (Gearhart and Mintz, 1972). Whether fusion with the satellite cells later, when they arrive, also contributes to multinucleation in amphibians may still be an open question. The cell protrusions interpreted by Youn and Malacinski (1981a) to be involved in cell fusion are instead probably functioning in movements of the cells; in any case, they were seen at tailbud stages, long before the myotubes became multinucleate. The types of muscle fibers differentiating in the larval and adult stages were reviewed by Radice and associates (1989), covering *Xenopus,* and also other amphibians. It should be remembered that *Xenopus* is not representative of the anurans in regard to myotome formation. As with most aspects of amphibian somite development, a number of routes to forming and differentiating the myotome have evolved. Some of these are described below.

The formation of muscles has been described and previous work on this subject from the nineteenth century onward was reviewed by Ryke (1953). According to Ryke, by stage 30 the dermatome has defined itself and the sclerotome has segregated off the ventral medial horn of the somite, leaving the myotome. Then the myotome forms a ventrolateral projection, the "*Urwirbelfortsätze,*" which moves ventrally to form the ventral somatic muscles; the remaining myotome forms the dorsal somatic muscles. The process of forming specific muscles in *Xenopus* is described in detail and summarized schematically in diagrams by Ryke (1953). These descriptions are thorough and detailed but they do infer origins from histological sections rather than from cell marking and tracing studies. More recently, Lynch (1990) used marked cells to trace the origin and development of the ventral abdominal musculature, as well as its innervation. Modern methods of cell tracing have been used to determine the origins and movements of avian myotome cells in great detail (see e.g., Denetclaw *et al.,* 1997), and these methods should be applied exhaustively to the amphibian somites as well.

D. Cellular Mechanisms of Myotome Formation and Development

1. Do the Myotome Cells Rotate as a Group or as Individuals?

The mechanism of myotomal cell rotation is not known. Hamilton (1969) was of the opinion that the somite rotated as a block of tissue, rather than cell by cell. However, Youn and Malacinski (1981a) argue against this possibility, noting that the entire somite does not rotate as a unit but instead shows the middle-to-dorsal and middle-to-ventral progressive pattern described above. To support their argument, they show that cells differ in their shapes and movements, depending where they are in the somitic mesoderm, and each cell following the general rule of turning longitudinally with the minimum of effort required at its particular location. Moreover, they observe large numbers of cell protrusions in their SEMs, putatively providing traction for such movements. The time-lapse observations of cell motility by Wilson *et al.* (1989) support this notion, showing that reorientation of cells occurs by local cell motility involving formation, extension, and retraction of protrusions characteristic of individually motile cells.

2. Questions of Traction

If the rotation is a matter of individual cell motility, as it appears to be, on what do the cells exert traction? Cells at the edges of the myotome could be crawling on the anterior surface of the segmental plate, the notochord, or the posterior surface of the previously formed somite. Perhaps it is more likely that they are crawling on extracellular matrix material in these spaces. However, the degree to which rotation requires traction on external surfaces has not been evaluated experimentally. Cells in the interior of the myotome must crawl on other cells or on extracellular matrix intercalated between the cells. A number of matrix molecules, matrix receptors, and cell adhesion molecules are expressed in the amphibian somite (see the discussion below), but experimental analysis of their role has not been done. Obviously, much work needs to be done on the cell motility and biomechanics of somite rotation.

3. Questions of Orientation

Myotome cell rotation occurs in an orderly fashion with respect to the adjacent structures of the embryo, but the cues that guide the direction of rotation are not known. McCaig (1986) presented evidence that the notochord provides some signal that acts at a distance to cause perpendicular orientation of myoblasts of *Xenopus* in culture, but no one has followed up on this report. The fact that the rotation process begins adjacent to the notochord is consistent with this structure being the primary cue for rotation. If this is true, how is the signal to rotate passed dorsally and ventrally to the rest of the somite: by further diffusion of a signaling molecule, by cell–cell contact, perhaps a stereotactic mechanism, or by matrix-mediated signaling? On the other hand, it has been reported that rotation, stretching (elongation) of myotome cells, and formation of myofibrils all take place in absence of a notochord (see Malacinski and Youn, 1982). If this is indeed the case, the reported notochord-induced alignment of myoblasts in culture is either wrong, irrelevant to the situation *in vivo,* or there are redundant mechanisms assuring cell rotation and other mechanisms fill in for the absence of the notochord. Again, this problem begs renewed attention.

4. The Mechanism and Regulation of Cell Elongation

Likewise, the mechanism of the cell elongation, its regulation, and its role in segmentation is not understood. First, it is not clear whether there is one elongation event, which then maintains the columnar shape through the remaining steps of somite morphogenesis, or whether there is a more or less permanent state of elongation that actually involves multiple regulatory and morphomechanical activities. The presomitic cells elongate mediolaterally as part of expressing MIB during convergent extension in the gastrula and early neurula stages. Then in the late neurula stages these cells also become elongated in the dorsoventral direction as they form the buttresslike array beneath the neural plate. These two elongation events appear to be different from one another. However, the second of these appears to be retained, for the most part, during the rotation movements. Second, the mechanism of this elongation, or of these elongations, is not understood in terms of the cytoskeleton, cell adhesion, and cell motility.

5. Is Rotation Essential for Segmentation and Muscle Differentiation?

Both recordings of living explants (Wilson *et al.,* 1989) and fixed preparations (Youn and Malacinski, 1981a) show that the movements of rotation and intersegmental furrow formation are intimately related in *Xenopus.* Although tightly linked to segmentation in normal development, rotation may not be essential for some sort of segmentation, defined as the formation of a boundary between blocks of tissue. First, not all the somitic cells that participate in segmentation also elongate and rotate. According to Youn and Malacinski (1981a), the dermatome also becomes segmented but these cells never elongate and thus there certainly is no evidence that they participate in rotation; instead the dermatome forms as and remains a thin sheet of cuboidal cells on the lateral surface of the somite. There is no evidence bearing on how the sclerotome segments, but rounded cells are not seen at the ventromedial surfaces of the somitic mesoderm during segmentation. The question of where sclerotome cells are located at segmentation will be discussed further below. Second, intersegmental furrows of some sort form in cultured open-faced explants of gastrula mesoderm (Shih and Keller, 1992c), but the cells do not rotate, either before or after this "segmentation." Thus it appears that an intersomitic furrow of some type can form in absence of rotation. This segmentation may be abnormal, however, and cell rotation may still be part of normal segmentation. Finally, if cell elongation and rotation is obligatory for normal segmentation in *Xenopus,* this is not a universal feature of the amphibians. Segmentation occurs without cell elongation and rotation in a number of species (see Section V.A–C). Segmentation also does not appear to be dependent on extension. In these explants, various amounts of convergent extension occur and so some somites will be broad and short and others longer and narrower (see also Elsdale and Davidson, 1983).

It does not appear that rotation must occur to express at least some muscle-specific proteins. In open-faced explants of the early gastrula, the myotome cell rotation does not occur, for unknown reasons (Wilson and Keller, 1991; Shih and Keller, 1992c), but the cells stain with the monoclonal antibody 12-101, a muscle marker (Kintner and Brockes, 1984). However, such explants have never been observed to twitch autonomously or in response to mechanical stimulation (R. Keller, unpublished observations). Whether or not they develop functional muscle is not known.

6. Tissue Interactions and Somite Differentiation

Many possibilities for analyzing the tissue interactions required for the development of the *Xenopus* somite have not been exploited. In the "Keller explants" of the dorsal IMZ (see Keller and Danilchik, 1988), the somitic mesoderm develops only in contact with notochord and the endodermal roof of the archenteron. Myotomal cells do elongate and rotate in such cases, and they stain with the 12-101 muscle-specific antibody without influence of neural or epidermal tissues. How far myotome development will go under these conditions is not known, and whether other components of the somite form under these conditions is also not known. These issues should be explored in *Xenopus* as well as in other amphibians. By using variants of the Wilson explants (see Wilson *et al.,* 1989; Keller *et al.,* 1989), interactions of nearly any combination of dorsal tissues at any stage can be achieved for analysis of patterning. The diversity of morphogenesis among the amphibians may be paralleled by diversity of somite patterning schemes.

V. Variations on a Theme: Somite and Myotome Formation in Other Anurans: *Bombina, Gastrotheca, Bufo, Pelobates,* and *Rana*

Somite formation varies in a number of aspects among the other anurans that have been studied in at least some detail. Using the work on *Xenopus* as a baseline, the major differences are summarized in Fig. 14. In the following discussion it should be kept in mind that differences in somite formation are being defined principally in terms of differences in myotome formation, since little is known about the dermatome and sclerotome.

A. *Bombina* and *Gastrotheca*

Somite differentiation was examined with histological sections in a comparative paper by Kielbowna (1981) on *Bombina, Pelobates,* and *Xenopus* (see also Kielbowna and Koscielski, 1979, 1981). *Bombina* has a very different way of accomplishing the myotomal arrangement seen in *Xenopus.* The presomitic cells are rotund rather than elongated at the outset, and they are not arrayed around a myocoel or "premyocoel," which is absent (Fig. 14, second row). Moreover, they do not appear to be organized into two layers, as is the case in *Xenopus.* Segmentation occurs among rounded, polymorphic cells. The cells then elongate parallel to the anterior–posterior axis and in doing so, they interdigitate, forming parallel arrays that comprise the full length of the somite (Fig. 14). (The term interdigitate is used here to describe this limited mixing of cells, rather than the term intercalation, which has already been applied to the repetitive intermixing of cells along the mediolateral axis during convergence and extension of both mesodermal and neural tissue; see Keller *et al.,* 1992). Note that in *Bombina,* segmentation is uncoupled from cell elongation; elongation still functions in myotome formation but only after segmentation. In *Bombina,* as in *Xenopus,* myotubes and a sarcotubular sys-

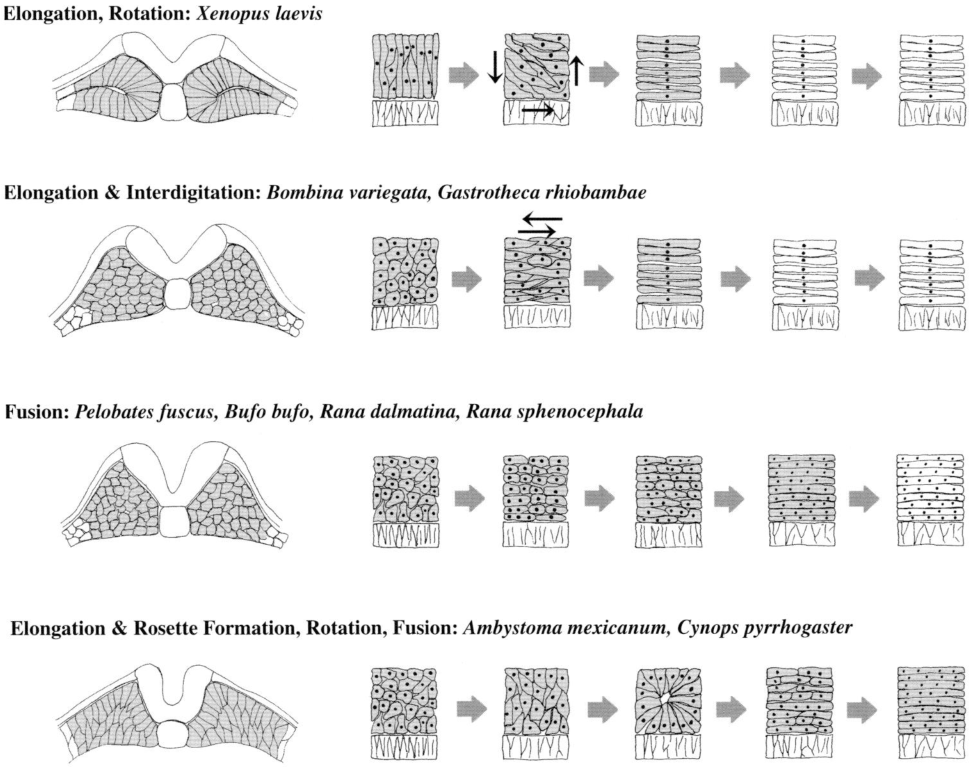

Figure 14 Variations of cell behavior comprising the process of segmentation and myotome formation are shown diagramatically. In the left column, cross-sectional views of the segmental plate are shown. In the series at the right, the sequence of cell behaviors bringing about myotome formation are diagrammed with anterior to the right. In *Xenopus,* the cells elongate mediolaterally prior to segmentation, and then, during segmentation they rotate, moving anteriorly next to the notochord, laterally anteriorly, and medially posteriorly to span the full length of the somite. In *Bombina* and *Gastrotheca,* the cells are initially rotund and polymorphic. When they do elongate, they also interdigitate, and span the full length of the somite. As in *Xenopus,* no fusion occurs and the cells remain mononucleate until stage 45. *Pelobates, Bufo,* and *Rana* have initially cuboidal cells, although some may elongate somewhat. These line up and fuse to form multinucleate myotomal cells. No myocoel is formed in any of these anurans. In the Urodeles *Ambystoma mexicanum* and *Cynops pyrrhogaster* myotome formation occurs by a number of steps. Initially polymorphic, generally rotund or cuboidal cells first elongate, but as they do so, they arrange themselves radially in all directions around a central point, forming a rosette. A myocoel forms at or near but not necessarily at the center point; then the cells rearrange, elongate, and fuse to form elongated, multinucleate cells, oriented parallel to the long axis of the embryo. Summarized from Brustis and Delbos (1976), Brustis (1979), Gatherer and Del Pino (1992), Kielbowna (1981), and Youn and Malacinski (1981a,b).

tem form in mononucleate myocytes. Multinucleation occurs only at later stages (stage 45), and it was thought to occur by cell fusion, resembling *Xenopus* in this regard. However, more recent work described above (Section IV.C) argues against fusion in *Xenopus* (Boudjelida and Muntz, 1987), and thus perhaps this subject should be revisited in *Bombina* as well.

Gastrotheca rhiobambae, a marsupial frog (broods its eggs in a dorsal pouch), shows a pattern of cell behavior during myotome formation similar to that seen in *Bombina.* The cells elongate and interdigitate, beginning from a previously unelongated state (Gatherer and Del Pino, 1992). The presomitic cells are arrayed in a weakly defined rosette, with cells appearing to be more epitheloid around the edges and rounded cells occupying the center (Gatherer and Del Pino, 1992). These "rosettes" are poorly defined compared to those in birds and *Ambystoma* (see below) and most closely resemble the arrangement in *Rana sphenocephala,* described below. In addition, this organization was only seen in anterior somites; caudal ones first appeared as cell aggregations with no sign of a rosette. No myocoel was observed in this species.

B. *Pelobates fuscus* and *Bufo bufo*

Pelobates fuscus, the European spadefoot toad, shows yet another version of myotome formation (Kielbowna, 1981). The cells initially have a rotund or cuboidal shapes and no myocoel is apparent (Fig. 14, third row). Segmentation occurs among these rounded cells, and after segmentation, the cells do not move. Instead, they line up in parallel rows and then fuse, forming multinucleate muscle cells spanning the full length of the somite (Fig. 14). Therefore myotube formation occurs in a multinucleate cell in *Pelobates* whereas it occurs in mononuclear cells in *Bombina* and *Xenopus.*

Based on histological and ultrastructural evidence, the common toad, *Bufo bufo,* appears to form its somites and myotomes in a fashion similar to *Pelobates* and *Rana* (Fig. 14, third row) (Brustis and Delbos, 1976; Brustis *et al.,* 1976; Brustis, 1979). Segmentation occurs among rotund cells. Myotome formation occurs as cells bounding the anterior and posterior borders of the somite elongate posteriorly and anteriorly, respectively, where they meet and fuse with the internal cells, which have become fusiform in the meantime.

Somite and myotome formation in the anuran *R. sphenocephala* has been described in detail with SEM by Youn and Malacinski (1981b), and it resembles that of *P. fuscus* and *B. bufo.* First, at least a fraction of the presomitic mesodermal cell population appears elongated mediolaterally, initially, and then they appear to radiate in all directions around a central focal point (Fig. 15A,B). However, they do not form the strong rosettes characteristic of somites of urodelean amphibians (see below) nor do they have a myocoel (Fig. 15A,B). The little elongation that is seen, occurs shortly before segmentation. The unsegmented somitic mesoderm shows no cell elongation but is composed of polymorphic cells in no distinct arrangment (Fig. 15C). With their elongation and orientation being less pronounced, any rotation of the cells that might occur is not obvious. Limited rotation does appear to occur during somite segmentation. It has been reported that just a few somites an-

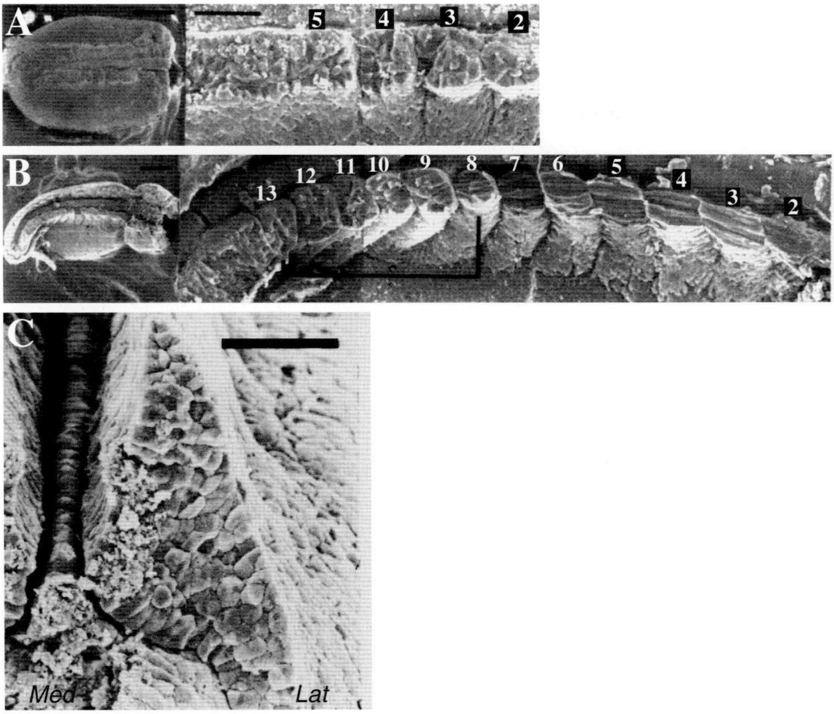

Figure 15 SEMs show dorsal views of frontal fractures through somites/somitic mesoderm at the mid neurula at stage 15 (A) and the late neurula at stage 18 (B) of *Rana sphenocephala.* The somites are numbered and the regions thought to be undergoing cell fusion are indicated at the right end of the bracket in B. The insets at the left show low magnification views of the whole embryo, for reference. A transverse fracture shows the cell morphology and arrangement in the segmental plate of *R. sphenocephala* (C). Compare with Figs. 10C and 11D for *Xenopus* and Fig. 16C and 18 for the urodele. Bars are 0.3 mm in low magnification micrographs of A and B, and 0.1 mm in all others. From Youn and Malacinski, 1981b.

terior to the last-formed somite, the myoblasts fuse to form a myotube that appears superficially equivalent to the myotomal cells spanning the somite in *Xenopus* (Youn and Malacinski, 1981a) (Fig. 15B). However, no direct evidence of fusion is offered other than external appearance of the cells in this study.

C. Selected Urodeles: Rosettes

Youn and Malacinski (1981b) examined somite formation in two species of urodeles, *Ambystoma mexicanum* and *P. waltl.* In both species, the paraxial mesoderm segments by rosette formation, with elongated cells radiating from a central point (Fig. 14, last row; Fig. 16A,B). Both species are said to be the same. Both differ from the well-known rosettes found in the bird in having more elongate cells in a

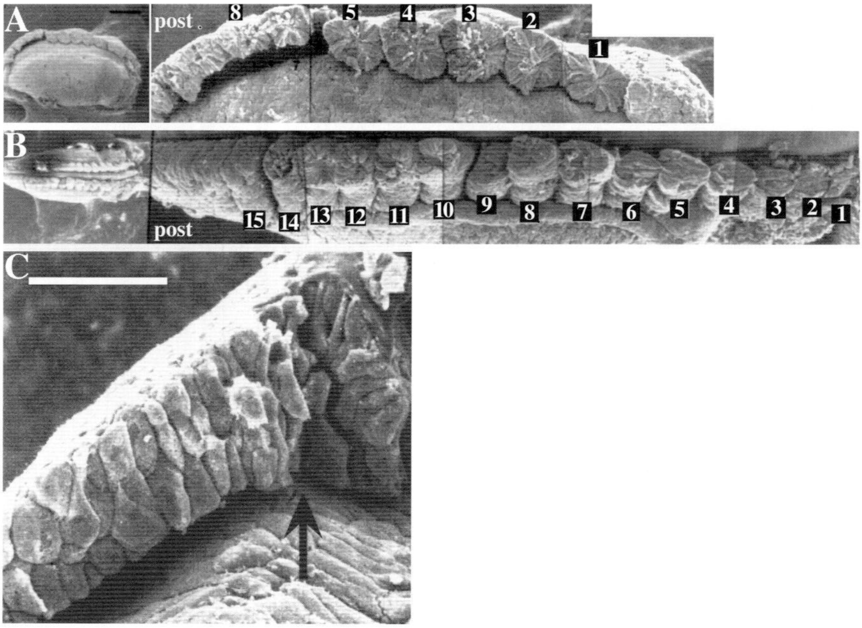

Figure 16 SEMs show segmentation and somite morphogenesis in *Ambystoma mexicanum,* a urodele, in a parasagittal fracture of stage 21–22 (early tailbud) (A) and a frontal fracture of stage 27–28, an advanced tailbud (B). A transverse fracture of the segmental plate shows the arrangement and morphology of the cells before segmentation (C). Bars = 300 and 100 μm, respectively, in the low and high magnification figures of A and B; bar = 50 μm in C. From Youn and Malacinski (1981b).

pronounced, radial array around a myocoel (Fig. 16A,B). In general, the cells do not show specialized cell shapes before segmentation (Fig. 16C), although those in the medial region did appear to be elongated somewhat in the mediolateral direction, and those farther laterally appeared to be elongated somewhat in the dorsoventral direction, and arranged in two layers (Fig. 16C). However, note that the inner ends of the cells of the two layers do not form a clear interface (a premyocoel), as in *Xenopus* (compare Fig. 10C with Fig. 16C), but rather interdigitate for about 10 to 20% of their lengths.

Rosette formation appears to involve rotation of the mediolaterally and dorsoventrally oriented cells to an anterior–posterior orientation (Fig. 17A–C). The cells at the corners of the incipient somite re-orient about 45°, often adopting a triangular morphology in doing so (Fig. 17B). Rotation to form a radial rosette appears to occur as the somite undergoes segmentation (see recently forming somite, Fig. 17C), much as the planar rotation occurred during segmentation in *Xenopus.* Unlike the other species discussed thus far, which either have a flattened discontinuity between layers of somitic mesoderm (premyocoel), or no myocoel at all, these urodeles have a pronounced myocoel (Fig. 17D). The myocoel of urodeles

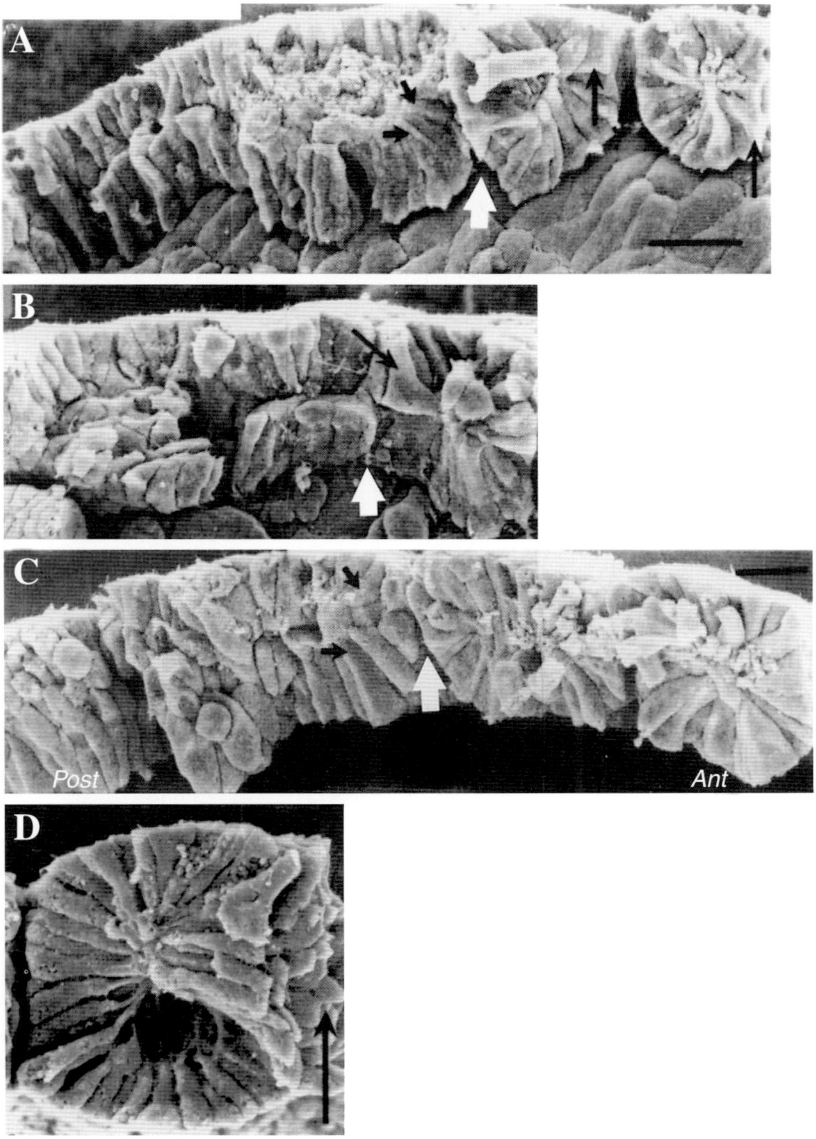

Figure 17 SEMs of parasagittal fractures show changes in cell shape and arrangement during rosette formation in *Pleurodeles* (A,B) and *Ambystoma* (C). Anterior is to the right. The short, thick white arrows show the most recently formed intersegmental furrows. Conically shaped cells thought to be undergoing rotation are indicated by short, black arrows. The triangular cells found at the corners of the rosettes are indicated by long black arrows. A fracture through an *Ambystoma* somite shows the myocoel (D). The arrow indicates the lateral-to-medial direction. The bars = 50 μm in A–C and 20 μm in D. From Youn and Malacinski (1981b).

is not related to formation of an interface between two layers, as is the formation of the premyocoel of *Xenopus*. It seems that in *Xenopus* the so-called premyocoel never makes a myocoel, and in *Ambystoma* the myocoel forms without a premyocoel. Also the myocoel of the urodeles forms after rosette formation and is not always at the center of a rosette (Fig. 17D). It appears that the radial arrangement of cells characteristic of a rosette occurs around a focal area, which may or may not be occupied by the myocoel when it develops. It should be kept in mind that these descriptions are based on interpretations of SEMs and that the movements of cells have not actually been observed directly.

Myotome formation was described only in *Ambystoma* in these studies. The anterior and posterior cells of the rosette elongate, extending toward one another, meeting at their inner ends, and fusing; of course, for cells at the central region of the rosette this requires little movement as their inner ends are close to one another at the outset (Fig. 18A–C). The triangular cells at the corners of the rosette lose this shape and likewise elongate toward one another to make a parallel array that then undergoe fusion (Fig. 18B). The anterior cells of the rosette are longer than the posterior cells, and Youn and Malacinski (1981b) describe the anterior cells as spreading more aggressively around the posterior cells, than vice versa, and suggested that this may be a step in the process of cell fusion. They also interpreted the signs of protrusive activity, such as filopodia, lamellipodia, and filolamellipodia (Fig. 18D), as evidence for cell fusion. However, these are locomotory organelles that may reflect active cell crawling instead. The evidence for fusion in this work appears to be whether or not cells span the entire length of the somite, the assumption being that if one does, a posterior and anterior cell in the rosette must have fused. In this period of stage 28 to 34, the myotome cells are supposed to become multinucleate based on histological evidence (Loeffler, 1969).

VI. The Dermatome and Sclerotome

The dermatome and sclerotome of the somites of amphibians have been ignored compared to the attention given the myotome, undoubtedly because the myotome is a large population of cells, dramatic in their behavior, whereas the other two components, particularly the sclerotome, are small, difficult to find, and less accessible to observation and experimentation.

A. The Dermatome

As noted above, the dermatome of *Xenopus* appears as a thin sheet of cells on the lateral surface of the somitic mesoderm (Fig. 10B,C, 11D), and it appears to be hoisted into this location by the columnarizing myotome cells, as they change shape (see Figure 5 of Hamilton, 1969; Youn and Malacinski, 1981a; also see

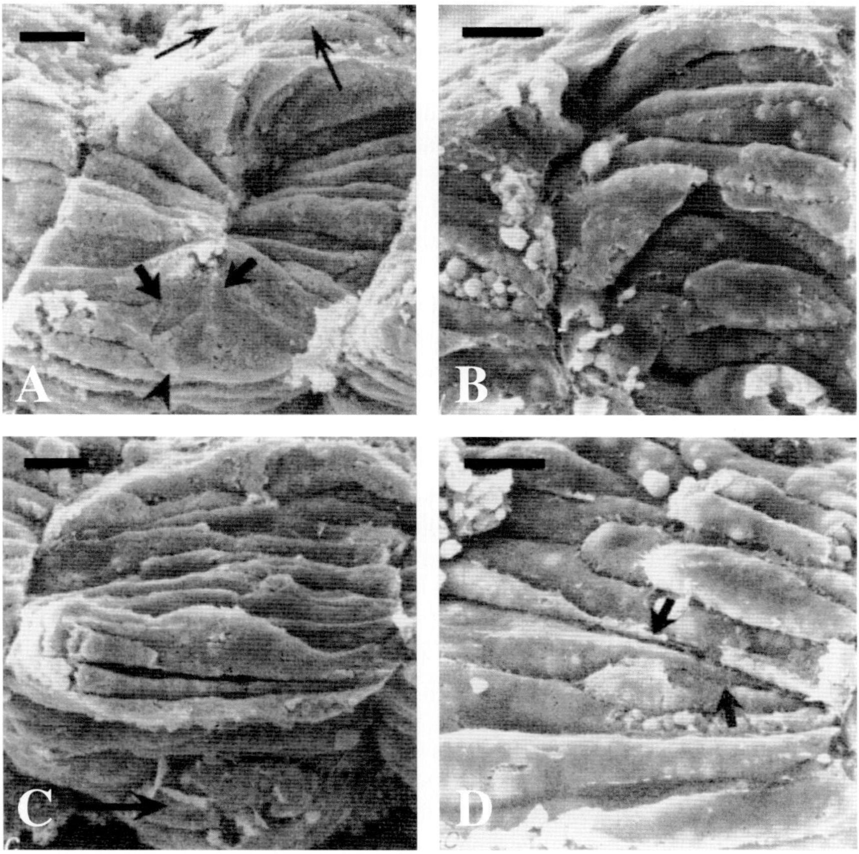

Figure 18 A dorsal view of a longitudinally fractured somite of *Ambystoma* (A) shows the adjacent ends of cells at the anterior and posterior parts of the somite applying protrusions on one another's surfaces, events thought to be in preparation for fusion (short arrows, arrowhead). The long arrows show what is believed to be the dermatome. Cells in the lateral region of the somite that were conical become oriented along the anterior–posterior axis (B). A more advanced state of fusion shows cells spanning the full length of the somite (C). The arrow at the bottom indicates cells thought to be the sclerotome. Cells in the anterior half of the somite develop long thin protrusions thought to be involved in fusion (arrows, D). Bars are 20 μm. From Youn and Malacinski (1981b).

Keller, 1976). Ryke (1953) describes the somite as first being divided into a lateral lamella of cuboidal cells, the dermatome, which is connected to the remaining sclero-myotome part of the somite at its dorsal extent, a description consistent with the above work. The dermatome of *Xenopus* is said to be segmented by some (Youn and Malacinski, 1981a) but the mechanism of segmentation has not been described, and others claim it is unsegmented (see Hamilton, 1969, especially Figure 5). Brustis and others (1976) describe the dermatome of *B. bufo* and *Rana dalmatina* as forming a single layer of cells histologically distinct from the myotome on the basis of having clearer cytoplasm, larger nuclei, and a rotund shape, acquired as a result of failing to elongate as the myotomal cells elongate. According to these investigators, it is segmented in these species, and in this regard they agree with the findings of Youn and Malacinski on *Xenopus.* In *Xenopus,* the dermatome moves off the lateral surface of the somite and becomes apposed to the epidermis at stage 29 or 30, an advanced tailbud stage in the second day of development (Nieuwkoop and Faber, 1967). Similar behavior occurs in *Bufo* and *Rana* (Brustis *et al.,* 1976).

No consistent and comprehensive picture of dermatome development emerges from these studies, and a number of issues remain clouded, even at the level of basic description and characterization. Is the dermatome segmented, and if so, how does it segment? Are there species differences in this regard? Where exactly do these cells originate and how do they come to form a sheet on the outer surface of the somite? How do the mechanisms of forming the dermatome in *Xenopus* and other amphibians differ? How do these cuboidal cells maintain their position while the myotomal cells rotate beneath them in *Xenopus?* What tissue interactions and factors regulate their differentiation and movement off the lateral surface of the myotome at later stages?

B. The Sclerotome

The sclerotome likewise is a small part of the *Xenopus* somite, appearing at their ventral-medial aspect as rounded, polymorphic cells, which segregate from the myotomal region at stage 30 (Ryke, 1953; Youn and Malacinski, 1981a) (stage 24 to 29–30 according to Nieuwkoop and Faber, 1967). In *B. bufo* and *R. dalmatina,* the sclerotome cell population forms on the medial surface of the myotome and appears to be larger and more distinct from the rest of the somite than its counterpart in *Xenopus* (Brustis *et al.,* 1976). Like the dermatome cells, the sclerotome cells appear different and distinct from the myotome largely as a result of failing to undergo the elongation that characterizes the myotomal cells (Brustis *et al.,* 1976). However, there are few convincing illustrations of the sclerotome or this process of their formation and segregation in any species. For example, in *Xenopus* the details of the origin of the sclerotome and what form it takes while awaiting dispersal from the myotome are not known. If the sclerotome forms by failing to participate in the rotation, then it should be visible on the medial surface of any freshly formed somite. No clearly identified population of this type is visible in most of the illustrations of *Xenopus.* So the question arises, where are the sclerotome cells at this stage? They could be a cryptic subpopulation of the elongated, putative "myotome" cells, and participate in elongation and rotation/segmentation, but later round up, rather than go on to become myotome. Or they could segment by some mechanism that does not require elongation, but then they should have shown up after segmentation as rotund cells in the proper location in at least some of the published illustrations. Turning to the urodele, in an elegant study in which he showed that the neural arches are formed from the sclerotome and not the trunk neural crest, Detwiler (1937) vital dyed the neural crest to distinguish them from sclerotome; he showed in histological sections that the sclerotome appears as a small population of cells adjacent to the notochord. One of the best illustrations of this population is shown by SEM in Fig. 18C. In summary, it is clear that much remains to be learned about the initial formation of amphibian sclerotome. A more convincing characterization of this important cell population in amphibians, including unambiguous identification with molecular markers, is badly needed.

In all species, the sclerotome cells are thought take up positions around the notochord and neural tube as scattered cells when they leave the somite in the tailbud stage. A general picture of how this occurs and how the vertebrae form from sclerotome emerges from a massive histological study on the development of the vertebral column of the urodele, focusing on *Triton vulgaris* (Mookerjee, 1930) and of the anuran, using embryos of *Rana temporaria, Bufo melanostictus, Bombina* and *X. laevis* (Mookerjee, 1931). To summarize vertebral development from Mookerjee's studies, in the mid tailbud stage the notochord becomes surrounded by a layer of cells, which initially is somewhat indefinite but later forms a distinct layer, the notochordal epithelium. About this time, sclerotome cells are liberated from the ventromedial part of the somite and scatter around both the neural tube and notochord. This is possible because the notochord separates from the underlying gut, leaving space between them; the dorsal aorta appears between the notochord and the gut. The sclerotome cells first scatter along the length of the notochord, and later they aggregate to form perichordal rings encircling the notochord and at the dorsolateral corners of notochord at the caudal end of the myotome. These later increase greatly in number, aggregate, and form the neural arches, extending up on both sides of the neural tube as well as ventrally at the caudal level of the myotome. The remainder of the vertebra is formed by one of two modes, depending on the species. In the "perichordal mode," additional cell aggregations, presumably sclerotome-derived, form chondrogenic masses beneath the notochord in the intermyotomal region; these encircle the notochord and eventually unite with the neural arches to form the centrum of the vertebra. As the vertebra develops at the caudal end of the myotomal segment, the notochord becomes constricted in this, the "intermyotomal" or future vertebral region. In the "epichordal mode," the notochord degenerates and the centrum of the vertebra forms from ventral extensions of the dorsal, neural arch elements.

One should keep in mind that these are anatomical studies that infer continuity of cell populations from histological analysis, and no direct cell tracing was done. Thus the exact course of the peregrinations of the sclerotome from the somite through the various complex anatomical structures attributed to them is not known from direct evidence. Also the pattern of proliferation that transforms a small population of cells into a large one has not been described very well. Although the sclerotome development in urodeles and anurans investigated thus far show general similarities, eventually substantial species differences arise in various aspects of notochord development, vertebral development, and rib formation, reflecting the differences in the anatomy of these animals. These differences are not well characterized beyond histological analysis. In addition, the sclerotome of the third and often forgotten order of amphibians, the legless Gymnophiona, is quite different from that of the commonly studied urodeles and anurans in both mass and morphology; these animals have large sclerotome as well as large numbers of somites (Wake, 1970; Wake and Wake, 1985). These facts argue that development of sclerotome should be investigated with modern methods and in a broader spectrum of species of amphibians than has been the case thus far.

C. Resegmentation

The segmental units eventually formed by the sclerotome, the vertebrae, are 180° out of phase with respect to the paraxial muscles formed by the myotome. One way to account for this development is the concept of "resegmentation" in which parts of the originally segmented somite units are reorganized to form secondary segmented units at the level of the intersomitic furrows that form the vertebrae (see Wake, 1970; also see Chap. 9 by Brand-Saberi and Christ, this volume). The mechanism of this "resegmentation," if it occurs at all in the commonly studied amphibians is not known. For it to occur, there should be some formal and consistent relationship between the original segmented units and the "resegmented" ones. There are several possibilities. The sclerotome cells from the posterior of one somite and those from the anterior of the next most caudal somite could then join to form a vertebral segment lying between the two myotomes (intermyotomal level). Alternatively, all the sclerotome at the myotomal level of one somite could move anteriorly into the intermyotomal region between that somite and the next and form a vertebra there. Finally, all sclerotome from one somite could move posteriorly into the intermyotomal region between that somite and the next most posterior one to form a vertebra offset a half segment posteriorly.

In the anuran and urodele amphibians, there is little evidence bearing on the problem of resegmentation (see Wake, 1970; Wake and Lawson, 1973). In the above work, the evidence bearing on how the sclerotome segments and segregates from the myotome is sporadic and not very satisfying. There are no cell tracing experiments that unequivocally show that sclerotome from a particular part of a somite forms a specific part of the vertebra. It could be that sclerotome cells from a number of myotomal levels mix indiscriminantly along the notochord and then aggregate in the intermyotomal regions with no relationship to the original segmental origin of the cells.

The work on avian embryos can serve as a model of what should be done in amphibians. Cell labeling experiments suggest that in birds, sclerotome cells from two adjacent somites form the vertebral anlage between the two somites (Ewan and Everett, 1992). More recently, chick–quail chimeras have been used in an elegant study showing the details of resegmentation in the avian embryo, including the contributions of adjacent somites to the vertebrae, ribs, and muscles (Huang *et al.,* 1996). Such a study done on a number of species of amphibian would at once clarify the development of the sclerotome and determine whether the variation that appears to be a common feature of other aspects of somitic mesoderm morphogenesis in amphibians is also found in the sclerotome. Of particular interest will be the legless caecelians (Gymnophiona), which have a very large population of sclerotome cells and large numbers of somites (Wake, 1970; Wake and Wake, 1985). In this regard they are more like the avian somite and very unlike the commonly studied anuran and urodele species described above. Finally, it will be informative to analyze sclerotome (and myotome) morphogenesis in species that develop directly into adults rather than going through a larval stage.

VII. Patterning of the Somitic Mesoderm by Adjacent Tissues

Amphibian embryos were used in the past for experimental analysis of the tissue interactions patterning the somitic mesoderm but have been eclipsed in this regard by other vertebrates in recent years. However, the variation in somite type and morphogenesis among experimentally accessible amphibians may offer opportunities to study underlying variations in patterning that might not be so readily available in other classes of vertebrates.

The early work uncovered the notion that tissue interactions are required for advanced sclerotome differentiation. If notochord development is suppressed completely by lithium treatment, no cartilage is found in the tail and trunk regions (Lehmann, 1935). However, if the notochord is removed by microsurgery at the early neural plate stage of *Ambystoma,* cartilage forms but in one large unsegmented mass in the anterior trunk and no neural arches are formed (Kitchen, 1949; also see Holtzer and Detwiler, 1953). These studies suggested that sclerotome in this species can segregate and make cartilage after the early neural plate stage without further interactions with the notochord, but it cannot segment without continued presence of the notochord. Thus it would seem that the notochord might induce sclerotome to form quite early, during or shortly after gastrulation. On the other hand, the only early removal of notochord in these types of studies was with lithium treatment, which may also have direct effects on the prospective somitic tissue.

Brustis (1978) examined the role of the notochord and neural tube in somite development, as well as the role of epidermis in dermatome development in two species with somewhat different results.The dorsal structures of the early neurula of *R. dalmatina* and *B. bufo* were excised, the endoderm of the roof of the archenteron was lifted up, and the notochord and neural tube were removed, leaving the explanted segmental plate and remaining adjacent tissues. Under these conditions, sclerotome cells appeared, segmented, and formed a mesenchymal cell population that segregated from the remainder of the somite. However, because of limited culture conditions, it was not determined how far the differentiation of the sclerotome could proceed without interactions with adjacent tissues. In similar experiments, removal of the dorsal epidermis from the early neurula failed to prevent formation and segregation of the dermatome from the myotome; however, lateral epidermis migrated over the somite in the course of the experiment. Brustis concluded that by the early neurula stage, formation and segregation of both sclerotome and dermatome did not require interactions of the somitic tissue with adjacent tissues in these species, but the question of their differentiation under these conditions was left unresolved.

Although the evidence shows that the ventral neural tube and notochord have a role in patterning the sclerotome of amphibians the details of when, where, and how these interactions occur, and what steps of sclerotome specification, segregation, movement, proliferation, and chondrogenesis are regulated is not known. The factors patterning somites are being identified and their functions characterized at a frenetic rate, particularly in birds and mammals (see Buffinger and Stockdale, 1994; Fan and Tessier-Lavingne, 1994; Fan *et al.,* 1995; Capdevila *et al.,* 1998; see Chap. 10 by Dockter and Chap. 13 by Ordahl *et al.,* this volume), while little of this type of work is being done in amphibians.

In this context, what can the amphibian system contribute? As described above, the amphibians offer a great range of diversity in somite type, in the way the somitic mesoderm is formed, in the process of segmentation, in the subsequent morphogenesis of the somite, and in the proportion of tissue allocated to its different components. This variation probably arises from variation in the parameters of the tissue interactions patterning somites. The nature of these variations could be probed in the amphibians, which allow interspecific grafting and recombination of tissues. Thus the design and evolution of systems of tissue interactions might be profitably approached by applying modern methods of description and analysis to selected amphibians. The relative lack of interest in using the amphibians for these types of studies thus far perhaps accounts for the fact that many of the molecular and immunological reagents and cell type markers so useful in other systems have not been developed for amphibians. Developing the appropriate reagents for amphibians that have diverse profiles of somitic development may uncover variations of tissue interactions which could not be investigated easily, or perhaps do not exist, in the vertebrate model systems currently dominating this area of research.

VIII. How do Cells Decide Where to Make an Intersomitic Furrow?

The question of how the somitic mesoderm of vertebrates decides where and when to make somites and how many to make has stimulated much thought (Cooke and

Zeeman, 1976; Meinhardt, 1986; Elsdale and Davidson, 1987; Davidson, 1988; Keynes and Stern, 1988; Tam and Trainor, 1994), and recently, some clues as to the molecular nature of the decision-making process has been uncovered with the finding that an autonomous, rhythmic expression of the mRNA of *c-hairy1,* an avian homolog of the *Drosophila* segmentation gene, may function as a molecular clock in vertebrate segmentation (Palmeirim *et al.,* 1997). In addition, homologs of the *Drosophila* neurogenic genes (*Notch, Delta,* and *RBP-Jϰ*) have been implicated in somitogenesis of vertebrates (reviewed by Pourquié in Chap. 4, this volume). Recent work on the amphibian in this regard, integrated with earlier experiments on patterning and morphogenesis, provides a new insight on how cell signaling and cell contact interactions might lead to somitogenesis.

A. Cell Signaling and Cell Decisions

Recent work has shown that the members of the Notch family of receptors and their ligands, the DSL proteins, which have been implicated in cell fate decisions mediated by cell local interactions in a number of systems are involved in vertebrate segmentation (see Conlon *et al.,* 1995), including amphibians (Jen *et al.,* 1997). A *Xenopus* homolog of *Drosophila Delta, X-Delta-2,* is expressed just posterior to the already segmented somites in a pattern of stripes (Jen *et al.,* 1997). These stripes are generated by progressive narrowing of an initially uniform pattern of expression such that is confined to the future anterior regions of the individual somites-to-be shortly before segmentation. A likely receptor, *Xenopus Notch-1,* is expressed uniformly in both the segmental plate and somites. Expression of an antimorphic form of *X-Delta-2* or a DNA-binding mutant of *Xenopus* Supressor of Hairless [XSu(H)^{DBM}], another component in the Notch signaling pathway, both affect the position of the intersegmental boundaries but not differentiation of the somitic tissue, at least when assayed in terms of myogenic markers (Jen *et al.,* 1997). As expected from the negative auto feedback regulation typical of this pathway, the normal decrease in expression of *X-Delta-2* in the posterior half of the somites did not occur but remained uniform under these conditions.

Jen and associates propose that a Notch/Delta-mediated inhibitory feedback mechanism of cell–cell interactions refines the boundaries of the somites. They note that the region in which this process is occurring is near the zone in which heat shock causes defects in somitogenesis. Elsdale and associates (Elsdale *et al.,* 1976; Pearson and Elsdale, 1979) using *Xenopus* and *Rana* embryos, and Armstrong and Graveson (1988), using axolotl embryos, showed that there was a region beginning three to five somites posterior to the last-formed somite in which heat shock would disrupt somitogenesis; the intervening tissue and the tissue posterior to this zone were resistant. Finally, these authors correlate the zone of Delta/Notch-mediated refinement of patterns of gene expression with the zone of somitomere formation (see Jacobson, 1988), and that the prospective segments defined by the sharpening boundaries of gene expression correspond to the approximate anterior–posterior length of the somitomeres (about 10 cells). The somitomeres are metameric, concentric arrangements of cells that precede segmentation and correspond to the somites in the trunk of vertebrate embryos (described in more detail in the next section).

B. Somitomeres

About 20 years ago, Steve Meier and associates used scanning electron microscopy and stereoimaging to reveal segmental, concentric arrangements of cells in the paraxial mesoderm from the head through the length of the axis, arrangements that were transient in the head but predicted the formation of the somites in the segmental plate (reviewed in Jacobson, 1988). These arrangements of cells, called somitomeres, were found in birds (Meier, 1981; Meier and Jacobson, 1982), mammals (Meier and Tam, 1982), teleost fish (Martindale *et al.,* 1987), reptiles (Packard and Meier, 1984), and amphibians, both urodeles (Jacobson and Meier, 1984, 1986) and anurans (*Xenopus*) (see Malacinski *et al.,* 1989). In the newt, *Taricha torosa,* Jacobson and Meier (1984) observed somitomeres formed about six somites ahead of the last segmented somite. As noted above, somitomeres extend through the zone of heat shock resistance and into the heat shock sensitive zone described by Elsdale and associates (Elsdale *et al.,* 1976; Pearson and Elsdale, 1979) and overlap the zone of Delta/Notch-mediated sharpening of segmental boundaries.

One possibility is that somitomere formation, or perhaps more likely, sharpening of its boundaries, is an early morphological response to the Delta/Notch negative feedback signaling. Part of the amplification of cell–cell differences in such a feedback system might be changes in cell–cell contact. Also, this study shows that the regulatory circuits controlling or refining pattern in somite segmentation are independent of those regulating muscle differentiation at this stage. This system appears to be conserved between *Xenopus* and mammals and thus may be common to all vertebrates (see Jen *et al.,* 1997), and thus common to all the amphibians. If so, this common segment-generating or segment-refining signaling system must eventually drive diverse downstream modules of cell behavior to actually accomplish segmentation, given the fact that segmentation occurs by a number of different cell behaviors in various species. The somitomeres, which are universal among vertebrates, may be transient cellular organizations that have been retained across all vertebrates as a universal morphological parallel to a universal cell-cell interaction pathway. In the midst of this diversification of design before and after segmentation, somitomeres may reflect something simple but important, the cellular equivalent of a football (American-style) "huddle," in which the cells transiently get together in groups to designate tasks, tasks which may be different depending on the species. If somitomeres represent early, universal morphogenic steps in somitogenesis involving an interplay of cell contact events and Delta/Notch-mediated signaling, the challenge will be to determine the nature of these interactions. Of course, it is not known what differentiatons of cell motility, cell adhesion, and cell contact behavior comprise the somitomeric phenotype. It is also not known how common morphogenic parameters of somitomeres, if indeed all somitomeres are the same in this regard, lead to different mechanisms of segmental boundary formation or somite organization.

C. A Progressive Anterior–Posterior Wave Regulating Morphogenic Properties?

It has been proposed that the anterior–posterior progression of segmentation is accompanied by related changes in nonsegmenting tissues, changes that could reflect the progress of a common anterior–posterior regulatory wave (Armstrong, 1989). Gillespie *et al.* (1985) present evidence for a "proteinaceous" presegmental wave, which they interpret to organize morphogenesis of the mesoderm in an anterior-to-posterior progression. Specifically, these authors describe a zone in which trypsin digestion will disrupt both somite formation and pronephric duct migration in the axolotl. The zone is about 4 somites-lengths long and travels about 3.5 somite-lengths ahead of the anterior–posterior progression of segmentation, a reasonable approximation of the distance ahead of segmentation that the heat-shock sensitive zone occurs. Earlier work argued for a local transient wave of adhesive changes in supporting pronephric duct migration just lateral to the segmental plate/somites (Poole and Steinberg, 1982). Interestingly, the waves of *X-Delta-2* expression linked to somitogenesis in *Xenopus* extend far lateral to the myogenic cells, into the lateral plate mesoderm (see Figure 3H of Jen *et al.,* 1997). However, the waves of *hairy* mRNA expression thought to constitute the molecular clock linked to avian segmentation occur only in the paraxial mesoderm (see Palmeirim *et al.,* 1997). In the end, it seems that the relationship between anterior–posterior progressions leading to metameric events in the paraxial tissue, and nonmetameric ones laterally, awaits resolution of the underlying control circuits. In addition, these putative progressions could be characterized better in both cell and molecular terms.

IX. Role of Morphomechanical Molecules in Segmentation, Somite Morphogenesis

Molecules involved in cell motility, adhesion, and regulation of cell shape should have major roles in controlling the cell behaviors underlying segmentation and somite morphogenesis. A survey of the work done in this area on amphibians shows promise, but the field is largely unexplored.

A. Extracellular Matrix and Matrix Receptors

The somites of *Xenopus* are surrounded by fibronectin and laminin, and cells adjacent to these surfaces express the β1 integrin subunit (Krotoski and Bronner-Fraser, 1990; Wedlich *et al.,* 1989). Experimental analysis in chick suggests that β1 integrins are necessary for somitic cell adhesion to axial tissues but not for their segmentation (Drake and Little, 1991; Drake *et al.,* 1992). Whether the matrix surrounding the somites/somitic mesoderm of the amphibian has a role in the internal morphogenesis of these tissues, a role in their relationship with other tissues, or have roles primarily in other processes, such as neural crest migration, is not known. Again, it is possible that species differences exist in this regard, given the range of cell behaviors during segmentation and somite morphogenesis.

Likewise, the role of the integrin family of matrix receptors in somite morphogenesis of amphibians has not been adequately studied. In the mammalian embryo, somitogenesis is accompanied by a complex regulation in space and time of several integrin subunits and extracellular matrix (ECM), the integrin α6 being particularly interesting (Bronner-Fraser *et al.,* 1992; Pow and Hendricks, 1995; Thorsteindottir *et al.,* 1995). Comparable analyses have not been done on the amphibian. There is evidence that the *Xenopus* somitic mesoderm will not tolerate expression of exogenous integrins. mRNAs encoding integrins α2 and α3 are expressed in the notochord but not the somitic mesoderm of *Xenopus* at gastrulation and neurulation; when α3 RNA is targeted to all the dorsal mesoderm, including somitic mesoderm, by injecting it into animal caps, the result is defects in somite segmentation, whereas targeting the RNA to nondorsal mesodermal regions (the vegetal region) does not affect somitogenesis (Meng *et al.,* 1997).

Segmentation is marked by the expression of several other molecules potentially involved in regulating adhesion and integrin-mediated signaling. ADAM 13, a member of the ADAMs family of membrane-anchored proteins having a disintegrin and metalloprotease domain, is expressed in the intersomitic furrows during or shortly after segmentation (Alfandari *et al.,* 1997). Since these molecules have potential adhesive and antiadhesive functions, their discovery in this context deserves further work. Segmentation is also characterized by the appearance of focal adhesion kinase (FAK), shown to be important in integrin-mediated signaling related to motility, in cells bounding the newly forming intersomitic furrows during rotation and segmentation (see Hens and DeSimone, 1995). The role of these molecules in controlling either cell adhesion or motility in the context of making an intersomitic furrow has not been tested experimentally. Expression of these molecules in this context is the most intriguing lead as to how the boundary actually forms. Therefore further experimental analysis should be done, particularly since it is possible to visualize normal and any resulting perturbed cell behavior during segmentation and rotation in *Xenopus* with videomicroscopy (Wilson *et al.,* 1989).

Tenascin, an extracellular matrix protein possibly playing some role in controlling cell movements, is expressed as mRNA from stage 14 by RNAase protection, and is found throughout the somite of *Xenopus* by stage 29–30 (tailbud) with *in situ* hybridization (Umbhauer *et al.,* 1992). In activin-induced animal caps, the protein appears in conjunction with muscle and is found in the muscle tissue next to the notochord and in the notochordal sheath (Umbhauer *et al.,* 1992). Epperlein and others (1988) found tenascin protein concentrated around the notochord, neural tube, and medial surface of the somites in both *Xenopus* and *A. mexicanum.* The fact that tenascin is associated with the surfaces of the major axial and paraxial tissues, suggests that it may be more likely to function in regulating neural crest migration and morphology, which it has been demonstrated to do *in vitro* (Epperlein *et al.,* 1988). The protein is expressed in somitic tissue after segmentation in both axolotl and *Xenopus.* What role it has in later development of the somites, if any, is not known.

SPARC (secreted protein, acidic, rich in cysteine/osteonectin/BM-40) is an ECM protein expressed in a number of places in vertebrate embryos, including the somites, and precocious expression or injection of anti-SPARC antibodies into the blastocoel of *Xenopus* results in head and axis defects, including defects in the somites (Purcell *et al.,* 1993; Damjanovski *et al.,* 1994, 1997). SPARC has a number of potential functions in morphogenesis (see Lane and Sage, 1994) but whether it has a direct role in somite morphogenesis and what that role might be is not known.

B. Cell Adhesion Molecules

There is evidence that cadherins play a role in modulating adhesion in the course of somite formation and morphogenesis. Defects in somite formation occur in mice with mutant N-cadherin, and these defects appear to involve lowered cell adhesion and disrupted epithelial organization of the somites (Radice *et al.*, 1997). In chick embryos, N-cadherin mediated cell adhesion appears to be involved in commitment of muscle progenitors to this phenotype and for their terminal differentiation (George-Weinstein *et al.,* 1997). In zebrafish, N-cadherin expression is high in epithelial somites and declines during myogenesis (Bitzur *et al.,* 1994). N-cadherin is expressed in *Xenopus* somitic mesoderm (Detrick *et al.,* 1990), but its role in somitogenesis has not been studied experimentally. Expression patterns suggest other cadherins that may be involved in somitogenesis. Cadherin 11 is correlated with somite segregation from the segmental plate in mouse, and is finally restricted to the sclerotome cells (Kimura *et al.,* 1995). M-cadherin was found associated with myotomal cells, initially uniformly on the cell surface and then localized at putative sites of cell–cell contact (Rose *et al.,* 1994). Likely candidates such as these should be explored in the amphibians, preferably using several species that differ in somite origin and morphogenesis.

N-CAM is also expressed in the somitic mesoderm of *Xenopus,* and overexpression of normal N-CAM unilaterally in *Xenopus* somitic mesoderm results in kinked axes, usually a sign of defective convergent extension, and disorganized myotome cells (Kintner, 1988). It would be useful to determine when these effects first appear. It is not known whether the primary effects of overexpression were early and on convergent extension or late and directly on somitogenesis. An early effect on convergent extension and early steps in morphogenesis might show up secondarily on somitogenesis. The role of N-CAM in somitogenesis should be pursued further in view of the fact that one can view cell behavior and contact interactions directly during somitogenesis, at least certain aspects of it, in cultured explants.

C. Cytoskeleton

The cell elongations, cell motility, and cell rearrangements occurring before, during, and after segmentation undoubtedly depend in large part on organized cytoskeleton to generate and transmit forces across the cells and tissue. Little of a specific nature is known about the functions of the cytoskeleton in these events, particularly the regulation of the cytoskeleton. There is a bit of information about FAK, described above, but beyond this, nothing is known. An initial step would be detailed characterization of the cytoskeletal organization of the somitic cells through development. Yolk amphibian material, particularly at later stages, has not been ideal for such studies, but at least some elements of the cytoskeleton can be visualized in later stages (see Lane and Keller, 1997). Cary and Klymkowsky (1994) have suggested that desmin, an intermediate filament protein eventually important in muscle fiber structural organization, may also be involved in early linkage of the somites to one another at the intermyotomal boundary. They found that desmin is concentrated in the medial and lateral tips of the myotomal cells prior to their segmentation and rotation, and after rotation, desmin remains at the same ends of the cells, which then bound the intermytomal septum. It is only after stage 30 that the desmin becomes associated with the sarcolemma and Z-discs in the interior of the developing muscle fiber.

D. Analysis of Morphomechanical Molecules: Advantages of the Amphibian System

Characterization and experimental manipulation of various components of the cell cytoskeleton, intercellular adhesion, and adhesion to matrix in amphibian somitogenesis should be particularly rewarding. The amphibian system has several advantages. First, the effects of such manipulations on cell motility can be visualized directly with videomicroscopy of cell behavior (Wilson *et al.,* 1989).

Second, the role of the cytoskeleton, ECM, matrix receptors, and cell adhesion molecules can be evaluated in biomechanical terms (see Moore *et al.,* 1995). These approaches can be very powerful in determining mechanism. The current experimental paradigm in developmental biology is to perturb the expression or function of a molecule and observe that some complex, multicellular morphogenic machine comes to a grinding halt. These sorts of experiments lead to conclusions about composition of the machinery underlying the process, a grocery list of necessary molecules, but rarely to the functional mechanisms, the diagrams of forces, and geometries of cell movements that are essential in understanding morphogenic mechanisms beyond the compositional level. The third advantage is that the large diversity of somite origin, morphogenesis, and differentiation that appears to be present across the spectrum of amphibian species offers a good opportunity for analysis of the design, integration, and evolution of components of a developmental system. Finally, it is now possible to use transgenic methods to study the function of these components in somitogenesis of amphibians (Kroll and Amaya, 1996; Amaya and Kroll, 1998).

X. Conclusions

1. The developmental pathway lead ing to the somites is not conservative but able to accommodate a variety of fate maps, patterns, and morphogenic movements both before, during, and after segmentation.

2. The diversity of pattern and morphogenesis in a number of available and experimentally approachable amphibians should be exploited to learn how the patterning and morphogenic processes involved in segmentation have become diversified and yet are tightly integrated in time and space.

3. The amphibian somite, particularly the sclerotome and dermatome, have not been adequately described in terms of their origin, patterning and subsequent development. Most needed are good cell tracing experiments and molecular markers for unambiguous identification, particularly of the sclerotome and dermatome.

4. Experiments pertaining to the patterning of the amphibian somite by adjacent tissues are some of the oldest in the field, but these issues have not been explored extensively in amphibians in a number of years. The available evidence suggests that in some aspects the amphibians may resemble birds in this regard, but the timing and details may be different. In addition, the large diversity of somite type within the amphibians imply differences in underlying patterning mechanisms, at least in degree, if not geometry and character.

5. Many aspects of the morphogenesis leading up to segmentation, the formation of intersegmental boundaries, and the subsequent morphogenesis of the somite can be observed and manipulated directly in explants of a number of amphibians. This, along with the traditional methods of molecular manipulation in the amphibian, the advent of transgenic methods to manipulate gene expression in

amphibians (Kroll and Amaya, 1996; Amaya and Kroll, 1998), and the development of high resolution recordings of fluorescently labeled cells, make the amphibian an ideal system to exploit in the study of the cellular and molecular basis of segmentation.

References

Adams, D., Keller, R., and Koehl, M. (1991). The mechanics of notochord elongation, straightening, and stiffening in the embryo of *Xenopus laevis*. *Development* **110,** 115–130.

Alfandari, D., Wolfsberg, T., White, J., and DeSimone, D. (1997). ADAM 13: A novel ADAM expressed in somitic mesoderm and neural crest cells during *Xenopus laevis* development. *Dev. Biol.* **182,** 314–330.

Amaya, E., and Kroll, K. (1998). A method for generating transgenic frog embryos. *In* "Methods in Molecular Biology: Molecular Embryology: Methods and Protocols" (P. Sharpe and I. Mason, Eds.), Humana Press, Totowa, New Jersey.

Ang, S.-L., and Rossant, J. (1994). HNF-3β is essential for node and notochord formation in mouse development. *Cell* **78,** 561–578.

Armstrong, J. B., and Graveson, A. C. (1988). Progressive patterning precedes somite segmentation in the Mexican axolotl (*Ambystoma mexicanum*). *Devel. Biol.* **126,** 1–6.

Armstrong, J. B. (1989). Morphogenetic waves during elongation. *In* "Developmental Biology of the Axolotl" (J. B. Armstrong and G. M. Malacinski, Eds.), Oxford Univ. Press, New York. pp. 73–82.

Bautzmann, H. (1926). Experimentelle Untersucheungen zur Abgrenzung des Organisationszentrums bei *Triton taeniatus*. *Wilhelm Roux' Arch. Entwicklungsmech. Org.* **108,** 283.

Bitzer, S., Kam, Z., and Geiger, B. (1994). Structure and distribution of N-cadherin in developing zebrafish embryos: Morphogenetic effects of ectopic overexpression. *Dev. Dyn.* **201,** 121–136.

Blackshaw, A., and Warner, A. (1976). Low resistence junctions between mesodermal cells during the development of trunk muscle. *J. Physiol.* **255,** 209–230.

Boudjelida, A., and Muntz, L. (1987). Multinucleation during myogenesis of the myotome of *Xenopus laevis:* A qualitative study. *Development* **101,** 583–590.

Bronner-Fraser, M., Artinger, M., Muschler, J., and Horwitz, A. (1992). Developmentally regulated expression of alpha 6 integrin in avian embryos. *Development* **115,** 197–211.

Brun, R., and Garson, J. (1984). Notochord formation in the Mexican salamander (*Ambystoma mexicanum*) is different from notochord formation in *Xenopus laevis*. *J. Exp. Zool.* **229,** 235–240.

Brustis, J.-J. (1978). Organisation precoce du dermatome et du sclerotome chez deuz Amphibiens Anoures *Rana dalmatina* Bon. et *Bufo bufo* L. *C. R. Acad. Sci. Paris* **287,** 1153–1155.

Brustis, J.-J. (1979). Aspects ultrastructuraux de l'organisation des somites et de la differenciation precoce des myotomes chez l'embryon du crapaud commun *Bufo bufo* L. *Arch. Biol.* **90,** 262–272.

Brustis, J.-J., and Delbos, M. (1976). Aspects ultrastructuraux de la formation des myoseptes chez l'embryon du crapaud commun, *Bufo bufo* L. (Amphibien anoure). *C. R. Acad. Sci. Paris* **281,** 1127–1129.

Brustis, J.-J., Landsmann, F., and Gipouloux, J.-D. (1976). Etude de la differenciation des somites chez les embryons de deux amphibiens Anoures: Crapaud commun (*Bufo bufo* L.) et Grenouille agile (*Rana dalmatina* Bon.). *Bull. Biol. Fr. Belg.*, 299–311.

Buffinger, N., and Stockdale, F. (1994). Myogenic specificaton in somites: Induction by axial structures. *Development* **120,** 1443–1452.

Capdevila, J., Tabin, C., and Johnson, R. L. (1998). Control of dorsoventral somite patterning by Wnt-1 and beta-catenin. *Dev. Biol.* **193,** 182–294.

Cary, R. B., and Klymkowsky, M. W. (1994). Desmin organization during the differentiation of the dorsal myotome in *Xenopus laevis*. *Differentiation* **56,** 31–38.

Chuang, H.-H. (1946). Defekt- und Vitalfärbungsversuche zur Analyse der Entwicklung der Kaudalen Rumpfabschnitte und des Schwanzes bei Urodelen. *Arch. Entwicklungsmech.* **143,** 19–21.

Clarke, J. D. W., Holder, N., Soffe, S. R., and Storm-Mathisen, J. (1991). Neuroanatomical and functional analysis of neural tube formation in notochordless *Xenopus* embryos: Laterality of the ventral spinal cord is lost. *Development* **112,** 499–516.

Cohen, A. (1938). Myotome fusion in the embryo of *Amblystoma punctatum* after treatment with lithium and other agents. *J. Exp. Zool.* **79,** 461–473.

Conlon, R. A., Reaume, A. G., and Rossant, J. (1995). Notch 1 is required for the coordinate segmentation of somites. *Development* **121,** 1533–1545.

Cooke, J., and Zeeman, E. (1976). A clock-and-wavefront model for control of the number of repeated structures during animal morphogenesis. *J. Theor. Biol.* **58,** 455–476.

Damjanovski, S., Malaval, L., and Ringuette, M. (1994). Transient expression of SPARC in the dorsal axis of early *Xenopus* embryos: Correlation with calcium-dependent adhesion and electrical coupling. *Int. J. Dev. Biol.* **38,** 439–446.

Damjanovski, S., Karp, X., Funk, S., Sage, E., and Ringuette, M. (1997). Ectopic expression of SPARC in *Xenopus* embryos interferes with tissue morphogenesis: Identification of a bioactive sequence in the C-terminal EF hand. *J. Histochem. Cytochem.* **45,** 634–655.

Davidson, D. (1988). Segmentation in frogs. *Development* **104**(Suppl.), 221 and 229.

Delarue, M., Sanchez, S., Johnson, K. E., Darribere, T., and Boucaut, J.-C. (1992). A fate map of superficial and deep circumblastoporal cells in the early gastrula of *Pleurodeles waltl. Development* **114,** 135–146.

Delarue, M., Johnson, K., and Boucaut, J.-C. (1996). Anteroposterior segregation of superficial and deep cells during gastrulation in *Pleurodeles waltl* and *Rana pipiens* embryos. *J. Exp. Zool.* **276,** 345–360.

Denetclaw, W. F., Christ, B., and Ordahl, C. (1997). Location and growth of epaxial myotome precursor cells. *Development* **124,** 1601–1610.

Detrick, R. J., Dickey, D., and Kintner, C. (1990). The effects of N-cadherin misexpression on morphogenesis in *Xenopus* embryos. *Neuron* **4,** 493–506.

Detwiler, S. R. (1937). Observations upon the migration of neural crest cells, and upon the development of the spinal ganglia and vertebral arches in *Amblystoma. Am. J. Anat.* **61,** 63–94.

Domingo, C., and Keller, R. (1995). Induction of notochord cell intercalation behavior and differentiation by progressive signals in the gastrula of *Xenopus laevis. Development* **121,** 3311–3321.

Drake, C. J., and Little, C. D. (1991). Integrins play an essential role in somite adhesion to the embryonic axis. *Dev. Biol.* **143,** 418–421.

Drake, C. J., Davis, L. A., Hungerford, J. E., and Little, C. D. (1992). Perturbation of beta 1 integrin mediated adhesions results in altered somite cell shape and behavior. *Dev. Biol.* **149,** 327–338.

Duprat, A., Romanovsky, A., Hurychova, D., and Macha, J. (1975). Immunofluorescence studies of amphibian myoblast differentiation. *J. Embryol. Exp. Morphol.* **34,** 113–123.

Elsdale, T., and Davidson, D. (1983). Somitogenesis in amphbian. IV. The dynamics of tail development. *J. Embryol. Exp. Morphol.* **76,** 157–176.

Elsdale, T. and Davidson, D. (1987). Timekeeping by frog embryos, in normal development and after heat shock. *Development* **99,** 41–49.

Elsdale, T., Pearson, M., and Whitehead, M. (1976). Abnormalities in somite segmentation following heat shock to *Xenopus* embryos. *J. Embrol. Exp. Morphol.* **35,** 625–635.

Epperlein, H. H., Halfter, W., and Tucker, R. P. (1988). The distribution of fibronectin and tenascin along migratory pathways of the neural crest in the trunk of amphibian embyros. *Development* **103,** 743–756.

Ewan, K. B. R., and Everett, A. W. (1992). Evidence for resegmentation in the formation of the vertebral column using the novel approach of retrovival-mediated gene transfer. *Exp. Cell Res.* **198,** 315–320.

Fan, C.-M., and Tessier-Lavigne, M. (1994). Patterning of mammalian somites by surface ectoderm and notochord: Evidence for sclerotome induction by a hedgehog homolog. *Cell* **79,** 1175–1186.

Fan, C.-M., Porter, J., Chiang, C., Chang, D., Beachy, P., and Tessier-Lavigne, M. (1995). Long-range sclerotome induction by sonic hedgehog: Direct role of the amino-terminal cleavage product and modulation by the cyclic AMP signaling pathway. *Cell* **81,** 457–465.

Gatherer, D. and Del Pino, E. (1992). Somitogenesis in the marsupial frog *Gastrotheca riobamdae. Int. J. Dev. Biol.* **36,** 283–291.

Gearhart, J. D., and Mintz, B. (1972). Clonal origins of somites and their muscle derivatives: Evidence from allophenic mice. *Dev. Biol.* **29,** 27–37.

George-Weinstein, M., Gerhart, J., Blitz, J., Simak, E., and Knudsen, K. A. (1997). N-cadherin promotes the commitment and differentiation of skeletal muscle precuror cells. *Dev., Biol.* **185,** 14–24.

Gillespie, L., Armstrong, J., and Steinberg, M. (1985). Experimental evidence for a proteinaceous presegmental wave required for morphogenesis of axolotl mesoderm. *Dev. Biol.* **10** (7), 220–226.

Halpern, M., Thisse, C., Ho, R., Thisse, B., Riggleman, B., Trevarrow, B., Weinberg, E., Postlethwait, J., and Kimmel, C. (1995). Cell-autonomous shift from axial to paraxial mesodermal development in zebrafish *floating head* mutants. *Development* **121,** 4257–4264.

Hamilton, L. (1969). The formation of somites in *Xenopus laevis. J. Embryol. Exp. Morphol.* **22,** 253–264.

Hens, M., and DeSimone, D. (1995). Molecular analysis and developmental expression of the focal adhesion kinase pp125^{FAK} in *Xenopus laevis. Dev. Biol.* **170,** 274–288.

Holtfreter, J. (1943). A study of the mechanics of gastrulation. Part I. *J. Exp. Zool.* **94,** 261–318.

Holtzer, H., and Detwiler, S. (1953). An experimental analysis of the development of the spinal column. *J. Exp. Zool.* **123,** 335–346.

Hoppler, S., and Moon, R. (1998). BMP-2/-4 and Wnt-8 cooperatively pattern the *Xenopus* mesoderm. *Mech. Dev.* **69,** 105–114.

Horb, M. E., and Thomsen, G. H. (1997). A vegetally localized T-box transcription factor in *Xenopus* eggs specifies mesoderm and endoderm and is essential for embryonic mesoderm formation. *Development* **124,** 1689–1698.

Huang, R., Zhi, Q., Neubuser, A., Muller, T.S., Brand-Saberi, B., Christ, B., and Wilting, J. (1996). Function of somite and somitocoele cells in the formation of the vertebral motion segment in avian embryos. *Acta Anat.* **155,** 231–241.

Hudson, C., Clements, R., Friday, V., Stott, D., and Woodland, H. (1997). Xsox17α and -β mediate endoderm formation in *Xenopus. Cell* **91,** 397–405.

Jacobson, A. (1988). Somitomeres: Mesodermal segments of vertebrate embryos. *Development* **104**(Suppl.), 209–220.

Jacobson, A., and Meier, S. (1984). Morphogenesis of the head of the newt: Mesodermal segments, neuromeres, and distribution of neural crest. *Dev. Biol.* **106,** 181–193.

Jacobson, A., and Meier, S. (1986). Somitomeres: The primordial body segments. *In* "Somites in Developing Embryos" (R. Bellairs, D. A. Ede, and J. Lash, Eds.), Plenum, New York.

Jen, W.-C., Wettstein, D., Chitnis, D., and Kintner, C. (1997). The Notch ligand, X-Delta-2, mediates segmentation of the paraxial mesoderm in *Xenopus* embryos. *Development* **124,** 1169–1178.

Jones, E., and Woodland, H. (1989). Spatial aspects of neural induction in *Xenopus laevis. Development* **107,** 785–791.

Jones, E. A., Abel, M. H. and Woodland, H. (1993). The possible role of mesodermal growth factors in the formation of endoderm in *Xenopus laevis. Roux's Arch. Dev. Biol.* **202,** 233–239.

Kao, K., and Elinson, R. (1988). The entire mesodermal mantle behaves as Spemann's Organizer in dorsoanterior enhanced *Xenopus laevis* embryos. *Dev. Biol.* **127,** 64–77.

Kao, K., Masui, Y. and Elinson, R. (1986). Lithium-induced respecification of pattern in *Xenopus laevis* embryos. *Nature* **322,** 371–373.

Keller, R. E. (1975). Vital dye mapping of the gastrula and neurula of *Xenopus laevis*. I. Prospective areas and morphogenetic movements of the superficial layer. *Dev. Biol.* **42,** 222–241.

Keller, R. E. (1976). Vital dye mapping of the gastrula and neurula of *Xenopus laevis*. II. Prospective areas and morphogenetic movements of the deep layer. *Dev. Biol.* **51,** 118–137.

Keller, R. E. (1986). The cellular basis of amphibian gastrulation. *In* "Developmental Biology: A Comprehensive Synthesis, Volume 2, The Cellular Basis of Morphogenesis." (L. Browder, Ed.), pp. 39–89 and 241–327. Plenum, New York.

Keller, R. (1991). Early embryonic development of *Xenopus laevis*. "Methods in Cell Biology, Volume 36, *Xenopus laevis:* Practical Uses in Cell and Molecular Biology" (B. Kay and B. Peng, Eds.) Chap. 5, pp. 61–113. Academic Press, San Diego.

Keller, R., and Danilchik, M. (1988). Regional expression, pattern and timing of convergence and extension during gastrulation of *Xenopus laevis*. *Development* **103,** 193–209.

Keller, R., and Jansa, S. (1992). *Xenopus* gastrulation without a blastocoel roof. *Dev. Dyn.* **195,** 162–176.

Keller, R. E., Cooper, M., Danilchik, M., Tibbetts, P., and Wilson, P. (1989). Cell intercalation during notochord development in *Xenopus laevis*. *J. Exp. Zool.* **251,** 134–154.

Keller, R., Domingo, C., and Shih, J. (1992). The patterning and functioning of protrusive activity during convergence and extension of the *Xenopus* organizer. *Development* (Suppl.), 81–91.

Keynes, R., and Stern, C. (1988). Mechanisms of vertebrate segmentation. *Development* **103,** 413–429.

Kielbowna, L. (1966). Cytological and cytophotometrical studies on myogenesis in *Xenpous laevis* (Daudin). *Zool. Pol.* **117,** 247–255.

Kielbowna, L., 1980. Two different types of myogenesis in *Xenopus laevis* (Daudin). *Zool. Pol.* **27,** 377–394.

Kielbowna, L. (1981). The formation of somites and early myotomal myogenesis in *Xenopus laevis, Bombina variegata* and *Pelobates fuscus*. *J. Embryol. Exp. Morphol.* **64,** 295–304.

Kielbowna, L., and Koscielski, B. (1979). Myotomal myogenesis in *Bombina variegata* L. *Roux's Arch. Dev. Biol.* **185,** 295–303.

Kimura, Y., Matsunami, H., Inoue, T., Shimamura, K., Uchida, N., Ueno, T., Miyazaki, T., and Takeichi, M. (1995). Cadherin-11 expressed in association with mesenchymal morphogenesis in the head, somite, and limb bud of early mouse embryos. *Dev. Biol.* **169,** 347–358.

Kintner, C. (1988). Effects of altered expression of the neural cell adhesion molecule, N-CAM, on early neural development in *Xenopus* embryos. *Neuron* **1,** 545–555.

Kintner, C., and Brockes, (1984). Monoclonal antibodies identify blastemal cells derived from differentiating muscle in newt limb regeneration, *Nature* **308,** 67–69.

Kitchen, I. C. (1949). The effects of notochordectomy in *Amblystoma mexicanum*. *J. Exp. Zool.* **112,** 393–411.

Kroll, K., and Amaya, E. (1998). Transgenic *Xenopus* embryos from sperm nuclear transplantations reveal FGF signaling requirements during gastrulation. *Development* **122,** 3173–3183.

Krotoski, D., and Bronner-Fraser, M. (1990). Distribution of integrins and their ligands in the trunk of *Xenopus laevis* during neural crest cell migration. *J. Exp. Zool.* **253,** 139–150.

Lane, M. C., and Keller, R. (1997). Microtubule disruption reveals that Spemann's organizer is subdivided into two domains by the vegetal alignment zone. *Development* **124,** 895–906.

Lane, T., and Sage, E. (1994). The biology of SPARC, a protein that modulates cell–matrix interactions. *Fed. Proc. Fed. Am. Soc. Exp. Biol.* **8,** 163–172.

Lehmann, F. E. (1935). Die Entwicklung von Rückenmark, Spinalganglien und Wirbelanlagen in Chordolosen Körperrregionen von Trintonlarven. *Rev. Suisse Zool.* **42,** 405–415.

Leyns, L., Bouwmeester, T., Kim, S.-H., Piccolo, S. and DeRobertis, E. (1997). Frzb-1 is a secreted antagonist of Wnt signaling expressed in the Spemann organizer. *Cell* **88,** 747–756.

Loeffler, C. A. (1969). Evidence for the fusion of myoblasts in amphibian embryos. I. Homoplastic transplantations of somatic material labeled with tritiated thymidine. *J. Morphol.* **128,** 403–426.

Lofberg, J. (1974). Apical surface topography of invaginating and non invaginating cells. A scanning-transmission study of amphibian neurulae. *Dev. Biol.* **36,** 311–329.

Lundmark, C. (1986). Role of bilateral zones of ingressing superficial cells during gastrulation of *Ambystoma mexicanum. J. Embryol. Exp. Morphol.* **97,** 47–62.

Lynch, K. (1990). Development and innervation of the abdominal muscle in embryonic *Xenopus laevis. Am. J. Anat.* **187,** 374–392.

McCaig, C. (1986). Myoblasts and notochord influence the orientation of somitic myoblasts from *Xenopus laevis. J. Exp. Morphol.* **93,** 121–131.

Malacinski, G. M., and Youn, B.-W. (1981). Neural plate morphogenesis and axial stretching in "notochord-defective" *Xenopus laevis* embryos. *Dev. Biol.* **88,** 352–357.

Malacinski, G. M., and Youn, B.-W. (1982). The structure of the anuran amphibian notochord and a re-evaluation of its presumed role in early embryogenesis. *Differentiation* **21,** 13–21.

Malacinski, G. M., Allis, C. D., and Chung, H.-M. (1974). Correction of developmental abnormalities resulting from localized ultraviolet irradiation of an amphibian egg. *J. Exp. Zool.* **189,** 249–254.

Malacinski, G., Neff, A., Radice, G., and Chung, H.-M. (1989). Amphibian somite development: Contrasts of morphogenetic and molecular differentiation patterns between the laboratory archetype species *Xenopus* (anuran) and axolotl (urodele). *Zool. Sci.* **6,** 1–14.

Martindale, M., Meier, S., and Jacobson, A. (1987). Mesodermal metamerism in the teleost, *Oryzias latipes. J. Morphol.* **193,** 241–252.

Meier, S. (1981). Development of the chick embryo mesoblast: Morphogenesis of the prechordal plate and cranial segments. *Dev. Biol.* **83,** 49–61.

Meier, S., and Jacobson, A. (1982). Experimental studies of the origin and expression of metameric pattern in the chick embryo. *J. Exp. Zool.* **219,** 217–232.

Meier, S., and Tam, P. (1982). Metameric pattern development in the embryonic axis of the mouse. I. Differentiation of the cranial segments. *Differentiation* **21,** 95–108.

Meinhardt, H. (1986). Hierarchical inductions of cell states: A model for segmentation in *Drosophila. J. Cell Sci.* **4**(Suppl.), 357–381.

Melby, A., Warga, R., and Kimmel, C. (1996). Specification of cell fates at the dorsal margin of the zebrafish gastrula. *Development* **122,** 2225–2237.

Meng, F., Whittaker, C. A., Ransom, D. G., and DeSimone, D. W. (1997). Cloning and characterization of cDNAs encoding the integrin alpha 2 and alpha 3 subunits from *Xenopus laevis. Mech. Dev.* **67,** 141–155.

Minsuk, S.B., and Keller, R. (1996). Dorsal mesoderm has a dual origin and forms by a novel mechanism in *Hymenochirus,* a relative of *Xenopus. Dev. Biol.* **174,** 92–103.

Minsuk, S., and Keller, R. (1997). Surface mesoderm in *Xenopus:* A revision of the stage 10 fate map. *Dev. Genes Evol.* **207,** 389–401.

Mookerjee, H. K. (1930). On the development of the vertebral column of Urodela. *Philos. Trans. R. Soc. London Ser. B* **218,** 415–446.

Mookerjee, H. K. (1931). On the development of the vertebral column of Anura. *Philos. Trans. Ro. Soc. London Ser. B* **219,** 165–196.

Moore, S., Keller, R., and Koehl, M. (1995). The dorsal involuting marginal zone stiffens anisotropically during its convergent extension in the gastrula of *Xenopus laevis. Development* **121,** 3131–3140.

Muntz, L. (1975). Myogenesis in the trunk and leg during development of the tadpole of *Xenopus laevis.* (Daudin 1802). *J. Embryol. Exp. Morphol.* **33,** 757–774.

Nieuwkoop, P. (1969a). The formation of the mesoderm in urodelean amphibians. I. Induction by the endoderm. Roux's Arch. Dev. Biol. **162,** 341–373.

Nieuwkoop, P. D. (1969b). The formation of mesoderm in urodelean amphibians. II The origin of dorsalventral polarity of the mesoderm. *Roux's Arch. Dev. Biol.* **162,** 298–315.

Nieuwkoop, P. and Faber, J. (1967). "Normal Table of *Xenopus laevis* (Daudin)," 2nd ed. North-Holland, Amsterdam.

Packard, D., and Meier, S. (1984). Morphological and experimental studies of the somitomeric organization of the segmental plate in snapping turtle embryos. *J. Embryol. Exp. Morphol.* **84,** 35–48.

Pasteels, J. (1942). New observations concerning the fate maps of presumptive areas of the young amphibian gastrula. (*Amblystoma* and *Discoglossus*). *J. Exp. Zool.* **89,** 255–282.

Palmeirim, I., Henrique, D., Ish-Horowicz, D., and Pourquie, O. (1997). Avian hairy gene expression identifies a molecular clock linked to vertebrate segmentation and somitogenesis. *Cell* **91,** 639–648.

Pearson, M., and Elsdale, T. (1979). Somitogenesis in amphibian embryos. I. Experimental evidence for an interaction between two temporal factors in the specification of somite pattern. *J. Embryol. Exp. Morphol.* **51,** 27–50.

Piccolo, S., Sasai, Y., Lu, B., and DeRobertis, E. (1996). Dorsoventral patterning in *Xenopus:* Inhibition of ventral signals by direct binding of chordin to BMP-4. *Cell* **86,** 589–598.

Poole, T., and Steinberg, M. (1982). Evidence for the guidance of pronephric duct migration by a craniocaudally travelling adhesion gradient. *Dev. Biol.* **92,** 144–158.

Pow, C. S., and Hendricks, A. G. (1995). Localization of integrin subunits alpha 6 and beta 1 during somitogenesis in the long-tailed macaque (*M. fascisularis*). *Cell Tissue Res.* **281,** 101–108.

Purcell, L., Gruia-Gray, J., Scanga, S., and Ringuette, M. (1993). Developmental anomalies of *Xenopus* embryos following microinjection of SPARC antibodies. *J. Exp. Zool.* **265,** 153–164.

Purcell, S., and Keller, R. (1993). A different type of amphibian mesoderm morphogenesis in *Ceratophrys ornata. Development* **117,** 307–317.

Radice, G., Neff, A., Shim, Y., Brustis, J. J., and Malacinski, G. (1989). Developmental histories in amphibian myogenesis. *Int. J. Dev. Biol.* **33,** 325–343.

Radice, G. L., Rayburn, H., Matsunami, H., Knudsen, K. A., Takeichi, M., and Hynes, R.O. (1997). Developmental defects in mouse embryos lacking N-cadherin. *Dev. Biol.* **181,** 64–78.

Rose, O., Rohwedel, J., Reinhardt, S., Bachmann, M., Cramer, M., Rotter, M., Wobus, A., and Starzinski-Powitz, A. (1994). Expression of M-cadherin protein in myogenic cells during prenatal mouse development and differentiation of embryonic stem cells in culture. *Dev. Dyn.* **201,** 245–259.

Ruffini, A. (1925). "Fisiogenia." F. Vallardi, Milano, Italy.

Ryke, P. A. J. (1953). The ontogenesis development of the somatic musculature of the trunk of the aglossal anuran *Xenopus laevis* (Daudin). *Acta Zool.* **34,** 1–70.

Sasai, Y., Lu, B., Piccoloo, S., and De Robertis, E. (1996). Endoderm induction by the organizer-secreted factors chordin and noggin in *Xenopus* animal caps. *EMBO J.* **15,** 4547–4555.

Scharf, S., and Gerhart, J. C. (1980). Determination of the dorsal–ventral axis in eggs of *Xenopus laevis:* Complete rescue of uv-impaired eggs by oblique orientation before first cleavage. *Dev. Biol.* **79,** 181–198.

Schoenwolf, G.C., Garcia-Martinez, V., and Dias, M. (1992). Mesoderm movement and fate during avian gastrulation and neurulation. *Dev. Dyn.* **193,** 235–248.

Schroeder, T. E. (1970). Neurulation in *Xenopus laevis.* An analysis and model based upon light and electron microscopy. *J. Embryol. Exp. Morphol.* **23,** 427–462.

Shih, J. and Fraser, S. (1996). Characterizing the zebrafish organizer: Microsurgical analysis at the early-shield stage. *Development* **122,** 1313–1322.

Shih, J., and Keller, R. (1992a). The epithelium of the dorsal marginal zone of *Xenopus* has organizer properties. *Development* **116,** 887–899.

Shih, J., and Keller, R. (1992b). Cell motility driving mediolateral intercalation in explants of *Xenopus laevis. Development* **116,** 901–914.

Shih, J., and Keller, R. (1992c). Patterns of cell motility in the organizer and dorsal mesoderm of *Xenopus. Development* **116,** 915–930.

Smith, J. C., Dale, L., and Slack, J. M. W. (1985). Cell lineage labels and region-specific markers in the analysis of inductive interactions. *J. Embryol. Exp. Morphol.* **89**(Suppl.), 317–331.

Smith, J. L., Gesteland, K. M., and Schoenwolf, G. C. (1994). Prospective fate map of the mouse primitive streak at 7.5 days of gestation. *Dev. Dyn.* **201,** 279–289.

Spemann, H. (1938). "Embryonic Development and Induction." Yale Univ. Press. New Haven, Connecticut.

Spemann, H., and Mangold, H. (1924). Über Induktion von Embryonalanlagen durch Implantation artfremder Organisatoren. *Roux' Arch. Entwicklungsmech.* **100,** 599–638.

Sudarwati, and Nieuwkoop, P. (1971). Mesoderm formation in the Anuran *Xenopus Laevis. Roux's Arch. Dev. Biol.*, 189–204.

Talbot, W., Trevarrow, W., Halpern, M., Melby, A., Farr, H., Postlethwaite, J., Jowett, T., Kimmel, C., and Kimmelman, D. (1995). Requirement for the homeobox gene *floating head* in zebrafish notochord development. *Nature* **378,** 150–157.

Thorsteindottir, S., Roelen, B. A., Freund, E., Gaspar, A. C., Sonnenberg, A., and Mummery, C. L. (1995). Expression patterns of laminin receptor splice variants alpha 6A, beta 1 and alpha 6 beta 1 suggest different roles in mouse development. *Dev. Dyn.* **204,** 240–258.

Umbhauer, M., Riou, J. F., Spring, J., Smith, J. C., and Boucaut, J. C. (1992). Expression of tenascin mRNA in mesoderm during *Xenopus laevis* embryogenesis: The potential role of mesoderm patterning in tenascin regionalization. *Development* **116,** 147–157.

Vogt, W. (1929). Gestaltanalyse am Amphibienkein mit örtlicher Vitalfärbung. II. Teil. Gastrulation und Mesodermbildung bei Urodelen und Anuren. *Wilhelm Roux' Arch. Entwicklungsmech. Org.* **120,** 384–706.

Wake, D. (1970). Aspects of vertebral evolution in the modern Amphibia. *Forma et Functio* **3,** 33–60.

Wake, D., and Lawson, R. (1973). Developmental and adult morphology of the vertebral column in the plethodontid salamander *Eurvcea bislineata* with comments on vertebral evolution in the Amphibia. *J. Morphol.* **139,** 251–300.

Wake, M., and Wake, D. (1985). Vertebral development in gymnophion amphibians: Resegmentation and homology. *Am. Zool.* **25,** 93a.

Wang, S., Krinks, M., Lin, K., Luyten, F. P., and Moos, M., Jr. (1997). Frzb, secreted protein expressed in the Spemann organizer, binds and inhibits Wnt-8. *Cell* **88,** 757–766.

Wedlich, D., Hacke, H., and Klein, G. (1989). The distribution of fibronectin and laminin in somitogenesis of *Xenopus.* Differentiation **40,** 77–83.

Wilson, P. A. (1990). The Development of the Axial Mesoderm in *Xenopus laevis.* Dissertation, University of California, Berkeley.

Wilson, P., and Keller, R. (1991). Cell rearrangement during gastrulation of *Xenopus:* direct observation of cultured explants. *Development* **112,** 289–300.

Wilson, P. A., Oster, G., and Keller, R. (1989). Cell rearrangement and segmentation in *Xenopus:* direct observation of cultured explants. *Development* **105,** 155–166.

Youn, B.-W., Keller, R., and Malacinski, G. (1980). An atlas of notochord and somite morphogenesis in several anuran and urodelean amphibians. *J. Embryol. Exp. Morphol.* **59,** 223–247.

Youn, B.-W., and Malacinski, G. (1981a). Somitogenesis in the amphibian *Xenopus laevis:* Scanning electron microscopic analysis of intrasomitic cellular arrangements during somite rotation. *J. Embryol. Exp. Morphol.* **64,** 23–46.

Youn, B.-W., and Malacinski, G. (1981b). Comparative analysis of amphibian somite morphogenesis: Cell rearrangement patterns during rosette formation and myoblast fusion. *J. Embryol. Exp. Morphol.* **66,** 1–26.

Zhang J., and King, M. L. (1996). *Xenopus* VegT RNA is localized to the vegetal cortex during oogenesis and encodes a novel T-box transcription factor involved in mesodermal patterning. *Development.* **122,** 4119–4129.

Zimmerman, L., DeJesus-Escobar, J. M., and Harland, R. M. (1996). The Spemann organizer signal noggin binds and inactivates bone morphogenetic protein-4. *Cell* **86,** 599–606.

8

Somitogenesis in Zebrafish

Scott A. Holley and *Christiane Nüsslein-Volhard*
Max-Planck-Institut für Entwicklungsbiologie
Tübingen, Germany

Both genetic and embryological studies in the zebrafish, *Danio rerio,* have contributed to our general understanding of how somites form and differentiate. In the zebrafish, mutants have been isolated that have specific effects on virtually every aspect of somite development. The *fss*-type mutants, defining 5 genes, affect somite segmentation and epithelialization. The *you*-type mutants, comprising 7 genes, and mutants in another 13 genes defective in notochord formation, have somites with abnormal pattern and morphology. Eighteen genes have been identified that are required for the differentiation and maintenance of the somitic musculature, and 2 genes have been identified that are involved in the development of motoneurons that innervate the somitic musculature. The true utility of the zebrafish lies in the ability to combine genetic analysis with embryological experimentation. Such analysis of somite segmentation suggests that homologues of both the *Drosophila* pair-rule and segment polarity genes, *her1* and *Sonic hedgehog,* respectively, are involved generating periodicity during somitogenesis. The *Sonic hedgehog* protein secreted from the notochord also induces the formation of specific muscle types including the slow muscle fibers which are initially induced in the medial somite and undergo a series of morphological transitions including migration through the somite to the lateral surface where they complete their differentiation. The role of the notochord in patterning the

Current Topics in Developmental Biology, Vol. 47

0070-2153/00 $30.00

somite is also demonstrated by its involvement in regulating the permissiveness of the somite to the extention of axons of primary motoneurons.

I. Introduction

Somites are repeated, epithelial structures within the mesoderm of the vertebrate embryo that give rise to the vertebrae and muscle of the trunk and tail (Fig. 1A). They are derived from the unsegmented, paraxial mesoderm located just lateral to the axial mesoderm or notochord. Somite formation (somitogenesis) occurs in an anterior to posterior sequence as clusters of cells in the mesenchymal, presomitic mesoderm undergo segmentation and form epithelial spheres (epithelialization) (Fig. 1C). The somites are then patterned by the adjacent notochord, neural tube, lateral plate mesoderm, and surface ectoderm to give rise to the appropriate, innervated muscle types and vertebrae.

Somite formation in the zebrafish embryo commences at 10.5 hr postfertilization with a somite pair being created approximately every 30 min in an anterior to posterior progression until 26–30 somite pairs are formed. Roughly, 7 somites form above the yolk cell, 10 above the yolk extension, and 9–13 posterior to the anus (Fig. 1A). The zebrafish somite gives rise to sclerotome and myotome, but it has not been determined if zebrafish somites have a dermomyotome as do somites of the mouse and chick (Fig. 1D) (Kimmel *et al.,* 1995). Moreover, in contrast to the mouse and chick where the sclerotome is the major component of the somite, the zebrafish somite is primarily myotome with only the ventral medial cells giving rise to sclerotome (Fig. 1D) (Morin-Kensicki and Eisen, 1997).

Using Nomarski optics, several structures can be distinguished within the somite *in vivo.* Both the transverse myosepta that form the anterior–posterior borders and the horizontal myoseptum which divides the somite into dorsal and ventral compartments can be distinguished (Fig. 1A,B). The dorsal compartments gives rise to epaxial muscle while the ventral compartment gives rise to hypaxial muscle (Kimmel *et al.,* 1995). Adjacent to the notochord, a specialized group of cuboidal, mesodermal cells, the adaxial cells, can be distinguished from the more lateral presomitic mesoderm prior to somite formation (Fig. 1C) (Thisse *et al.,* 1993). At this time, these cells also express the myogenic gene *MyoD* which is later expressed in the posterior half of each somite (Fig. 2I) (Weinberg *et al.,* 1996). Three to six of the 20 or so adaxial cells in each somite will become Engrailed-expressing muscle pioneers which are the first cells within the somite to develop a myotomal morphology (Fig. 3C) (Felsenfeld *et al.,* 1991; Hatta *et al.,* 1991; Ekker *et al.,* 1992). The muscle pioneers elongate in the middle of the somite near the future horizontal myoseptum which also is thought to be a derivative of the adaxial cells (Waterman, 1969; Devoto *et al.,* 1996). In mutants with a loss of or reduction of the adaxial cells and their derivatives, the somites develop a characteristic U shape

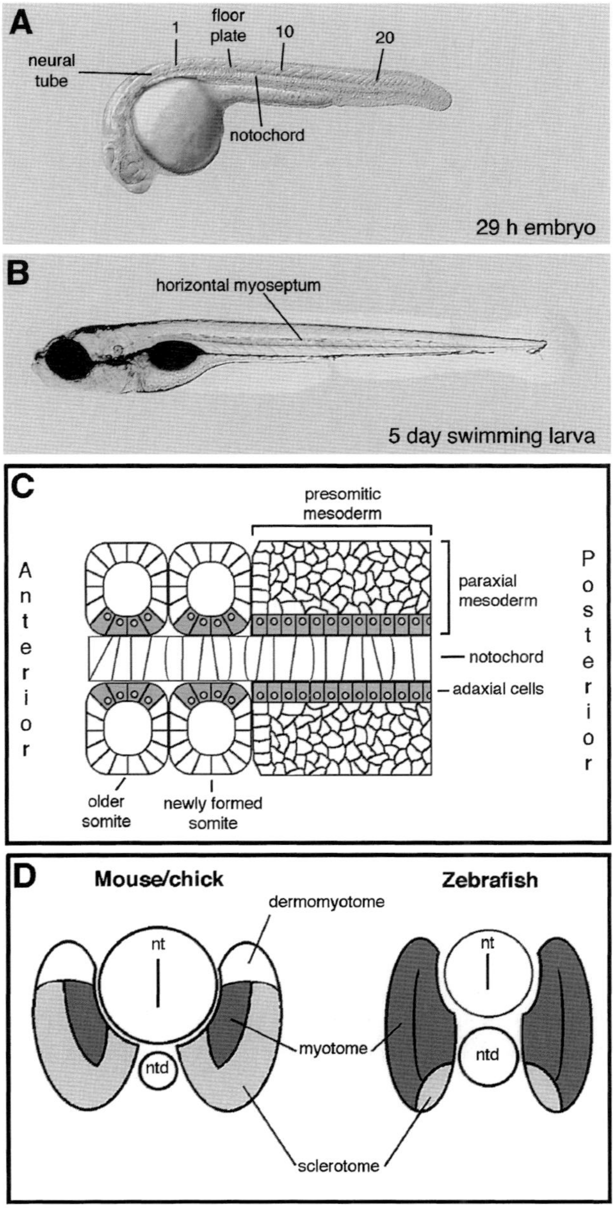

Figure 1 Features of zebrafish somites. In all figures, unless otherwise specified, anterior is left and posterior is right while dorsal is up and ventral is down. The somites are numbered in (A). (B) The horizontal myoseptum can be seen along with pigment cells that are aligned along it. (C) A schematic horizontal section illustrating the transition from mesenchymal presomitic mesoderm to epithelial somites. Additionally, the distinctive organization of the adaxial cells is shown (in gray). (D) Schematic transverse sections showing the differences between the mouse/chick somite and zebrafish somite. Note the relative differences in the sizes of the sclerotome and myotome. Neural tube (nt); notochord (ntd). Panels A and B were adapted from Haffter *et al.,* 1996.

rather than the V shape found in wild-type embryos (Halpern *et al.,* 1993; Odenthal *et al.,* 1996; van Eeden *et al.,* 1996).

Here, we review our current understanding of somite formation in the zebrafish. This includes both a review of embryological analysis of somite formation and differentiation as well as a review of the results of two large-scale genetic screens that identified 42 genes that are neccessary for the formation of the paraxial mesoderm, segmentation of the paraxial mesoderm, patterning of the somites, differentiation of the myotome and development of motoneurons that innervate the somites (Tables I–IV).

II. The Zebrafish as a Model Organism

The ability to perform both embryological and genetic analysis makes the zebrafish, *D. rerio,* a powerful tool for the study of pattern formation and morphogenesis during vertebrate development. Zebrafish eggs are fertilized externally, and, due to their transparency, living embryos can be observed with a dissecting or compound microscope. One pair of fish can produce up to 200 transparent, synchronously developing embryos per week. The embryos can be monitored throughout somitogenesis and formation of the notochord, brain, and eyes which all take place within 24 hr after fertilization (Kimmel *et al.,* 1995). The ability to easily examine the morphology of zebrafish embryos facilitates the isolation of mutants defective in embryonic and larval development as evidenced by the completion of two large-scale genetic screens (Haffter *et al.,* 1996; Driever *et al.,* 1996). The accessiblity of zebrafish embryos also permits embryological manipulation. Cell transplantation experiments can be used to produce genetic mosaics which facilitate the analysis of specific mutants. Visualization of individual cells can be enhanced by the use of vital dyes that label either cell membranes or nuclei, allowing for the analysis of cell lineage. A potentially more powerful tool is green fluorescent protein (GFP) which, as has been demonstrated for GATA-1 and GATA-2, can be expressed under the control of tissue specific enhancers either transiently in embryos injected with the DNA construct or as a stable transgene to allow for the *in vivo* observation of gene expression and identification of specific cell types (Long *et al.,* 1997; Meng *et al.,* 1997).

For positional cloning, however, the zebrafish still lags behind other systems. The zebrafish has 25 chromosomes and a rather large genome of approximately 1.7×10^9 bp (Postlethwait *et al.,* 1994). Moreover, the zebrafish lacks a large number of visible adult markers and balancer chromosomes. However, the large clutch size is enormously helpful for the generation of mutant embryos for recombination mapping of both mutations and molecular polymorphisms. Mapping projects underway in several laboratories using polymerase chain reaction (PCR) based polymorphisms promises to expand the current genetic map (Johnson *et al.,* 1996;

Knapik *et al.,* 1996; Postlethwait *et al.,* 1994). Already, using these markers, the zebrafish mutant *one-eyed pinhead* has been positionally cloned and shown to encode an epidermal growth factor (EGF) related ligand (Zhang *et al.,* 1998). Another PCR-based technique that as been successfully used in plants, AFLP (amplified fragment length polymorphism), is currently being used to isolate polymorphisms that are closely linked (within 0.5 cM) to specific mutations (Vos *et al.,* 1995; Simons *et al.,* 1997). This combination of an expanding genetic map and the ability to generate closely linked markers will facilitate positional cloning in the zebrafish.

George Streisinger first realized the potential of and developed genetic analysis in the zebrafish (Walker and Streisinger, 1983). He and his colleagues at the University of Oregon used gamma rays to induce mutations. Among the early mutants isolated are *spadetail* (*spt*), *no tail* (*ntl*), and *fibrils unbundeled* (*fub*) (Felsenfeld *et al.,* 1991; Halpern *et al.,* 1993; Kimmel *et al.,* 1989). *spadetail* embryos fail to form trunk somites because paraxial mesoderm precursor cells move into the tail during gastrulation rather than converging to the dorsal side as in wild-type embryos (Ho and Kane, 1990; Kimmel *et al.,* 1989). The *no tail* phenotype is due to a mutation in the homologue of the mouse *Brachyury* (*T*) gene (Schulte-Merker *et al.,* 1994). *no tail* embryos lack both the notochord and the tail. The somites that do form in these embryos lack the adaxial cells and their derivatives, the muscle pioneers and horizontal myoseptum (Halpern *et al.,* 1993). *fibrils unbundeled* embryos fail to form striated muscle within the somites. Histological analysis indicates that muscle differentiation is initiated but that the myofibrils fail to organize properly (Felsenfeld *et al.,* 1991; Granato *et al.,* 1996).

More recently, using the chemical mutagen ethylnitrosourea (ENU), two large-scale genetic screens were performed (Driever *et al.,* 1996; Haffter *et al.,* 1996). Five classes of mutants were isolated that have specific effects on the somites. (1) The *fused somites* (*fss*) type mutants which define 5 genes affect the formation of the somite boundaries and anterior–posterior patterning within each somite (Table I) (van Eeden *et al.,* 1996). (2) Reflecting the role of the notochord in somite patterning, mutants in 12 complementation groups were isolated that primarily affect the notochord and also affect somite development. These mutants can be divided into two groups (Table II). Mutants in 5 genes, including *no tail,* are defective in early notochord differentiation and also have somites lacking the horizontal myoseptum. Mutants in an additional 7 genes lack a fully differentiated notochord and have abnormal, dense somites (Table II) (Odenthal *et al.,* 1996; Stemple *et al.,* 1996). (3) The *you*-type mutants which include the zebrafish *sonic hedgehog* gene and 6 other genes have normal notochord but have somites that lack the horizontal myoseptum (Table III) (van Eeden *et al.,* 1996; Schauerte *et al.,* 1998). (4) An additional class of mutants comprising 18 genes affects the differentiation and maintenance of the somitic musculature. (5) Finally, mutants in 2 genes were isolated that have specific defects in the development of primary motoneurons (Table IV) (Granato *et al.,* 1996).

III. Segmentation of the Paraxial Mesoderm

A. Background

Several lines of evidence indicate that segmentation of the paraxial mesoderm takes place in the presomitic mesoderm before morphological signs of segments are evident. First, in both *Xenopus* and zebrafish, following a short heat shock, one to two abnormal somites form, but only after several normal somites are created (Elsdale *et al.,* 1976; Kimmel *et al.,* 1991). Second, as is discussed below, cell labeling experiments in zebrafish indicate that cells become restricted to individual somites while in the presomitic mesoderm (Müller *et al.,* 1996). Furthermore, it has been shown in the chick that anterior–posterior polarity of the somite is established early and is maintained independent of its orientation with respect to the environment: newly formed somites that are rotated along their anterior–posterior axis form vertebrae with inverted polarity (Aoyama and Asamoto, 1988). Additional grafting experiments suggest that somite borders form only when anterior and posterior somite compartments are placed adjacent to each other (Stern and Keynes, 1987).

In the mouse, gene knockout experiments have identified a requirement for several genes during segmentation and epithelialization of the paraxial mesoderm. Embryos mutant for *Mesp2,* a bHLH transcription factor expressed in a single stripe in the anterior presomitic mesoderm, display severe defects including extensive fusion of vertebrae and loss of pattern within each somite. *Mesp2* mutant embryos also show aberrant expression of other genes that are differentially expressed in the presomitic mesoderm such as *Notch1, Notch2,* and *FGFR1.* Interestingly, however, in *Mesp2* embryos some delayed, irregular segmentation is observed within the dermomyotome. Additionally, in these embryos, histological sections indicate that the cervical region exhibits more residual segmentation than the thoracolumbar region (Saga *et al.,* 1997). Thus, the more anterior region of the embryos is less severely affected than the posterior end. Another gene required for somite formation in the mouse is *paraxis,* a basic helix–loop–helix (bHLH) transcription factor expressed in the anterior presomitic mesoderm and in the somites. *paraxis* mutant embryos fail to form epithelial somites but exhibit signs of segmentation and segment polarity (Burgess *et al.,* 1996).

Knockouts of several components of the *Notch* pathway, including *Notch1, Delta 1, RBP-J* (*suppressor of hairless*), and *presenilin* 1 (a transmembrane protein homologous to *SEL-12* which facilitates signaling through *Notch/Lin-12*), produce embryos with defects in segmentation and anterior–posterior pattern (Conlon *et al.,* 1995; Hrabé Angelis *et al.,* 1997; Oka *et al.,* 1995; Wong *et al.,* 1997). The phenotypes of these knockouts are, in general, less severe than the *Mesp2* knockout in that the residual segmentation is more extensive and, typically, the vertebrae are not fused. Another *Notch* pathway gene, *lunatic fringe* is expressed in a dynamic, striped pattern in the presomitic mesoderm in the mouse embryo (Johnson *et al.,* 1997).

In the zebrafish, the *Notch* homologues *Notch1a, Notch5,* and *Notch6* have complex expression patterns that include the presomitic mesoderm and the developing somites (Bierkamp and Campos-Ortega, 1993; Westin and Lardelli, 1997). *Notch1a* is expressed at high levels in the tailbud and presomitic mesoderm and at low levels in the posterior of each somite (Bierkamp and Campos-Ortega, 1993). Interestingly, the expression of *Notch1a* in the tailbud and presomitic mesoderm is strongly reduced in *wit* embryos which exhibit both a neurogenic-like phenotype similar to *Notch* mutants in *Drosophila* and a segmentation phenotype in the posterior somites similar to the *Notch* knockout mouse (Conlon *et al.,* 1995; Heitzler and Simpson, 1991; Jiang *et al.,* 1996; Lehmann *et al.,* 1983). *Notch5* is expressed in a single stripe in the presomitic mesoderm and in the posterior of young somites while *Notch6* is expressed uniformly in the presomitic mesoderm and in the anterior of developing somites. An additional *Notch* homologue, *Notch1b,* is not expressed in the presomitic mesoderm but is expressed in the posterior of each somite (Westin and Lardelli, 1997). The *Notch* ligand homologues, *DeltaD* and *DeltaC,* are expressed in the tailbud and in a pair of bilateral stripes in the anterior presomitic mesoderm that likely correspond to the primordia of adjacent somites. After somite formation, *DeltaD* is expressed in the anterior and *DeltaC* is expressed in the posterior of each each somite (Dornseifer *et al.,* 1997; Haddon *et al.,* 1998). Similar to data from *Xenopus* experiments, misexpression of *DeltaD* via mRNA injection leads to defects in somite formation as indicated by *MyoD* expression (Dornseifer *et al.,* 1997; Jen *et al.,* 1997).

Other genes that may be involved in segmentation of the paraxial mesoderm are *snail1* and *her1. snail1* is expressed in the tailbud and, like *MyoD,* is expressed in the adaxial cells, in the presomitic mesoderm, and in a posterior stripe in each developing somite (Hammerschmidt and Nüsslein-Volhard, 1993; Thisse *et al.,* 1993; Weinberg *et al.,* 1996). The *snail1* expression differs from *MyoD* in that the adaxial expression fades in each somite as it forms while the *MyoD* expression remains (Hammerschmidt and Nüsslein-Volhard, 1993; Thisse *et al.,* 1993). Additionally, *snail1* expression can be observed in two faint stripes in the anterior presomitic mesoderm (Thisse *et al.,* 1993; Weinberg *et al.,* 1996). These stripes likely correspond to the anlage of individual somites and suggests that *snail1* may be involved in segmentation of the paraxial mesoderm. Another gene, *her1* is unique, as is discussed below, in that it is expressed only in the anlage of the odd number somites beginning with somite five (Müller *et al.,* 1996).

B. The *fss*-Type Mutants

In the zebrafish, the *fss*-type mutants defining five genes have defects in both the segmentation and epithelialization of the somitic mesoderm (Table I). Unlike

Table I Mutations Affecting Somite Segmentation and Epithelialization

Gene	Number of alleles	Other phenotypes	References[a]
fused somites (*fss*)	2	None	A
beamter (*bea*)	3	None	A
deadly seven (*des*)	10	None	A
after eight (*aei*)	4	None	A
white tail (*wit*)	1	Brain, spinal cord	A, B

[a] References: A, van Eeden *et al.*, 1996; B, Jiang *et al.*, 1996.

spadetail embryos, the paraxial mesoderm forms normally in these mutants, but epithelial somites fail to form properly. *fss* affects the formation of all somites while, in *beamter* (*bea*) embryos, the first three to four somites form but the remainder do not. In *deadly seven* (*des*), *after eight* (*aei*) and *white tail* (*wit*) embryos, the first seven to nine somites form but the more posterior presomitic mesoderm does not segment or epithelialize (Fig. 2A–F) (van Eeden *et al.,* 1996). In each of these mutants, in the unsegmented regions, the segmental expression of *MyoD* is lost, and *MyoD* is expressed throughout the somitic mesoderm (Fig. 2 I,J) (van Eeden *et al.,* 1996). Surprisingly, each of these mutants, except *wit* which exhibits the additional *Notch*-like neurogenic phenotype, is homozygous viable. Segmentation clearly persists in these embryos as the vertebrae, although clearly malformed, are not fused (Fig. 2 G,H) (van Eeden *et al.,* 1996). This residual segmentation is also observed in the myotome as irregular somite boundaries form later during embryogenesis (Fig. 2 K,L) (van Eeden *et al.,* 1996). All possible double mutant combinations between *fss, bea, des,* and *aei* have been made, and none of the doubles exhibit a morphological phenotype stronger than the most severe single mutant indicating that neither the formation of anterior somite borders in *bea, des,* and *aei* nor the late-forming irregular somite borders are due to genetic redundancy within the *fss*-type genes (van Eeden *et al.,* 1996, 1998). Similarly,

Figure 2 Phenotypes of the *fss*-type mutants. A–D are images of live, 11–13 somite stage embryos. (A) Somites can be clearly seen in wild-type embryos at this stage. However, all somite borders are absent in *fss* (B) and only one or two irregular somites formed in the *bea* mutant embryo (C). The first seven somites can be seen in the *des* embryo (D). (E,F) Dorsal views of wild-type (E) and *bea* (F) embryos at the five to six somite stage. While organized somite boundaries can be seen in wild-type embryos, only irregular somite boundaries can be observed in the *bea* embryo. (G,H) Skeletal stainings of wild-type and *fss* larvae, respectively. Both the neural and hemal arches are formed irregularly in *fss* embryos, but the vertebrae are not fused. (I,J) *MyoD* expression in wild-type and *fss* embryos, respectively. The segmented pattern of expression is lost in *fss* embryos. (K,L) Regular somite boundaries can be seen within the birefringence in the wild-type myotome (K) and irregular somite boundaries can be seen in the myotome of *fss* larvae. Additionally, note that there is no significant reduction in the length of the *fss* embryo. Both photos are of 3.5-day-old larvae. All panels were adapted from van Eeden *et al.,* 1996.

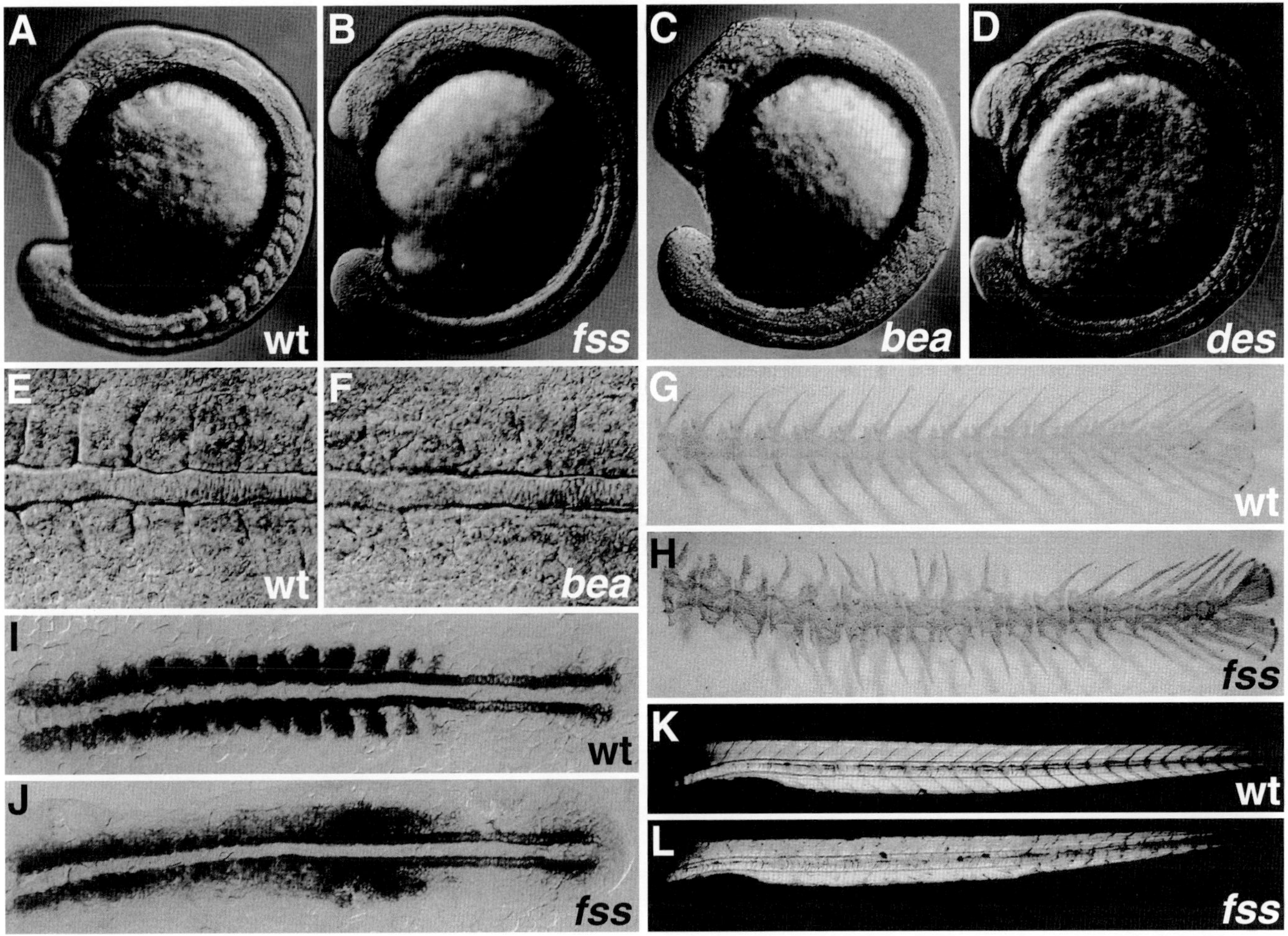
A
wt
B
fss
C
bea
D
des
E
wt
F
bea
G
wt
H
fss
I
wt
J
fss
K
wt
L
fss

analysis of embryos from homozygous mutant females indicates that formation of these borders is not due to a maternal contribution (van Eeden *et al.,* 1996).

In the *fss*-type mutants, the formation of the irregular somite borders later during embryogenesis correlates temporally with the rearrangement and stacking of the adaxial cells (van Eeden *et al.,* 1996). The adaxial cells initially are large cuboidal cells found adjacent to the notochord (Thisse *et al.,* 1993). In wild-type embryos, after formation of the somite borders, some of these cells elongate to span the anterior–posterior length of the somite, and the cells stack such that the nuclei align medially along the anterior–posterior midline of the somite (Waterman, 1969). The stacking of the adaxial cells occurs in *fss* mutants in the absence of somite borders and, moreover, irregular somite boundaries appear to form where the edges of two adjacent stacks meet (van Eeden *et al.,* 1996).

Adaxial cells and their derivatives are missing in mutants that have no notochord such as *no tail* and in the *you*-type mutants, such as *you too* (*yot*), which have a morphologically normal notochord. *yot/fss* double mutant embryos lack all somite boundaries and all evidence of polarity within the myotome. The vertebrae could not be analyzed because *yot* embryos die before the vertebrae are formed; thus, it is not known whether this double mutant lacks segmentation within all somite derivatives. Nonetheless, this result indicates that the activity of the *you*-type genes is required for the formation of irregular boundaries within the myotome of *fss*- embryos and is consistent with the hypothesis that the adaxial cells are directly responsible for the formation of these boundaries (van Eeden *et al.,* 1998).

C. Conservation of Segmentation Mechanisms between Insects and Vertebrates?

Segmentation is best understood in *Drosophila* where a genetic hierarchy controlling this process has been defined. In this hierarchy, the gap genes act in broad regions spanning the anlage of several segments to regulate the expression of the pair-rule genes. These genes are required in portions of every other segment to establish the correct expression of the segment polarity genes which pattern the region within each segment (reviewed in Pankratz and Jäckle, 1990). A significant difference between the *Drosophila* mode of segmentation and somitogenesis in vertebrates is that *Drosophila* is a long-germ band insect, meaning that all segments are created simultaneously, whereas somitogenesis is a sequential process in which each bilateral somite pair segments from the presomitic mesoderm in an anterior to posterior sequence. Somitogenesis is more similar to segmentation in the short-germ band insects such as *Tribolium* which form segments in a sequential fashion as well.

Until recently, it was believed that segmentation in vertebrates evolved independently of segmentation in invertebrates, but three reports describing the expression of an *engrailed* homologue in *Amphioxus* and *hairy* homologues in both the chick and zebrafish have lead to a reexamination of this hypothesis. *Amphioxus engrailed* is expressed in the posterior part of the first eight somites, reminiscent of *Drosophila engrailed* which is a segment polarity gene expressed in a portion of each segment (Holland *et al.,* 1997). *c-hairy1,* a chick homologue of the *Drosophila* pair-rule gene *hairy,* is expressed in cyclic waves whose temporal periodicity corresponds to the time that it takes for one somite pair to form. Interestingly, these waves of expression do not appear to be directed by cell displacement or by an intercellular signal and, thus, appear to be created by a "clock" within each cell that coordinates expression of this gene (Palmeirim *et al.,* 1997). *her1,* a more distantly related *hairy* homologue in the zebrafish, is expressed in a pattern more reminiscent of the pair-rule genes. This expression pattern is pair-rule-like in that *her1* is expressed in the tailbud and in two or three stripes demarcating the anlage of alternating somites within the presomitic mesoderm. This expression pattern is dynamic in that the most anterior stripe fades just before morphological somites can be distinguished, and, at the same time, new stripes of expression "bud" from the tailbud. Elegant cell labeling experiments demonstrated that the stripes of *her1* expression correspond to the odd number somites beginning with the fifth somite while the intervening nonexpressing stripes correspond to the even number somites (Müller *et al.,* 1996). This expression pattern is even more similar to the expression of the *Tribolium hairy* homologue which is also expressed in stripes within the region where segmentation is taking place (Sommer and Tautz, 1993).

The activity of the *fss*-type genes is required for the proper expression of *her1.* Interestingly, however, *fss* which has the most severe morphological phenotype has the smallest effect on *her1* expression. In *fss* embryos, the pair-rule pattern of expression is established, but the anteriormost stripe fades prematurely. Conversely, in *wit, aei,* and *des* embryos, the pair-rule expression pattern is lost by the 10 somite stage. In these embryos, only a disorganized expression in the anterior presomitic mesoderm is observed. In *bea* embryos, *her1* is expressed at a uniform level throughout the tailbud and presomitic mesoderm. Thus, *bea, wit, aei,* and *des* are required for the establishment of the pair-rule expression pattern of *her1,* and *fss* is necessary for the maintenance of this expression pattern (van Eeden *et al.,* 1998). Linkage analysis, however, indicated that *her1* does not cosegregate with any of the *fss*-type mutations (van Eeden *et al.,* 1998). Thus, it remains unclear what effect a mutation in *her1* would have on somite formation.

Although it has been demonstrated in recent years that organisms with diverse body plans utilize orthologous genetic mechanisms during development, it remains unclear how closely related the segmentation mechanism of *Drosophila* is with the segmentation mechanism(s) employed during somitogenesis. Given that the *fss*-type mutations affect both the segmentation of the paraxial mesoderm and the pair-rule pattern of *her1,* it is possible that some pair-rule-like mechanism functions during somite formation (van Eeden *et al.,* 1998). However, analysis of *c-hairy* suggests a different or additional role for *Drosophila* pair-rule homologues in creating or controling periodicity during avian somitogenesis (Palmeirim *et al.,*

1997). Analysis of the *fss/yot* double mutant in zebrafish suggests that a homologue of the *Drosophila* segment polarity gene *hedgehog* is involved in confering periodicity to somite derivatives. This correlation is extended by the observation that, as in *Drosophila, hedgehog* activity is required for the proper segmental expression of *engrailed* (van Eeden *et al.,* 1996; Schauerte *et al.,* 1998). However, the important distinction regarding *hedgehog* signaling is that while *hh* has a segmented expression pattern in *Drosophila,* in zebrafish, *shh* does not. In summary, while it appears that homologues of both the *Drosophila* pair-rule and segment polarity genes are involved in segmenting the paraxial mesoderm in vertebrates, it is likely that their specific functions differ significantly from the roles that they play during *Drosophila* embryogenesis.

D. Segmentation Summary

The somites and the rhombomeres of the hindbrain are the most obvious manifestations of segmentation within the zebrafish embryo, and it is not clear whether these domains are patterned by the same mechanism. Analysis of zebrafish mutants indicates that segmentation of the somites and the hindbrain involves both distinct and shared genes. None of the *fss*-type mutants have an effect on the segmentation of the rhombomeres indicating that their function is required only in the somitic mesoderm (van Eeden *et al.,* 1996). Similarly, embryos mutant for *valentino* are defective in the formation of rhombomeres five and six but have normal somites (Moens *et al.,* 1996). The fact that some genes are required for both processes is demontrated by the zebrafish mutant *choker* which exhibits both somite and hindbrain defects (Kelsh *et al.,* 1996; van Eeden *et al.,* 1996).

During somitogenesis alone, there must be at least three different classes of genes required for the segmentation of the paraxial mesoderm. *fss* somites represents a class of genes necessary for the formation of all somites. A second group of genes are needed for development of the posterior somites but not the most anterior somites. *bea* is not required for the formation of the first three to five somites, while *wit, des,* and *aei* are not necessary for the formation of the first seven to nine somites. It should be noted that no mutants were isolated in which only the most anterior somitic mesoderm was unsegmented. The existence of an additional mode of segmentation within the somitic mesoderm is demonstrated by the observation that some segmental pattern exists in *fss* somite embryos. This mode of segmentation requires signaling through the *you*-type genes (van Eeden *et al.,* 1998).

It is interesting to note that each of the mouse knockouts, like the *fss*-type mutants, have some residual segmentation. This residual segmentation differs within each mutant in that the *Notch1* mouse exhibits more segmentation than the *Mesp2* mouse (Conlon *et al.,* 1995; Saga *et al.,* 1997). In the zebrafish, the elimination of *yot,* a member of the *you*-type genes, eliminates this residual segmentation in the *fss* background. Another member of the *you*-type mutants is *sonic you* which encodes the zebrafish homologue of *Sonic hedgehog* (Schauerte *et al.,* 1998). In this regard, it would be interesting to know the phenotype of a mutant combination in the mouse similar to *fss/yot* such as *Sonic hedgehog/Mesp2.*

Another similarity between the mouse and fish mutants is that, in both organisms, the anterior region is less severely affected than more posterior regions (Hrabé Angelis *et al.,* 1997; Saga *et al.,* 1997; van Eeden *et al.,* 1996). In the zebrafish, the more anterior somites do show some differences in their development. For instance, the first six somites appear to form more rapidly (three per hour) than the remainder (two per hour) (Hanneman and Westerfield, 1989). Additionally, the first five to seven somites exhibit a more synchronous development than the more posterior somites. In the first five somites, the adaxial cells rearrange simultaneously while in the more posterior somites the adaxial undergo this change in each successive somite as it matures (van Eeden *et al.,* 1996). This phenomenon is observed with Engrailed expression and refinement of *snail1* expression. Both of these processes occur synchronously in the first seven somites and in an anterior to posterior sequence within remainder of the somitic mesoderm (Hammerschmidt and Nüsslein-Volhard, 1993; Hatta *et al.,* 1991).

Analysis of both the *fss*-type mutants and the mouse knockouts has also helped to elucidate the relationships between segmentation, epithelialization and the differentiation of somitic cell lineages. Epithelialization appears to be dependent on segmentation occuring in the presomitic mesoderm. However, analysis of *paraxis* embryos and the *fss*-type embyros indicates that some segmentation can occur in the absence of epithelialization. Differentiation of the mytome, dermomyotome, and sclerotome proceeds in the absence of or reduction of both segmentation and epithelialization. That the somite-derived tissues develop in the abscence of somite epithelialization is clearly demonstrated by the fact that four of the five *fss*-type mutants are viable and fertile (van Eeden *et al.,* 1996). Interestingly, in *Xenopus,* epithelial somites are never formed. The presomitic mesoderm in *Xenopus* consists of mononucleated myotomal cells which express *MyoD* in a uniform, unsegmented pattern (Hamilton, 1969; Hopwood *et al.,* 1992). MyoD expression becomes segmented as balls of these myotomal cells bud off of the anterior segmental plate, rotate 90° and develop into mature musculature (Hamilton, 1969; reviewed in Jen *et al.,* 1997).

IV. Somite Patterning and Differentiation

A. Somite Patterning

After segmentation, the somite is patterned into dermomyotome, myotome, and sclerotome by interactions with the surrounding tissues. In the chick, it has been demonstrated that both the neural tube and the surface ectoderm have dorsalizing effects on somite pattern. Candidate dorsalizing molecules include members of the Bmp and Wnt families. Bmp4, Wnt1, Wnt3, and Wnt3A are all expressed in the roof plate of the neural tube while Bmp7 and Wnt4 are expressed in a more broad domain in the dorsal neural tube. Additionally, Bmp4, Wnt3, Wnt4, Wnt6, and Wnt7B are expressed in the surface ectoderm (Hollyday *et al.,* 1995; Parr *et al.,* 1993; Pourquié *et al.,* 1996). Signals from the floor plate and notochord are involved in ventralizing the somite as removal of both of these structures in the chick embryo results in a dorsalization of the somite (Dietrich *et al.,* 1997).

The ventralizing activity of the floor plate and notochord appears to be encoded primarily by members of the *hedgehog* family. *shh* is expressed in the floor plate and notochord in mouse, chick, and zebrafish. In both the mouse and chick, *shh* is a potent inducer of sclerotome, a weak inducer of myotome and antagonist dermomyotome formation. In the chick, *shh* synergizes with Wnt family members to specify myotome, indicating that myotome specification may depend on both dorsalizing and ventralizing signals, an idea supported by grafting experiments (Dietrich *et al.,* 1997; Münsterberg and Lassar, 1995; Stern *et al.,* 1995). Misexpression of a constitutively active form of the antagonist of *hh* signaling, cAMP-dependent protein kinase A (PKA), represses the formation of *shh*-inducible structures while a dominant–negative form of PKA mimics the effects of *shh* (Concordet *et al.,* 1996; Fan *et al.,* 1995; Hammerschmidt *et al.,* 1996; Johnson *et al.,* 1994; Weinberg *et al.,* 1996). *patched,* the putative *hh* receptor, is expressed in the sclerotome of the mouse and chick (Goodrich *et al.,* 1996; Marigo and Tabin, 1996). Similarly, in the zebrafish, *patched* is expressed in the adaxial cells and later in the developing somite, and this expression can be induced by misexpression of *shh* or dominant–negative PKA and repressed by an activated PKA (Concordet *et al.,* 1996). Loss-of-function data from the mouse indicate a requirement of *shh* in somite patterning. In the *shh* knockout, both *Pax1* expression, which marks the sclerotome, and *myf5* expression, which marks the myotome, are reduced while *Pax3* expression in the dermomyotome is expanded (Chiang *et al.,* 1996).

In the zebrafish, there are three *hh* family members that are expressed along the axial midline. *shh* is expressed in both the floor plate and the notochord (Krauss *et al.,* 1993). *tiggy-winkle hedgehog* (*twhh*) is restricted to the floor plate and ventral midline of the neural tube while *echidna hedgehog* (*ehh*) is expressed in the notochord (Currie and Ingham, 1996; Ekker *et al.,* 1995). Misexpression of *shh* leads to an increase in *MyoD* expression and in the number of Engrailed-expressing muscle pioneers (Concordet *et al.,* 1996; Hammerschmidt *et al.,* 1996; Weinberg *et al.,* 1996). A second study found that induction of additional muscle pioneers required misexpression of *ehh* in conjunction with *shh* (Currie and Ingham, 1996). Nevertheless, the myotome inducing ability of *hh* signaling is conserved. Additional gain-of-function experiments, however, underscore the differences between the zebrafish somite and that of the mouse and chick.

As mentioned previously, the sclerotome in the zebrafish is relatively small.

Both the bHLH protein Twist, and *Pax9,* as in the mouse, are expressed in the zebrafish sclerotome (Hammerschmidt *et al.,* 1996; Neubuser *et al.,* 1995; Nornes *et al.,* 1996; Morin-Kensicki and Eisen, 1997). While overexpression of *shh* induces both myotome and sclerotome formation in the mouse and chick, in the zebrafish, *shh* induces myotome at the expense of sclerotome (Fan and Tessier-Lavigne, 1994; Hammerschmidt *et al.,* 1996; Johnson *et al.,* 1994). The loss-of-function data in the mouse and zebrafish are not so different in that both *shh* mutants have sclerotome. However, the sclerotome is more strongly reduced in the mouse mutant. Thus, *shh* may be required for specifying only a portion of the sclerotome, and this portion may be relatively large in the mouse. The zebrafish sclerotome, which is relatively small in size, may be composed primarily of sclerotome that is specified independently of *shh.* Perhaps further analysis of the role of *hh* signaling in somite patterning may reveal how the relative differences in the sizes of myotome and sclerotome between the zebrafish and mouse/chick is determined.

B. The Notochord Mutants

Analysis of a number of notochord mutants in the zebrafish again demonstrates the role of the notochord in patterning the paraxial mesoderm. In mutants lacking the notochord, the adaxial cells, the muscle pioneers, and the horizontal mysoseptum are all absent. In these mutants, transplanted wild-type cells that populate notochord can rescue the formation of each of these structures (Halpern *et al.,* 1993; Odenthal *et al.,* 1996). Embryos mutant for *floating head* (*flh*), the zebrafish *Xnot* homologue, are defective in the specification of the chordomesoderm and, thus, have no notochord and no tail (Halpern *et al.,* 1995; Melby *et al.,* 1996; Talbot *et al.,* 1995). In *flh* embryos, the expression of *brachyury* in the axial mesoderm is confined to the posteriormost region while the trunk somites are fused below the neural tube (Halpern *et al.,* 1995; Odenthal *et al.,* 1996; Talbot *et al.,* 1995). A similar phenotype is seen in *bozozok* (*boz*) embryos (Sonica-Krezel *et al.,* 1996; Stemple *et al.,* 1996). In *momo* (*mom*) embryos, *brachyury* expression is confined to the posterior axial mesoderm resulting in the loss of notochord in the trunk region. In both *flh* and *mom, MyoD* is expressed in the adaxial cells indicating that a differentiated notochord is not required for this expression, but, in regions lacking notochord, Engrailed expression in the muscle pioneers is absent (Halpern *et al.,* 1995; Odenthal *et al.,* 1996).

As mentioned previously, the *no tail* (*ntl*) phenotype is due to a mutation in the zebrafish homologue of the mouse gene *Brachyury* (*T*) (Halpern *et al.,* 1993; Schulte-Merker *et al.,* 1994). *no tail* embryos lack both the notochord and the tail. The somites that do form in these embryos lack most *MyoD* expression in the adaxial cells, muscle pioneers, and the horizontal myoseptum (Halpern *et al.,* 1995; Odenthal *et al.,* 1996). *doc* mutant embryos like *ntl,* have notochord precursor

Table II Genes Required for Notochord and Somite Development

Gene	Number of alleles	Other loss-of-function phenotypes	References[a]
Group 1. Phenotypes: defects in early notochord formation and loss of horizontal myoseptum			
floating head (*flh*)	4	Floorplate, motility	A, B, C
bozozok (*boz*)	1	Floorplate, cyclopia	D, E
no tail (*ntl*)	7	No tail formed	C, F
momo (*mom*)	1	—	C
doc	3	—	C
dino (*din*)	2	Ventralizing	C, G
Group 2. Phenotypes: defects in late notochord development, dense abnormal somites			
sleepy (*sly*)	19	Disorganized brain, retinotectal pathfinding	C, D
grumpy (*gup*)	12	Disorganized brain, retinotectal pathfinding	C, D
bashful (*bal*)	26	Disorganized brain, retinotectal pathfinding	C, D
dopey (*dop*)	5	Day 2 embryo degeneration	C, D
sneezy (*sny*)	6	Day 2 embryo degeneration	C, D
happy (*hap*)	6	—	C
gno	1	None	D

[a]References: A, Halpern *et al.*, 1995; B, Talbot *et al.*, 1995; C, Odenthal *et al.*, 1996; D, Stemple *et al.*, 1996; E, Solnica-Krezel *et al.*, 1996; G, Hammerschmidt *et al.*, 1996.

cells but differentiated notochord cells can be seen only in the tail region. While these embryos exhibit a slight reduction in the adaxial *MyoD* expression, they do not express Engrailed in the region lacking notochord cells (Odenthal *et al.*, 1996).

An additional class of mutants defining seven genes has been isolated in which the notochord is undifferentiated and the somites have a disorganized morphology (Table II). Mutants in *sleepy, bashful, grumpy, sneezy, dopey, happy,* and *gno* have horizontal myosepta but have disorganized, dense somites (Odenthal *et al.*, 1996; Stemple *et al.*, 1996; van Eeden *et al.*, 1996).

C. The *you*-Type Mutants

While the notochord mutants demonstrate the importance of the notochord in somite patterning, the somite phenotypes in these mutants are likely indirect and due to a loss or reduction of *shh* expression. For instance, *doc* acts cell autonomously in the notochord, but is required for the proper expression of Engrailed within the somites. Indeed, the only Engrailed expression observed in the somites of *doc* embryos is found in the posterior of the embryo where a few vacuolated notochord cells develop and express *shh* (Odenthal *et al.*, 1996). Interestingly, a group of mu-

Table III Genes Required for Somite Patterning, but Not Notochord Development: The *you*-Type Genes

Gene	Number of alleles	Other loss-of-function phenotypes	References[a]
you	3	Circulation, motility	A
you-too (*yot*)	2	Spinal cord, circulation, motility, retinotectal projection	A
sonic-you (*syu*)	4	Motility, circulation, pectoral fins	A, B
chameleon (*con*)	4	Spinal cord, circulation, motility, retinotectal projection, pectoral fins	A
u-boot (*ubo*)	1	Fins, motility, melanophores	A
choker (*cho*)	1	Pigment pattern, hindbrain	A, C
iguana (*igu*)	2	Floor plate, brain, jaw	D, C

[a]References: A, van Eeden *et al.,* 1996; B, Schauerte *et al.,* 1998; C, Kelsh *et al.,*1996; Haffter *et al.,* 1996; Brand *et al.,* 1996.

tants, the *you*-type group, share the somite phenotype with the notochord mutants but have normal notochords. Thus, the *you*-type genes may be more directly involved in mediating and interpreting the signal(s) from the notochord to the somite (van Eeden *et al.,* 1996).

The six genes of the *you*-type group are required for formation of the horizontal myoseptum (Table III, Fig. 3) (van Eeden *et al.,* 1996). Thus, the mutants, *you* (*you*), *you-too* (*yot*), *sonic-you* (*syu*), *chameleon* (*con*), *u-boot* (*ubo*), *choker* (*cho*), and *iguana* (*igu*) as their group name indicates, have U-shaped somites (Brand *et al.,* 1996; Haffter *et al.,* 1996; van Eeden *et al.,* 1996). *yot* is the strongest of the *you*-type mutants in that no adaxial cells form; thus, no adaxial *MyoD* expression is observed (Fig. 3E,F), no somitic Engrailed expression is seen, and no horizontal myosepta are formed (Fig. 3A,B) (van Eeden *et al.,* 1996). Additionally, the more lateral expression of *MyoD* is affected in that full expression in the medial-anterior domain is never achieved and expression is not maintained in older somites (van Eeden *et al.,* 1996). The phenotypes of *syu, con,* and *you* mutant embryos are less severe in that some *MyoD* expression is seen in the adaxial cells (Schauerte *et al.,* 1998; van Eeden *et al.,* 1996). Likewise, in *con* and *you* embryos, the number of Engrailed expressing cells is reduced but a few remain in each segment while, in the most severe *syu* allele, all somitic Engrailed expression is lost (Fig. 3C,D) (van Eeden *et al.,* 1996; Schauerte *et al.,* 1998). In *ubo* embryos, *MyoD* expression is unaffected, and, interestingly, a larger number of cells express Engrailed at reduced levels (van Eeden, 1997; van Eeden *et al.,* 1996). *cho* embryos, initially studied for their pigment cell phenotype, have reduced horizontal myosepta but appear to have morphologically normal muscle pioneers (Kelsh *et al.,* 1996; van Eeden *et al.,* 1996). *igu* embryos are interesting in that they only

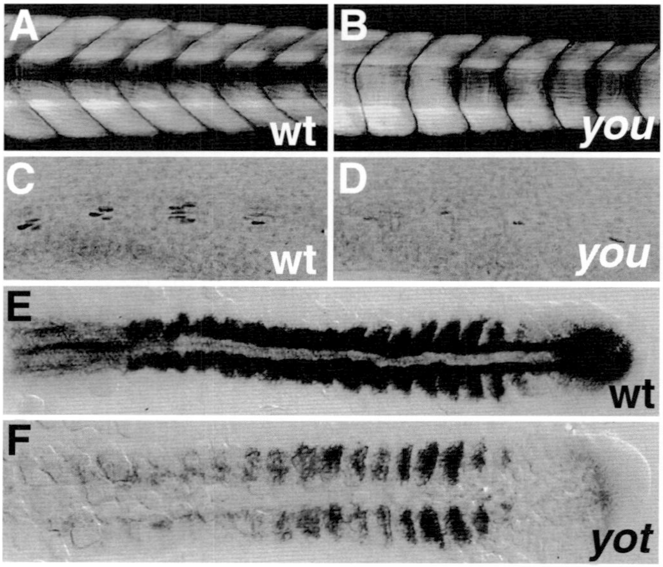

Figure 3 Phenotypes of the *you*-type mutants. (A,B) Dark field images of wild-type and *yot* somites, respectively. (A) In wild-type, the horizontal myoseptum divides the somite into dorsal and ventral compartments, and, in addition, the somite has a characteristic chevron shape. (B) The horizontal myoseptum is absent in *yot* embryos and the somites are "u" shaped. Engrailed expression in the muscle pioneers of wild-type (C) and *you* (D) embryos. The number of muscle pioneers is reduced in *you* embryos. *MyoD* expression retains its segmentally expressed pattern in both wild-type (E) and *yot* (F) embryos. However, the longitudinal expression in the adaxial cells is absent in *yot* embryos. All panels were adapted from van Eeden *et al.*, 1996.

lack the horizontal myoseptum in the posterior somites (Brand *et al.*, 1996; Haffter *et al.*, 1996).

Given that *sonic-you* is the zebrafish homologue of *shh*, it is possible that the *you*-type genes may encode additional components of the *hh* signaling pathway (Schauerte *et al.*, 1998). Cell transplantation experiments indicate that *ubo* and *yot* functions are required in the somite, and thus these genes may be involved in receiving or directly responding to the *hh* signal (van Eeden *et al.*, 1996). Indeed, injection of *shh* or a dominant negative PKA cannot induce additional muscle pioneers in *yot* embryos but can do so in *ubo, con,* and *you.* Thus, *yot* must act within the somite downstream of both *shh* and PKA (Schauerte *et al.*, 1998).

D. Muscle Differentiation Mutants

Eighteen complementation groups have been defined in the zebrafish that are required for the development or maintenance of the somitic musculature (Table IV). Most of these mutants were detected by their reduced motility and, on more de-

Table IV Genes Required for Differentiation, Maintenance, or Innervation of the Somitic Musculature

Gene	Number of alleles	Loss-of-function phenotypes	References[a]
Genes required for muscle differentiation or maintenance			
sloth (*slo*)	2	Reduced muscle striation,	A
frozen (*fro*)	1	immotile, defective in early muscle development	
fibrils unbundeled (*fub*)	3	Reduced muscle striation, immotile, defective in myofibril organization?	A, B
turtle (*tur*)	22	Reduced muscle striation, reduced motility	A
buzz-off (*buf*)	5		
faulpelz (*fap*)	2		
slow motion (*slw*)	1		
schnecke (*sne*)	1		
hermes (*hem*)	2		
duesentrieb (*dus*)	1		
mach two (*mah*)	1		
slop (*slp*)	1	Reduced muscle striation, reduced motility, heart	A
jam	1		
slinky (*sky*)	1		
sapje (*sap*)	2	Somite degeneration	A
softy (*sof*)	3		
schwammerl (*sml*)	2		
runzel (*ruz*)	1		
Genes specifically required for motoneuron development			
unplugged (*unp*)	1	Motility defects, CaP axon growth abnormal	A
diwanka (*diw*)	3	"Accordion" motility defect, defective CaP, RoP, and MiP motoneuron development	A

[a] References: A, Granato *et al.*, 1996; B, Felsenfeld *et al.*, 1991.

tailed analysis, were designated as muscle mutants. Defects in the musculature can be seen as loss of birefringence which is normally created when polarized light passes through the parallel organization of the myofibrils. Embryos mutant for *fibrils unbundeled* (*fub*), *sloth* (*slo*), or *frozen* (*fro*) are immotile and exhibit a loss of birefringence (Fig. 4A,B) (Felsenfeld *et al.*, 1991; Granato *et al.*, 1996). *snail1* expression in the adaxial cells and myotome is normal in these embryos indicating that early specification of muscle cell fates is normal (Granato *et al.*, 1996). Histological sections indicate that *fub, slo,* and *fro* lack myofibril differentiation. In *slo* and *fro,* the myotomal nuclei remain round whereas in wild-type and in *fub*

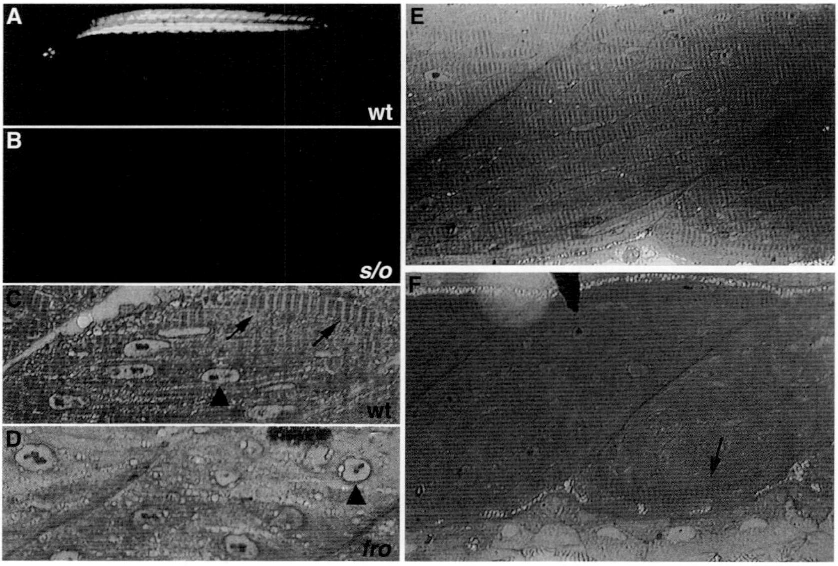

Figure 4 Phenotypes of the muscle mutants. (A,B) Dark-field images of 4-day-old larvae. While birefringence is intense in the wild-type larvae (C), birefringence is lost in *slo* larvae indicating a loss of myofibril organization. (C,D) Sagital–lateral sections through the myotome of 36-hr embryos. (C) In wild-type, organized muscle fibers can be seen (arrows) and nuclei lose their round morphology and elongate (arrowhead). In contrast, in *fro* embryos (D), organized muscle fibers are not seen and the nuclei do not elongate (arrowhead). (E,F) Optical sections of the myotome of 3-day-old larvae. Muscle fibers are rarely observed in *dus* larvae (arrow) (F) compared with wild-type (E). All panels were adapted from Granato *et al.,* 1996.

the nuclei are elongated (Fig. 4C,D) (Granato *et al.,* 1996). This indicates that *slo* and *fro* functions are required early in muscle cell differentiation (Granato *et al.,* 1996). Meanwhile, *fub* appears to function later during the organization of the myofibrils (Felsenfeld *et al.,* 1991).

Embryos mutant for *turtle* (*tur*), *buzz-off* (*buf*), *faulpelz* (*fap*), *slow motion* (*slw*), *schnecke* (*sne*), *hermes* (*hem*), *duesentrieb* (*dus*), *mach two* (*mah*), *slop* (*slp*), *jam,* or *slinky* (*sky*) show movement but have a slow escape response. Mutants for each of these genes also exhibit reduced birefringence. In most of these embryos, the muscle fibers appeared essentially normal when examined under a compound microscope. However, *slp, dus,* and *buf* embryos have obvious defects. *slp* embryos have roughly 80% of the muscle fibers of a wild-type embryo, whereas *dus* embryos have only a few recognizable fibers in each somite (Fig. 4E,F) (Granato *et al.,* 1996). *buf* embryos are distinct in that the total number of muscle fibers appear normal, but the arrangement of the sarcomeres is less ordered than in wild-type embryos (Granato *et al.,* 1996).

Embryos mutant for *sapje* (*sap*), *softy* (*sof*), *schwammerl* (*sml*), or *runzel* (*ruz*) behave normally for the first 96 hr, but at this time they exhibit slower motility and reduced birefringence as the somitic muscle begins to degenerate. In *sap* and *sof* embryos the degeneration is distributed among individual dorsal and ventral myotomes, whereas in *ruz* embryos the degeneration encompasses the entire somitic mesoderm. Strikingly, in each of these mutants, the jaw and heart muscles remain intact indicating that the defects are specific to the somitic muscle (Granato *et al.,* 1996).

E. Slow and Fast Muscle Differentiation

Within the axial musculature of a vertebrate exist two to three different muscle fiber types. Slow muscle fibers are specialized for contractions of low-force and long duration while fast muscle fibers are specialized for high-force contractions of short duration. Slow muscle fibers are smaller, darker in color, contain more lipid and mitochondria, and are more heavily vascularized than fast muscle fibers which are large and pale in color (Hoyle, 1983). A third intermediate fiber type also exists. In the zebrafish, the three muscle fiber types develop at distinct times and are localized to specific regions of the myotome. The slow muscle is localized to the lateral periphery in a wedge centered on the horizontal myoseptum. Fast muscle constitutes the majority of the somitic musculature, whereas the intermediate muscle is located laterally just medial to the slow muscle (van Raamsdonk *et al.,* 1982). Ultrastructural analysis of the developing zebrafish myotome implied that the medial muscle forms first giving rise to fast muscle while the lateral muscle formed last giving rise to the slow muscle (Waterman, 1969). Conversely, histochemical analysis suggested that slow muscle cells form initially in the medial region and that some of these cells later differentiated into fast muscle (van Raamsdonk *et al.,* 1978).

At a first approximation, it might appear that the different muscle types in the zebrafish embryo might be specified by signals determining the medial–lateral pattern of the somite with lateral signals promoting slow muscle fate and medial signals promoting fast muscle formation. Medial–lateral patterning of the dermomyotome in the chick is likely to be patterned in this way. The medial, central, and lateral regions of the dermomyotome give rise, respectively, to the epaxial muscle, the dermis, and the hypaxial muscle. Noggin, the BMP inhibitor (Holley *et al.,* 1996; Zimmerman *et al.,* 1996), is expressed in the dorsal somite and may antagonize BMP-4 expressed in both the neural tube and lateral plate mesoderm to pattern the dermomyotome along the medial lateral axis (Hirsinger *et al.,* 1997; Marcelle *et al.,* 1997; Pourquié *et al.,* 1996).

In part, slow and fast muscle are specified by signals differing along the medial–lateral axis. Slow muscle cells are specified by *shh* emanating from the notochord. Embryos that lack notochord, and thus *shh* expression, such as *boz* and *ntl* exhibit a loss of slow but not fast muscle, whereas overexpression of *shh* promotes the formation of slow muscle at the expense of fast muscle (Blagden *et al.,* 1997). As previously described by van Raamsdonk *et al.* (1978), the slow muscle cells are the first to express muscle specific markers (Devoto *et al.,* 1996; Blagden *et al.,* 1997). Likewise, as Waterman observed, these initially developing muscle cells are the medial, adaxial cells (Blagden *et al.,* 1997; Devoto *et al.,* 1996; Waterman, 1969). Resolution of these data comes from cell labeling experiments which demonstrate that the adaxial cells undergo elongation to span the anterior–posterior length of each somite and then migrate laterally through the somite to give rise to the slow muscle at the lateral surface of the myotome (Devoto *et al.,* 1996). Indeed, this can be inferred from transverse sections immunolabeled with an antibody specific to slow muscle which show a band of slow muscle cells migrating from the medial to the lateral edge of the myotome (Fig. 5) (Blagden *et al.,* 1997; Devoto *et al.,* 1996).

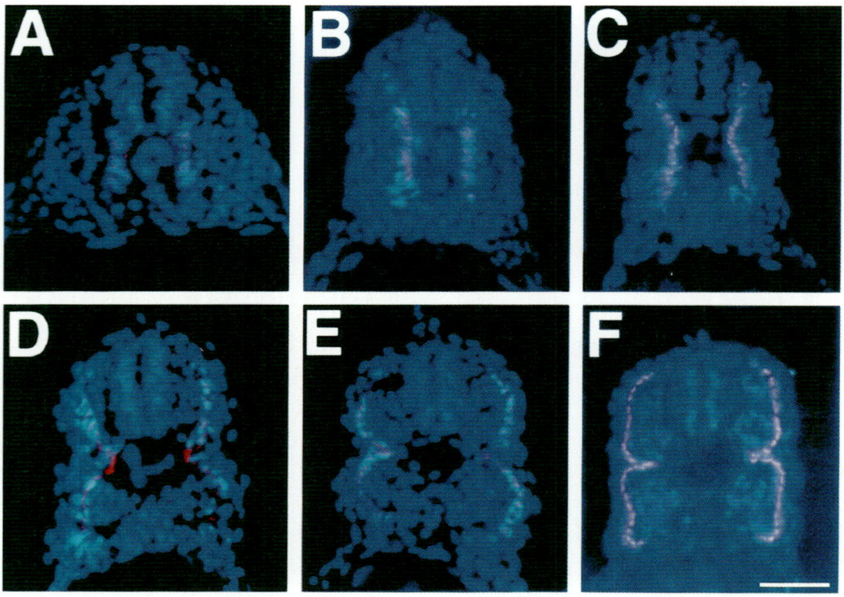

Figure 5 Radial migration of adaxial cells from the medial to lateral somite. All panels are transverse sections through the caudal trunk immunolabeled with F59 antibody, which marks the adaxial/slow muscle cells, and counterlabeled with Hoechst to mark all nuclei. (A) At 17 hr postfertilization (hpf), the adaxial cells are located adjacent to the notochord. (B) At 18.5 hpf, the adaxial cells begin to move dorsally and ventrally. This is also when the adaxial cells extend along the anterior–posterior axis to span entire individual somites. (C) At 20.5 hpf, the adaxial cells have nearly completed their dorsal–ventral migration. (D) At 21.5 hpf, the adaxial cells begin to migrate radially to the lateral somite. (E) At 23 hpf, most of the adaxial cells have reached the lateral surface of the myotome. (F) At 24 hpf, the adaxial cells are now lateral; however, some adaxial cells along the horizontal myoseptum remain attatched to the notochord creating a medial involution of these cells. All panels were adapted from Devoto *et al.,* 1996.

It is not clear whether the Engrailed expressing muscle pioneers, which also undergo this elongation and migration in the region that gives rise to the horizontal myoseptum, ultimately form a different subset of muscle fibers. The muscle pioneers do behave differently than the other adaxial cells in that their medial surface remains adjacent to the notochord as the rest of the cell flattens and migrates to the lateral surface. While the other adaxial cells lose contact with the notochord shortly after beginning migration, the muscle pioneers do not lose this connection until after 48 hr (Devoto *et al.,* 1996; Waterman, 1969).

IV. Innervation of the Somitic Musculature

A. Analysis of Primary Motoneurons

During zebrafish development, three or four identified primary motoneurons extend axonal projections in stereotypical pathways to innervate specific regions of the myotome (Eisen *et al.,* 1986, 1990; Myers *et al.,* 1986). The three best studied primary motoneurons are named according to position along the spinal cord: the caudal primary (CaP), middle primary (MiP), and rostral primary (RoP) (Fig. 6A).

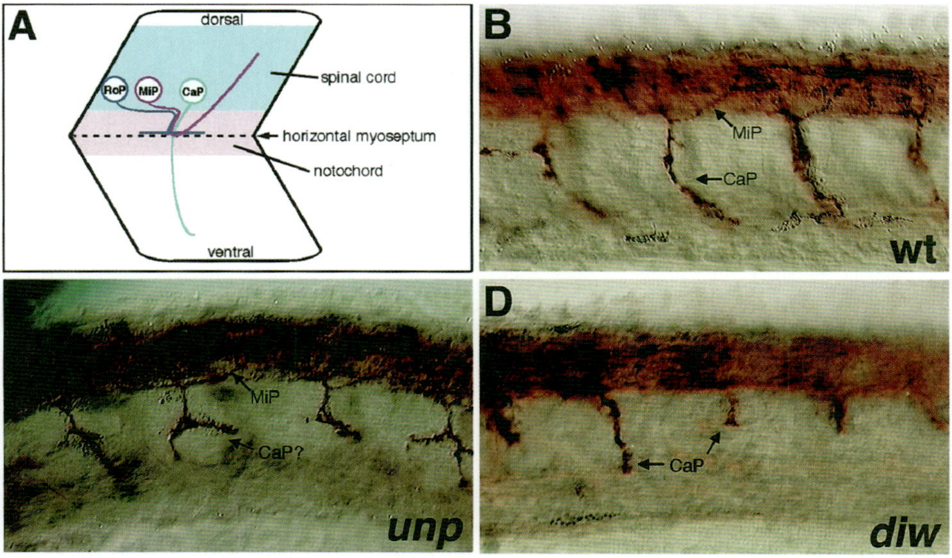

Figure 6 Phenotypes of mutants defective in motoneuron development. (A) Schematic of the three primary motoneurons and their stereotypic axonal projections. (B–D) Both the CaP and MiP axons can be visualized by using the zpn-1 antibody. (B) Wild-type embryo. (C) In *unp* embryos, the MiP axons seem normal, but what are likely CaP axons bifurcate and extend abnormally. (D) In *diw* embryos, MiP axons are absent while CaP axons appear much shorter than in wild-type embryos. All panels were adapted from Granato *et al.,* 1996.

The CaP growth cone is the first to leave the spinal cord and extends ventrally along the medial surface of the myotome until it reaches the horizontal myoseptum where growth pauses before the axon extends into the ventral myotome. The MiP growth cone initially extends caudally and then exits the ventral root to follow the path of the CaP to the horizontal myoseptum. There after a short pause, the growth cone extends dorsally and caudally. Finally, the RoP intially follows the path of the MiP to the horizontal myoseptum where it begins to grow laterally within the horizontal myoseptum and forms branches that extend from the medial to lateral surfaces of the myotome (Eisen *et al.,* 1986).

Analysis of the CaP axon indicates that its cell-specific pathway is partly regulated by signals from the notochord that make the dorsal myotome nonpermissive to CaP growth cones. By performing heterochronic, somite transplantations, it was demonstrated that, at 16 hr postfertilization, a time before the growth cones leave the spinal cord, both the dorsal and ventral myotome are permissive for extension of the CaP growth cone. Later, at 19 hr postfertilization, a time when the CaP growth cone leaves the horizontal myoseptum along its specific path, the dorsal myotome becomes nonpermissive for axonal outgrowth. When somites were removed at 16 hr postfertilization, and allowed to develop in culture for 3 hr prior to transplantation, then the dorsal myotome remained permissive for CaP axon outgrowth indicating that extrinsic signals are required to regulate the permissiveness of the dorsal myotome. The requirement for the notochord in this signaling process was demonstrated by transplanting somites derived from *flh* embryos, which lack notochord, into wild-type embryos. The dorsal regions of these somites remained permissive indicating that the notochord is required to change the permissiveness of the dorsal myotome (Beattie and Eisen, 1997). It must be noted, however, that the CaP axon extends ventrally, albeit abnormally, in *flh* embryos indicating that other factors must be involved in determining this axonal growth pathway (Beattie and Eisen, 1997; Talbot *et al.,* 1995).

B. Mutants Defective in Primary Motoneuron Development

As mentioned above, CaP motoneurons in *flh* embryos extend ventrally but do so at a reduced frequency and do not project as far as in wild-type embryos (Talbot *et al.,* 1995). Similar effects are seen in the *you*-type mutants. In *yot* embryos, most of the primary motoneurons run along the spinal cord while a few CaP axons extend ventrally. This phenotype is observed at a lower frequency in *you, syu,* and *con* embryos (Schauerte *et al.,* 1998; van Eeden *et al.,* 1996). The early axonal tracts in the *you*-type mutants appear normal indicating that the axon defect is specific to the motoneurons (van Eeden *et al.,* 1996). Given that *syu* encodes the zebrafish homologue of *shh,* it is likely that the defects in motoneuron axon growth in the *you*-type mutants are due to altered signaling from the notochord and that *shh* is involved in regulating the permissiveness of the myotome to CaP growth cones (Beattie and Eisen, 1997; Schauerte *et al.,* 1998). In the *fss*-type mutants, the CaP axons are irregularly spaced reflecting a loss of pattern along the anterior–posterior axis, and sometimes truncate prematurely (van Eeden *et al.,* 1996).

Mutants for *unplugged* (*unp*) and *diwanka* (*dwi*) have more specific effects on the motoneurons. *unp* mutant embryos are almost completely immotile at 24 hr but recover, and, by day 6 can swim short distances. These embryos have motoneurons, likely CaP motoneurons, that extend axons into the ventral myotome, but these axons have abnormal morphology and branching (Fig. 6B,C) (Granato *et al.,* 1996). MiP motoneuron, commissural interneurons, and Rohon–Beard neurons appear normal in *unp* embryos indicating that either CaP motoneurons or their ventral axonal pathway are specifically affected. *dwi* embryos belong to the "accordion" class of motility mutants. When touched, these embryos contract such that the anterior–posterior body axis shortens by 5 to 10%. This phenotype suggests that somites on both sides of the body contract simultaneously and, thus, lack the inhibitory mechanism which prevents motoneurons on one side of the body from firing while the opposing muscles are contracted (Fetcho, 1992; Granato *et al.,* 1996; Roberts, 1989). *dwi* mutant embryos have truncated CaP motoneurons and lack RoP and MiP motoneurons (Fig. 6 B,D; Granato *et al.,* 1996). Commissural interneurons and Rohon–Beard mechanosensory neurons are normal in *dwi* embryos indicating that *dwi* specifically affects the development of motoneurons. No other mutants in the "accordion" class exhibit this motoneuron phenotype (Granato *et al.,* 1996).

V. Conclusions

The 1990s have seen a rapid advancement in our understanding of the patterning and differentiation of the paraxial mesoderm. We now have some knowledge of how the surrounding tissues affect somite pattern, and, in many cases, the molecules involved in these interactions have been elucidated. A more sophisticated understanding of the temporal and functional relation between the various signals awaits further inquiry. Our knowledge of the mechanisms of segmentation and epithelialization is less detailed. Clearly, the *Notch* pathway is involved in segmentation of the paraxial mesoderm, but how this pathway functions during somitogenesis is not known. The *Notch* pathway may provide a molecular entry point into two of the more interesting questions concerning somite formation. First, how is the segmentation process governed spatially and temporally and how does this relate to species-specific somite number? Second, how is the repetitive, positional information interpreted to produce the mesenchymal to epithelial transition during somite formation? The zebrafish is a relative newcomer to the study of somitogenesis, but the ability to both genetically and embryologically manipulate the zebrafish embryo makes this organism particularly well suited for the future analysis of such complex processes.

Beyond telling us how elements of the vertebrate trunk and tail are formed, analysis of the genetic regulatory network governing somitogenesis may inform us as to how such regulatory networks evolve. It is the complexity of somite formation that engenders this possibility given that it is likely that acquisition of new gene function follows gene duplication. For example, in the zebrafish, as many as four *Notch* and three *hedgehog* family members may be involved in somitogenesis. To what extent do the functions of these genes overlap? What functions are unique to individual family members? Once we understand the answers to such questions regarding the various *Notch's, hh's, Wnt's,* etc., in each of the different organisms in which somite formation is analyzed, then, perhaps, we can make some general conclusions about changes in regulatory hierarchies and correlate these changes with changes in morphology. In this regard, the multiorganismal approach to studying development may result in a synergy that both accelerates the understanding of developmental mechanisms and provides insight into the relationship between evolution and development.

Acknowledgments

We thank Freek van Eeden, Michael Granato, Steve Devoto, and Pascal Haffter for the use of their figures. We thank Siegfried Roth and Henry Roehl for comments on the manuscript. S. A. H. is supported by a fellowship from the Cancer Research Fund of the Damon Runyon-Walter Winchel Foundation.

References

Aoyama and Asamoto (1988). *Development* **104,** 15–28.

Beattie, C. E., and Eisen, J. S. (1997). Notocord alters the permissiveness of myotome for pathfinding by an identified motoneuron in embryonic zebrafish. *Development* **124,** 713–720.

Bierkamp, C., and Campos-Ortega, J. A. (1993). A zebrafish homologue of the *Drosophila* neurogenic gene Notch and its pattern of transcription during early embryogenesis. *Mech. Dev.* **43,** 87–100.

Blagden, C. S., Currie, P. D., Ingham, P. W., and Hughes, S. M. (1997). Notochord induction of zebrafish slow muscle mediated by Sonic hedgehog. *Genes Dev.* **11,** 2163–2175.

Brand, M., Heisenberg, C.-P., Warga, R., Plegri, F., Karlstrom, R. O., Beuchle, D., Picker, A., Jiang, Y.-J., Furutani-Seiki, M., van Eeden, F. J. M., Granato, M., Haffter, P., Hammerschmidt, M., Kane, D., Kelsh, R., Mullins, M., Odenthal, J., and Nüsslein-Volhard, C. (1996). Mutations affecting development of the midline and general body shape during zebrafish embryogenesis. *Development* **123,** 129–142.

Burgess, R., Rawls, A., Brown, D., Bradley, A., and Olson, E. N. (1996). Requirement of the *paraxis* gene for somite formation and musculoskeletal patterning. *Nature* **384,** 570–573.

Chiang, C., Litingtung, Y., Lee, E., Young, K. E., Corden, J. L., Westphal, H., and Beachy, P. A. (1996). Cyclopia and defective axial patterning in mice lacking *Sonic hedgehog* gene function. *Nature* **383,** 407–413.

Concordet, J.-P., Lewis, K. E., Moore, J. W., Goodrich, L. V., Johnson, R. L., Scott, M. P., and Ingham, P. W. (1996). Spatial regulation of a zebrafish patched homologue reflects the roles of sonic hedgehog and protein kinase A in neural tube and somite patterning. *Development* **122,** 2835–2846.

Conlon, R. A., Reaume, A. G., and Rossant, J. (1995). Notch1 is required for the coordinate segmentation of somites. *Development* **121,** 1533–1545.

Currie, P. D., and Ingham, P. W. (1996). Induction of a specific muscle cell type by a hedgehog-like protein in zebrafish. *Nature* **382,** 452–455.

Devoto, S., Melançon, E., Eisen, J. S., and Westerfield, M. (1996). Identification of separate slow and fast muscle precursor cells *in vivo,* prior to somite formation. *Development* **122,** 3373–3380.

Dietrich, S., Schubert, F. R., and Lumsden, A. (1997). Control of dorsoventral pattern in the chick paraxial mesoderm. *Development* **124,** 3895–3908.

Dornseifer, P., Takke, C., and Campos-Ortega, J. A. (1997). Overexpression of a zebrafish homologue of the *Drosophila* neurogenic gene *Delta* perturbs differentiation of primary neurons and somite development. *Mech. Dev.* **63,** 159–171.

Driever, W., Solnica-Krezel, L., Schier, A. F., Neuhauss, S. C. F., Malicki, J., Stemple, D. L., Stainier, D. Y. R., Zwartkruis, F., Abdelilah, S., Rangini, Z., Beleak, J., and Boggs, C. (1996). A genetic screen for mutations affecting embryogenesis in zebrafish. *Development* **123,** 37–46.

Eisen, J. S., Myers, P. Z., and Westerfield, M. (1986). Pathway selection by growth cones of identified motoneurons in live zebra fish embryos. *Nature* **320,** 269–271.

Eisen, J. S., Pike, S. H., and Romancier, B. (1990). An identified motoneuron with variable fates in embryonic zebrafish. *J. Neurosci.* **10,** 34–43.

Ekker, M., Wegner, J., Akimenko, M. A., and Westerfield, M. (1992). Coordinate embryonic expression of three zebrafish engrailed genes. *Development* **116,** 1001–1010.

Ekker, S. C., Ungar, A. R., Greenstein, P., von Kessler, D. P., Porter, J. A., Moon, R. T., and Beachy, P. A. (1995). Patterning activities of vertebrate hedgehog proteins in the developing eye and brain. *Curr. Biol.* **5,** 944–955.

Elsdale, T., Pearson, M., and Whitehead, M. (1976). Abnormalities in somite segmentation following heat shock to *Xenopus* embryos. *J. Embyol. Exp. Morphol.* **35,** 625–635.

Fan, C. M., and Tessier-Lavigne, M. (1994). Patterning of mammalian somites by surface ectoderm and notochord: Evidence for sclerotome induction by a hedgehog homolog. *Cell* **79,** 1175–1186.

Fan, C. M., Porter, J. A., Chiang, D. T., Beachy, P. A., and Tessier-Lavigne, M. (1995). Long-range sclerotome induction by sonic hedgehog: Direct role of the amino-terminal cleavage product and modulation by the cyclic AMP signaling pathway. *Cell* **81,** 457–465.

Felsenfeld, A. L., Curry, M., and Kimmel, C. B. (1991). The fub-1 mutation blocks initial myofibril formation in zebrafish muscle pioneer cells. *Dev. Biol.* **148,** 23–30.

Fetcho, J. R. (1992). The spinal motor system in early vertebrates and some of its evolutionary changes. *Brain, Behav. Evol.* **40,** 82–97.

Goodrich, L. V., Johnson, R. L., Milenkovic, L., McMahon, J. A., and Scott, M. P. (1996). Conservation of the hedgehog/patched signaling pathway from flies to mice: Induction of a mouse patched gene by Hedgehog. *Genes Dev.* **10,** 301–312.

Granato, M., van Eeden, F. J. M., Schach, U., Trowe, T., Brand, M., Furutani-Seiki, M., Haffter, P., Hammerschmidt, M., Heisenberg, C.-P., Jiang, Y.-J., Kane, D. A., Kelsh, R. N., Mullins, M. C., Odenthal, J., and Nüsslein-Volhard, C. (1996). Genes controlling and mediating locomotion behavior of the zebrafish embryo and larva. *Development* **123,** 399–413.

Haddon, C., Smithers, L., Schneider-Maunoury, S., Coche, T., Henrique, D., and Lewis, J. (1998). Multiple *delta* genes and lateral inhibition in zebrafish primary neurogenesis. *Development* **125,** 359–370.

Haffter, P., Granato, M., Brand, M., Mullins, M. C., Hammerschmidt, M., Kane, D. A., Odenthal, J., van Eeden, F. J. M., Jiang, Y.-J., Heisenberg, C.-P., Kelsh, R. N., Furantani-Seiki, M., Vogelsang, E., Beuchle, D., Schach, U., Fabian, C., and Nüsslein-Volhard (1996). The identification of genes with unique and essential functions in the development of the zebrafish, *Danio rerio. Development* **123,** 1–36.

Halpern, M. E., Ho, R. K., Walker, C., and Kimmel, C. B. (1993). Induction of muscle pioneers and floorplate is distinguished by the zebrafish no tail mutation. *Cell* **75,** 99–111.

Halpern, M. E., Thisse, C., Ho, R. K., Thisse, B., Riggleman, B., Trevarrow, B., Weinberg, E. S.,

Postlethwait, J. H., and Kimmel, C. B. (1995). Cell-autonomous respecification of axial mesoderm in zebrafish floating head mutants. *Development* **121,** 4257–4264.

Hamilton, L. (1969). The formation of somites in *Xenopus. J. Embyol. Exp. Morphol.* **22,** 253–264.

Hammerschmidt, M., and Nüsslein-Volhard, C. (1993). The expression of a zebrafish gene homologous to *Drosophila* snail suggests a conserved function in invertebrate and vertebrate gastrulation. *Development* **119,** 1107–1118.

Hammerschmidt, M., Bitgood, M. J., and McMahon, A. P. (1996). Protein kinase A is a common negative regulator of Hedgehog signaling in the vertebrate embryo. *Genes Dev.* **10,** 647–658.

Hanneman, E., and Westerfield, M. (1989). Early expression of acetyl-choline-sterase activity in functionally distinct neurons of the zebrafish. *J. Comp. Neurol.* **284,** 350–361.

Hatta, K., Bremiller, R., Westerfield, M., and Kimmel, C. B. (1991). Diversity of expression of engrailed-like antigens in zebrafish. *Development* **112,** 821–832.

Heitzler, P., and Simpson, P. (1991). The choice of cell fate in the epidermis of *Drosophila. Cell* **64,** 1083–1092.

Hirsinger, E., Duprez, D., Jouve, C., Malapert, P., Cooke, J., and Pourquié, O. (1997). Noggin acts downstream of Wnt and Sonic Hedgehog to antagonize BMP4 in avian somite patterning. *Development* **124,** 4605–4614.

Ho, R. K., and Kane, D. A. (1990). Cell-autonomous action of zebrafish spt-1 mutation in specific mesodermal precursors. *Nature* **348,** 728–730.

Holland, L. Z., Kene, M., Williams, N. A., and Holland, N. D. (1997). Sequence and embryonic expression of the amphioxus engrailed gene (AmphiEn): The metameric pattern of transcription resembles that of its segment-polarity homolog in *Drosophila. Development* **124,** 1723–1732.

Holley, S. A., Neul, J. L., Attisano, L., Wrana, J. L., Sasai, Y., O'Connor, M. B., De Robertis, E. M., and Ferguson, E. L. (1996). The *Xenopus* dorsalizing factor Noggin ventralizes *Drosophila* embryos by preventing DPP from activating its receptor. *Cell* **86,** 607–617.

Hollyday, M., McMahon, J. A., and McMahon, A. P. (1995). Wnt expression patterns in chick embryo nervous system. *Mech. Dev.* **52,** 9–25.

Hopwood, N. D., Pluck, A., Gurdon, J. B., and Dilworth, S. M. (1992). Expression of XMyoD protein in early *Xenopus laevis* embryos. *Development* **114,** 31–38.

Hoyle, G. (1983). Muscles and their neural control. *In* "Muscles and Their Neural Control," pp. 263–311. Wiley, New York.

Hrabé Angelis, M., McIntyre, J., and Gossler, A. (1997). Maintenance of somite borders in mice requires the *Delta* homologue *Dll1. Nature* **386,** 717–721.

Jen, W.-C., Wettstein, D., Turner, D., Chitnis, A., and Kintner, C. (1997). The Notch ligand, X-Delta-2, mediates segmentation of the paraxial mesoderm in *Xenopus* embryos. *Development* **124,** 1169–1178.

Jiang, Y.-J., Brand, M., Heisenberg, C.-P., Beuchle, D., Furutani-Seiki, M., Kelsh, R. N., Warga, R. M., Granato, M., Haffter, P., Hammerschmidt, M., Kane, D. A., Mullins, M. C., Odenthal, J., van Eeden, F. J. M., and Nüsslein-Volhard, C. (1996). Mutations affecting neurogenesis and brain morphology in the zebrafish, *Danio rerio. Development* **123,** 205–216.

Johnson, R. L., Laufer, E., Riddle, R. D., and Tabin, C. (1994). Ectopic expression of Sonic hedgehog alters dorsal–ventral patterning of somites. *Cell* **79,** 1165–1173.

Johnson, S. H., Rauskolb, C., Wilson, R., Prabhakaran, B., Irvine, K. D., and Vogt, T. F. (1997). A family of mammalian *fringe* genes implicated in boundary determination and the Notch pathway. *Development* **124,** 2245–2254.

Johnson, S. L., Gates, M. A., Johnson, M., Talbot, W. S., Horne, S., Baik, K., Rude, S., Wong, J. R., and Postlethwait, J. H. (1996). Centromere-linkage analysis and consolidation of the zebrafish genetic map. *Genetics* **142,** 1277–1288.

Kelsh, R. N., Brand, M., Jiang, Y.-J., Heisenberg, C.-P., Lin, S., Haffter, P., Odenthal, J., Mullins, M. C., van Eeden, F. J. M., Furutani-Seiki, M., Granato, M., Hammerschmidt, M., Kane, D. A., Warga, R. M., Beuchle, D., Vogelsang, L., and Nüsslein-Volhard, C. (1996). Zebrafish pigmentation and the processes of neural crest development. *Development* **123,** 369–389.

Kimmel, C. B., Kane, D. A., Walker, C., Warga, R. M., and Rothman, M. B. (1989). A mutation that changes cell movement and cell fate in the zebrafish embryo. *Nature* **337,** 358–362.

Kimmel, C. B., Schilling, T. F., and Hatta, K. (1991). Patterning of body segments of the zebrafish embryo. *Curr. Top. Dev. Biol.* **25,** 77–110.

Kimmel, C. B., Ballard, W. M., Kimmel, S. R., Ullmann, B., and Schilling, T. F. (1995). Stages of embryonic development of the zebrafish. *Dev. Dyn.* **203,** 253–310.

Knapik, E. W., Goodman, A., Atkinson, O. S., Roberts, C. T., Shiozawa, M., Sim, C. U., Weksler-Zangen, S., Trolliet, M. R., Futrell, C., Innes, B. A., Koike, G., McLaughlin, M. G., Pierre, L., Simon, J. S., Vilallonga, E., Roy, M., Chiang, P.-W., Fishman, M. C., Driever, W., and Jacob, H. J. (1996). A reference cross DNA panel for zebrafish (*Danio rerio*) anchored with simple sequence length polymorphisms. *Development* **123,** 451–460.

Krauss, S., Concordet, J. P., and Ingham, P. W. (1993). A functionally conserved homology of the *Drosophila* segment polarity gene: shh is expressed in tissues with polarizing activities in zebrafish embryos. *Cell* **75,** 1431–1444.

Lehmann, R., Jiménez, F., Dietrich, U., and Campos-Ortega, J. A. (1983). On the phenotype and development of mutants of early neurogenesis in *Drosophila melanogaster. Roux's Arch. Dev. Biol.* **192,** 62–74.

Long, Q., Meng, A., Wang, H., Jessen, J. R., Farrell, M. J., and Lin, S. (1997). GATA-1 expression pattern can be recapitulated in living transgenic zebrafish using GFP reporter gene. *Development* **124,** 4105–4111.

Marcelle, C., Stark, M. R., and Bronner-Fraser, M. (1997). Coordinate actions of BMPs, Wnts, Shh, and Noggin mediate patterning of the dorsal somite. *Development* **124,** 3955–3963.

Marigo, V., and Tabin, C. J. (1996). Regulation of patched by sonic hedgehog in the developing neural tube. *Proc. Natl. Acad. Sci. U.S.A.* **93,** 9346–9351.

Melby, A. E., Warga, R. M., and Kimmel, C. B. (1996). Specification of cell fates at the dorsal margin of the zebrafish gastrula. *Development* **122,** 2225–2237.

Meng, A., Tang, H., Ong, B. A., Farrell, M. J., and Lin, S. (1997). Promoter analysis in living zebrafish embryos identifies a cis-acting motif required for neuronal expression. *Proc. Natl. Acad. Sci. U.S.A.* **94,** 6267–6272.

Moens, C. B., Yan, Y.-L., Appel, B., Force, A. G., and Kimel, C. B. (1996). *valentino:* A zebrafish gene required for normal hindbrain segmentation. *Development* **122,** 3981–3990.

Morin-Kensicki, E. M., and Eisen, J. S. (1997). Sclerotome development and peripheral nervous system segmentation in embryonic zebrafish. *Development* **124,** 159–167.

Müller, M., Weiszäcker, E., and Campos-Ortega, J. A. (1996). Expression domains of a zebrafish homologue of the *Drosophila* pair-rule gene *hairy* correspond to primordia of alternating somites. *Development* **122,** 2071–2078.

Münsterberg, A., and Lassar, A. (1995). Combinatorial signals from the neural tube, floor plate and notochord induce myogenic bHLH gene expression in the somite. *Development* **121,** 651–660.

Myers, P. Z., Eisen, J. S., and Westerfield, M. (1986). Development and axonal outgrowth of identified motoneurons in the zebrafish. *J. Neurosci.* **6,** 2278–2289.

Neubuser, A., Koseki, H., and Balling, R. (1995). Characterization and developmental expression of Pax9, a paired-box-containing gene related to Pax1. *Dev. Biol.* **170,** 701–716.

Nornes, S., Mikkola, I., Krauss, S., Delghand, M., Perander, M., and Johansen, T. (1996). Zebrafish Pax9 encodes two proteins with distinct C-terminal transactivating domains of different potency negatively regulated by adjacent N-terminal sequences. *J. Biol. Chem.* **271,** 26914–26923.

Odenthal, J., Haffter, P., Vogelsang, E., Brand, M., van Eeden, F. J. M., Furutani-Seiki, M., Granato, M., Hammerschmidt, M., Heisenberg, C.-P., Jiang, Y.-J., Kane, D. A., Kelsh, R. N., Mullins, M. C., Warga, R. M., Allende, M. L., Weinberg, E. S., and Nüsslein-Volhard, C. (1996). Mutations affecting the formation of the notochord in the zebrafish, *Danio rerio. Development* **123,** 103–115.

Oka, C., Nakano, T., Wakeham, A., de la Pompa, J. L., Mori, C., Sakai, T., Okazaki, S., Kawaichi, M., Shiota, K., Mak, T. W., and Honjo, T. (1995). Disruption of the mouse RBP-J Kappa results in early embryonic death. *Development* **121,** 3291–3301.

Palmeirim, I., Henrique, D., Ish-Horowicz, D., and Pourquié, O. (1997). Avian *hairy* gene expression identifies a molecular clock linked to vertebrate segmentation and somitogenesis. *Cell* **91,** 639–648.

Pankratz, M. J., and Jäckle, H. (1990). Making stripes in the *Drosophila* embryo. *Trends Genet.* **6,** 287–292.

Parr, B. A., Shea, M. J., Vassileva, G., and McMahon, A. P. (1993). Mouse *Wnt* genes exhibit discrete domains of expression in the early embryonic CNS and limb buds. *Development* **119,** 247–261.

Postlethwait, J. H., Johnson, S. L., Midson, C. N., Talbot, W. S., Gates, M., Ballinger, E. W., Africa, D., Andrews, R., Carl, T., Eisen, J. S., Horne, S., Kimmel, C. B., Hutchinson, M., Johnson, M., and Rodriguez, A. (1994). A genetic linkage map for the zebrafish. *Science* **264,** 699–703.

Pourquié, O., Fan, C. M., Coltey, M., Hirsinger, E., Watanabe, Y., Bréant, C., Francis-West, P., Brickell, P., Tessier-Lavigne, M., and Le Douarin, N. M. (1996). Lateral and axial signals involved in avian somite patterning: A role for BMP4. *Cell* **84,** 461–471.

Roberts, A. (1989). The neurons that control axial movements in a frog embryo. *Am. Zool.* **29,** 53–63.

Saga, Y., Hata, N., Koseki, H., and Taketo, M. M. (1997). *Mesp2:* A novel mouse gene expressed in the presegmented mesoderm and essential for segmentation initiation. *Genes Dev.* **11,** 1827–1839.

Schauerte, H. E., van Eeden, F. J. M., Fricke, C., Odenthal, J., Strähle, U., and Haffter, P. (1998). Sonic Hedgehog is not required for floor plate induction in the zebrafish. *Development* **125,** 2983–2993.

Schulte-Merker, S., van Eeden, F. J. M., Halpern, M. E., Kimmel, C. B., and Nüsslein-Volhard, C. (1994). *no tail* (*ntl*) is the zebrafish homologue of the mouse *T* (*Brachyury*) gene. *Development* **120,** 1009–1015.

Simons, G., van der Lee, T., Diergaarde, P., van Daelen, R., Groenendijk, J., Frijters, A., Büschges, R., Hollricher, K., Töpsch, S., Schulze-Lefert, P., Salamini, F., Zabeau, M., and Vos, P. (1997). AFLP-Based fine mapping of the *Mlo* gene to a 30-kb DNA segment of the barley genome. *Genomics* **44,** 61–70.

Sommer, R. J., and Tautz, D. (1993). Involvement of an orthologue of the *Drosophila* pair-rule gene hairy in segment formation of the short germ-band insect of *Tribolium* (Coleoptera). *Nature* **361,** 448–450.

Sonica-Krezel, L., Stemple, D. L., Mountcastle-Shah, E., Rangini, Z., Neuhauss, S. C. F., Malicki, J., Schier, A. F., Stainier, D. Y. R., Zwartkruis, F., Abdelilah, S., and Driever, W. (1996). Mutations affecting cell fates and cellular rearrangements during gastrulation in zebrafish. *Development* **123,** 67–80.

Stemple, D. L., Solnica-Krezel, L., Zwartkruis, F., Neuhauss, S. C. F., Schier, A. F., Maliki, J., Stainier, D. Y. R., Abdelilah, S., Rangini, Z., Mountcastle-Shah, E., and Driever, W. (1996). Mutations affecting development of the notochord in zebrafish. *Development* **123,** 117–128.

Stern, C. D., and Keynes, R. J. (1987). Interactions between somite cells: The formation and maintenance of segment boundaries in the chick embryo. *Development* **99,** 261–272.

Stern, H. M., Brown, A. M., and Hauschka, S. D. (1995). Myogenesis in paraxial mesoderm: Preferential induction by dorsal neural tube and by cells expressing Wnt-1. *Development* **121,** 3675–3686.

Talbot, W. S., Trevarrow, B., Halpern, M. E., Melby, A. E., Farr, G., Postlethwait, J. H., Jowett, T., Kimmel, C. B., and Kimelman, D. (1995). A homeobox gene essential for zebrafish notochord development. *Nature* **378,** 150–157.

Thisse, C., Thisse, B., Schilling, T. F., and Postlethwait, J. H. (1993). Structure of the zebrafish *snail1* gene and its expression in wild-type, spadetail and no tail mutant embryos. *Development* **119,** 1203–1215.

van Eeden, F. J. M. (1997). Genetic analysis of somite formation and patterning in the zebrafish, *Danio rerio.* Ph. D. Dissertation, Eberhard-Karls Universität Tübingen.

van Eeden, F. J. M., Granato, M., Schach, U., Brand, M., Furutani-Seiki, M., Haffter, P., Hammerschmidt, M., Heisenberg, C.-P., Jiang, Y.-J., Kane, D. A., Kelsh, R. N., Mullins, M. C., Odenthal,

J., Warga, R. M., Allende, M. L., Weinberg, E. S., and Nüsslein-Volhard, C. (1996). Mutations affecting somite formation and patterning in the zebrafish *Danio rerio. Development* **123,** 153–164.

van Eeden, F. J. M., Holley, S. A., Haffter, P., Campos-Ortega, J., and Nüsslein-Volhard, C. (1998). Zebrafish segmentation and pair-rule patterning. *Developmental Genetics* **23,** 65–76.

van Raamsdonk, W., Pool, C. W., and te Kronnie, G. (1978). Differentiation of muscle fiber types in the teleost *Brachydanio rerio. Anat. Embryol. Berl.* **153,** 137–155.

van Raamsdonk, W., van't Veer, L., Veeken, K., te Kronnie, T., and de Jager, S. (1982). Fiber type differentiation in fish. *Mol. Physiol.* **2,** 31–47.

Vos, P., Hogers, R., Bleeker, M., Reijans, M., van de Lee, T., Hornes, M., Frijters, A., Pot, J., Peleman, J., Kuiper, M., and Zabeau, M. (1995). AFLP: A new technique for DNA fingerprinting. *Nucleic Acids Res.* **23,** 4407–4414.

Walker, C., and Streisinger, G. (1983). Induction of mutations by gamma-rays in pregonial germ cells of zebrafish embryos. *Genetics* **103,** 125–136.

Waterman, R. E. (1969). Development of the lateral musculature in the teleost, *Brachydanio rerio:* A fine structural study. *Am. J. Anat.* **125,** 457–493.

Weinberg, E. S., Allende, M. L., Kelly, C. S., Abdelhamid, A., Murakami, T., Anderman, P., Doerre, O. G., Grunwald, D. J., and Riggleman, B. (1996). Developmental regulation of zebrafish *MyoD* in wild-type, *no tail* and *spadetail* embryos. *Development* **122,** 271–280.

Westin, J., and Lardelli, M. (1997). Three novel *Notch* genes in zebrafish: Implications for vertebrate *Notch* gene evolution and function. *Dev. Genes Evol.* **207,** 51–63.

Wong, P. C., Zheng, H., Chen, H., Becher, M. W., Sirinathsinghji, D. J. S., Trumbauer, M. E., Chen, H. Y., Price, D. L., Van der Ploeg, L. H. T., and Sisodia, S. S. (1997). Presenilin 1 is required for *Notch1* and *Dll1* expression in the paraxial mesoderm. *Nature* **387,** 288–292.

Zhang, J., Talbot, W. S., and Schier, A. F. (1998). Positional cloning identifies zebrafish *one-eyed pinhead* as a permissive EGF-related ligand required during gastrulation. *Cell* **92,** 241–251.

Zimmerman, L., Dejesusescobar, J., and Harland, R. (1996). The Spemann organizer signal Noggin binds and inactivates bone morphogenetic protein-4. *Cell* **86,** 599–606.

9

Rostrocaudal Differences within the Somites Confer Segmental Pattern to Trunk Neural Crest Migration

Marianne Bronner-Fraser
Division of Biology and Beckman Institute
California Institute of Technology
Pasadena, California 91125

I. Introduction

Neural crest cells are one of the most migratory and pleiotropic of embryonic cells types. These cells originate within the central nervous system, but emigrate from the neural tube shortly after its closure. They subsequently migrate along well-characterized pathways to populate numerous and diverse derivatives ranging from sensory and autonomic ganglia of the peripheral nervous system to the cartilage and bone of the face.

Initially, premigratory neural crest cells in the midline of the dorsal neural tube are columnar epithelial cells that appear morphologically similar to other neuroepithelial cells. Preceding their migration, these midline cells undergo changes in cell–cell adhesion, including a decrease in N-cadherin expression and a concomitant increase in cad-7 expression (Nakagawa and Takeichi, 1995). Once neural crest cells leave the neural tube, they initially enter a cell-free space filled

Current Topics in Developmental Biology, Vol. 47

0070-2153/00 $30.00

with extracellular matrix molecules. Subsequently, they interact with and migrate through adjacent tissues. The neural tube, epidermis, somites, notochord, and dorsal aorta are structures on or near neural crest migratory pathways. Thus, it is likely that interactions with these tissues or their associated extracellular matrices influence the migration of neural crest cells.

In the trunk region of the amniote embryo, neural crest cells migrate away from the neural tube and follow two primary pathways: a dorsolateral pathway between the ectoderm and somite; a ventral pathway through the rostral half of each somite (Fig. 1A) (Rickmann *et al.,* 1985; Bronner-Fraser, 1986). The cells following the dorsolateral stream give rise to melanocytes. The cells of the ventral stream localize in the sensory and sympathetic ganglia, as well as adrenomedullary sites, where they differentiate into neuronal, supportive, or adrenal chromaffin cells.

Development of the neural crest is intimately related to development of the somites. The somites bud off from the segmental plate with one pair of somites forming approximately every 90 min in the chick. Formation of the somites precedes neural crest emigration from the neural tube. The first neural crest cells to leave the neural tube do so about three to five somites rostral to the most recently formed somites. At the level of somite V, the epithelial somites begin to undergo an epithelial to mesenchymal conversion. The dorsal portion of the somites remains epithelial, forming the dermomyotome (precursors to muscle and dermis), and the ventral part becomes mesenchymal, forming the sclerotome (precursor to vertebrae). Concomitant with this epithelial to mesenchymal conversion, neural crest cells first enter the rostralmost part of the somite. They subsequently migrate ventrally and intermingle with cells of the rostral half of the sclerotome. This pattern is reiterated throughout the brachial and truck somites.

II. Rostrocaudal Polarity of the Somites Influences Trunk Neural Crest Migration

A hallmark of the developing peripheral nervous system is its inherent segmentation. After neural crest cells migrate from the neural tube and through the somites, they condense to form segmentally arranged sensory and sympathetic ganglia. For each somite, a single sensory and sympathetic ganglion forms. This exquisite and

---

Figure 1 Neural crest migration through the rostral half of the sclerotome is controlled by information within the somites. Longitudinal sections through the truck region of a 2.5-day chick embryo stained with the neural crest marker, HNK-1. (A) A normal embryo illustrating the segmental migration of neural crest cells through the rostral (r) half of each somitic sclerotome, but their absence from the caudal (c) half sclerotome. (B) An embryo 1 day after 180° inversion of the segmental plate. The somites appear morphologically normal but their rostrocaudal polarity is reversed. Neural crest cells now migrate to the portion of the somite that was originally rostral, but is inverted with respect to the rest of the embryo. The small somite (arrow) indicates the border of the rotation. nt, Neural tube. (Modified from Bronner-Fraser and Stern, 1991.)

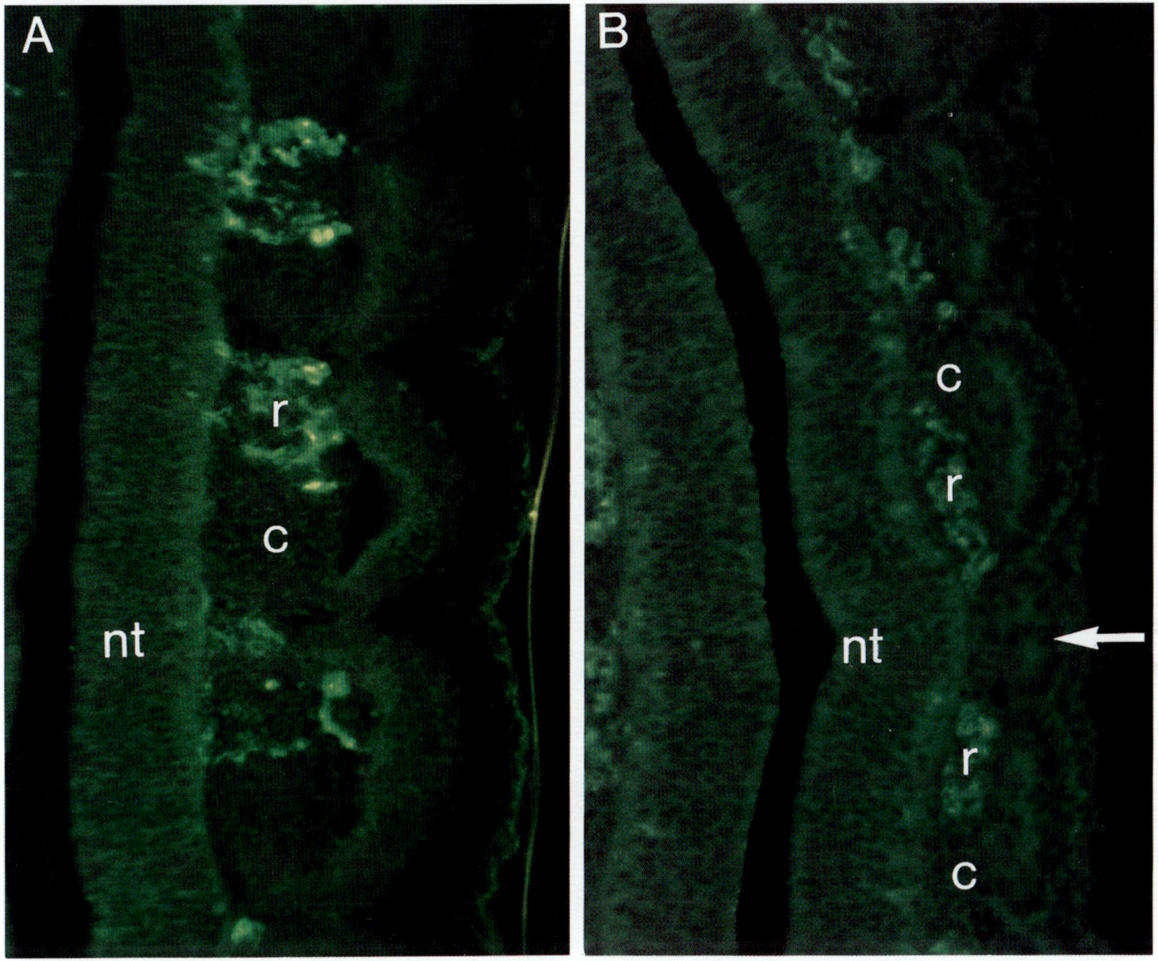
A
r
c
nt
B
c
r
nt
r
c

reproducible pattern suggests the presence of some inherent segmental information in the embryo that is responsible for segmental migration and gangliogenesis of neural crest cells. This could be manifested in the form of inherent cues within the neural tube that direct neural crest cells to migrate in a metameric pattern. Alternatively, the tissues through which neural crest cells migrate may contain patterning information that results in segmental migration. For example, there may be permissive cues in the rostral half of the sclerotome and/or inhibitory cues in the caudal half of the sclerotome.

The relationship between neural crest cells and their surrounding tissues has been explored by manipulating the neural tube and/or somites in a series of grafting experiments. Removal of the somites results in the failure of segmentation of neural crest-derived ganglia, suggesting that somites are necessary for the segmentation of the neural crest. On the other hand, inversion of the segmental plate in the rostrocaudal dimension reverses the pattern of neural crest migration (Bronner-Fraser and Stern, 1991), such that neural crest cells migrate through the caudal (original rostral) halves of the rotated sclerotomes (Fig. 1B). Motor axons, which also traverse the rostral sclerotome, exhibit similar behavior to neural crest cells after segmental plate rotation. This suggests that the information necessary to guide neural crest cells and motor axons is intrinsic to the somites. Furthermore, these experiments show that the rostrocaudal polarity of the somites is already established at the segmental plate stage. In contrast, the dorsoventral polarity of the somites is subject to interactions with adjacent tissues such as the ectoderm and ventral neural tube/notochord (Fan and Tessier-Lavigne, 1994; Munsterberg *et al.*, 1995; Munsterberg and Lassar, 1995).

Other experimental manipulations demonstrate that the caudal half somite is inhibitory whereas the rostral half somite is permissive for neural crest migration and motor axon guidance. Construction of somites with all rostral or all caudal half sclerotomes results in the absence of segmentation or the absence of migration, respectively, for both neural crest cells and motor axons (Stern and Keynes, 1987). Taken together, these findings demonstrate the segmental migration of both neural crest cells and motor axons is due to cues inherent to the somite. These cues may be caused by attractive cues in the rostral half sclerotome, inhibitory cues in the caudal half sclerotome, or a combination of both. The finding that somites containing all caudal half sclerotomes cannot support neural crest migration strongly suggests that some of the guidance cues are inhibitory.

III. Patterns of Extracellular and Cell Surface Molecules within the Somites

As a first step in identifying candidate molecules that may be involved in neural crest cell migration, it is necessary to look for those expressed along neural crest pathways as well as differentially within the somites. Numerous glycoproteins, proteoglycans, and glycosaminoglycans have been described within the somites through which neural crest cells and axons navigate, including fibronectin (Newgreen and Thiery, 1980), laminin (Krotoski *et al.*, 1986), tenascin (Tan *et al.*, 1987), and various collagens (Duband and Thiery, 1987; Perris *et al.*, 1991a). Of the proteoglycans present within the embryo, heparan sulfate proteoglycans (Perris *et al.*, 1991b) appear on neural crest cell pathways, while some forms of chondroitin sulfate proteoglycans are generally present in regions from which neural crest cells are absent (Tan *et al.*, 1987; Perris *et al.*, 1991b), such as the perinotochordal space.

Of particular interest are molecules that are differentially distributed within the somites. Peanut lectin binding molecules (Stern and Keynes, 1987; Norris *et al.*, 1989), chondroitin sulfate proteoglycans (Oakley and Tosney, 1991), and T-cadherin (Ranscht and Bronner-Fraser, 1991) are all present in the caudal half of the sclerotome both during and following neural crest cell migration and motor axon elongation. In addition, ephrins have recently been identified in the caudal half sclerotome (Krull *et al.*, 1997; Wang and Anderson, 1997; see below). Molecules distributed selectively in the rostral portion of the sclerotome include butyrlcholinesterase (Layer *et al.*, 1988) and an antigen that appears to be a nerve growth factor (NGF) receptor (Tanaka *et al.*, 1989). Other matrix components including fibronectin, laminin, and collagen IV remain uniformly distributed throughout the sclerotome. Hence, differential distribution of permissive and nonpermissive extracellular matrix (ECM) molecules, together with changes in the ability of cells to interact with the matrix, may determine the segmented migratory pattern of neural crest cells in the trunk.

Neural crest cells themselves may alter the extracellular matrix through which they migrate. For example, collagen type III is initially distributed uniformly through the somite. At late stages of migration, however, it is confined to the caudal half sclerotome and is reorganized by neural crest cells (Fig. 2) (Perris *et al.*, 1991a). A similar reorganization is observed with tenascin (Stern *et al.*, 1989) and some proteoglycans (Perris *et al.*, 1991b). These findings highlight the fact that it is important to consider the timing of a molecule's distribution before assigning a causative role in an event like cell migration. Furthermore, one cannot tell by distribution alone whether the localization of a particular molecule reflects a consequence or cause of a migratory event.

Recently, some of the cell surface receptors on neural crest cells have been identified. For example, integrins mediate neural crest cell binding to fibronectin and laminin substrates (Lallier and Bronner-Fraser, 1991). Addition of antibodies against chick β_1 integrin (Bronner-Fraser, 1985) to neural crest cells *in vitro* causes rapid rounding and detachment of cells on fibronectin or laminin substrates. These experiments suggest that integrins may function as important receptors for trunk neural crest migration.

IV. Eph-Family Receptor Tyrosine Kinases and Their Ligands

Eph receptor tyrosine kinases and their ligands display intriguing patterns of expression in the developing nervous system. Their distributions are consistent with

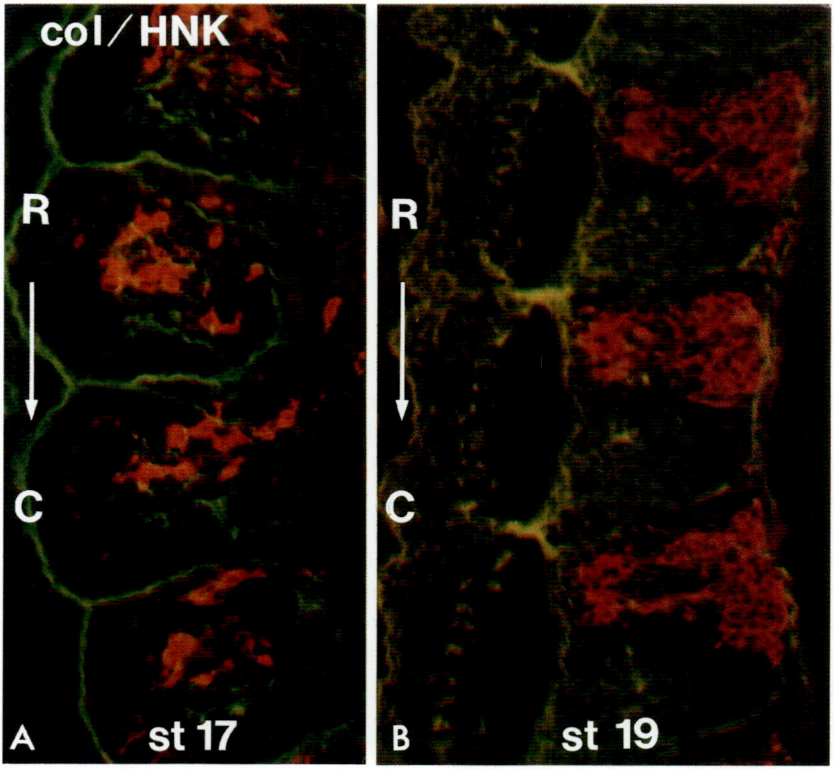

Figure 2 Some extracellular matrix molecules become reorganized during the course of neural crest migration. Longitudinal sections through the trunk showing the distribution of collagen III (green) relative to neural crest cells (red). (A) During early stages of neural crest migration (stage 17), neural crest cells migrate through the rostral half of the somite, and collagen is distributed uniformly throughout the sclerotome. (B) During advanced stages of neural crest migration (stage 19), neural crest cells continue to migrate through the rostral half of the sclerotome; collagen type III is now present only in the caudal half sclerotome. R → C indicates rostral to caudal. (Modified from Perris *et al.,* 1991a.)

potentially important roles in early neural patterning (Brambilla and Klein, 1995; Becker *et al.,* 1994; Flenniken *et al.,* 1996; Gale *et al.,* 1996a; Henkemeyer *et al.,* 1994). Eph receptors are functionally divided into two subclasses, Eph A receptors, which interact primarily with a Glycosylphosphatidylinositol-linked subclass of ligands, and Eph B receptors, which interact primarily with a transmembrane subclass of ligands (Gale *et al.,* 1996). The ligands require membrane attachment and multimerization to effectively activate receptor tyrosine phosphorylation (Davis *et al.,* 1994). Eph receptor–ligand interactions have been implicated in axonal patterning events, such as the topographic organization of the retinotectal projection (Cheng *et al.,* 1995; Drescher *et al.,* 1995; Holash and Pasquale, 1995; Kenny *et al.,*

1995). Other family members are expressed in the developing forebrain and hindbrain (Macdonald *et al.,* 1994; Henkemeyer *et al.,* 1994; Becker *et al.,* 1994; Bergemann *et al.,* 1995; Irving *et al.,* 1996) and appear to play a role in their regional organization (Xu *et al.,* 1995, 1996).

Because they are potential mediators of inhibitory interactions, members of the Eph family could mediate exclusion of neural crest cells from the caudal sclerotome. To determine whether certain Eph receptors and their transmembrane ligands exhibit the proper spatiotemporal distribution to be involved in guiding neural crest cells, avian embryos were stained with receptor–Fc or ligand–Fc fusion proteins (Davis *et al.,* 1994; Gale *et al.,* 1996a). Because ligand–receptor interactions among members of the subclass are known to be highly promiscuous, the ephrin-B–Fc has the potential to bind to and detect all members of the corresponding receptor subclass. Similarly, a representative receptor–Fc fusion protein will bind to and detect all members of the ligand subclass. Although staining with these reagents cannot be used to identify the distribution of single family members, it has the significant advantage of demonstrating the distribution of an entire subclass of relevant receptors or ligands (Gale *et al.,* 1996a).

Staining with ephrin-B1–Fc reveals the presence of Eph-family receptors localized on neural crest cells as they emigrate from the neural tube and travel through the rostral sclerotome (Fig. 3A). Additional labeling was observed on rostral sclerotomal cells and the dermomyotome. The cognate Eph-family ligands are present in the caudal half of the somite, from which neural crest cells are normally absent (Fig. 3B). The ligands first appear in the young epithelial somites, before their overt differentiation into sclerotome and dermomyotome, and clearly preceding the entry of the first neural crest cells (Krull *et al.,* 1997).

To identify specific ligands and receptors, we have used *in situ* hybridization with avian-specific probes. The receptor tyrosine kinase EphB3 was observed in the rostral regions of mature somites, including both neural crest and sclerotomal cells, whereas ephrin B1 ligand was found in the caudal sclerotome (Krull *et al.,* 1997). In the developing rat, Wang and Anderson (1997) find that EphB1 is the primary receptor expressed by neural crest and that the caudal sclerotome expresses both ephrin-B1 and ephrin-B2. These data suggest the interesting possibility that different species may use different individual family members to subserve the same function. However, there appears to be broad conservation within the subclass.

Because ephrins and their receptors are expressed from early times in somite formation through the development of the dermomyotome and sclerotome, it is interesting to speculate that Eph receptor–ligand interactions may play a variety of roles as a function of developmental time. One possibility is that these molecules guide several different interactions within the developing somite, playing a role first in the budding of somites from the segmental plate, later in organizing the rostrocaudal polarity within the somites, and, finally, in guiding the migration of the neural crest cells through the somites.

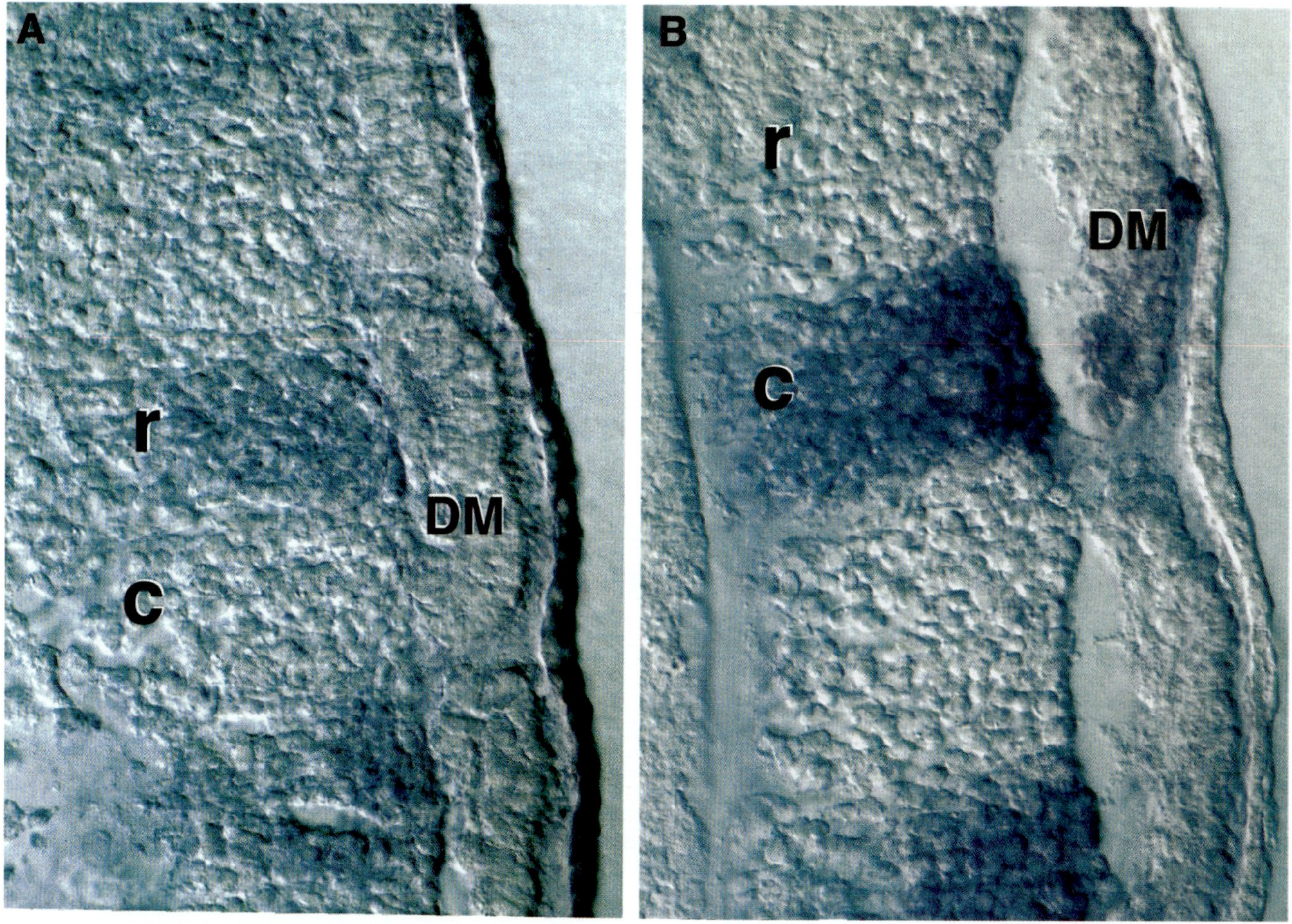
A
r
c
DM
B
r
c
DM

The above findings demonstrate that Eph B3 receptor and the cognate ephrin B1 ligand are good candidates for inhibiting neural crest migration. An *in vitro* assay was used to test whether ephrins are sufficient to inhibit neural crest movement. Migrating neural crest cells, plated on a patterned two-dimensional substrate composed of alternating lanes of multimeric ephrin B1 versus fibronectin alone, avoided lanes containing ephrin B1 and instead migrated on lanes composed of fibronectin alone (Fig. 4A; Krull *et al.*, 1997). This demonstrates that the ligand is sufficient to inhibit neural crest migration. Interestingly, when soluble ligand was added to the alternating lanes, neural crest cells migrated on both ephrin- and fibronectin-containing lanes (Fig. 4B). Furthermore, ligand presented in dimeric form (ephrin-B1–Fc) both competitively inhibited and caused low levels of signal transduction, whereas monomeric ligand (ephrin B1-myc) functioned only as a competitive inhibitor. These *in vitro* results demonstrate: (1) that substrate bound ligand can directly inhibit migrating neural crest cells; (2) that ligand must be presented discontinuously to cause a segmental pattern of migration; and (3) that soluble ligand can be used as a competitive inhibitor of substrate-bound ligand action. The findings demonstrate that Eph receptors and their cognate ligands have the correct spatiotemporal distribution to impact neural crest migration and, in culture, have the ability to block their migration. It remained necessary to test the functional significance of these molecules in living tissues. This has been facilitated by techniques making it possible to both monitor and perturb neural crest cells as they migrate through intact somites.

V. Dynamic Analysis of Trunk Neural Crest Migration

Attempts to define the molecular bases of the cues that affect segmental migration of neural crest cells have been hampered by the lack of an accessible assay system that permits neural crest migration to be observed directly *in vivo*. To examine dynamic aspects of trunk neural crest migration over time and to perturb candidate guiding molecules, a novel explant system was developed that allows visualization of neural crest migration in a three-dimensional living preparation (Krull *et al.*, 1995). Trunk regions of the embryo, placed in explant culture, continue to develop fairly normally for 2 days or more. In particular, the somites develop normally, forming a dermomyotome and sclerotome. Furthermore, these somites express appropriate manifestations of rostrocaudal polarity, including the expression of

Figure 3 Eph receptors and ligands have reciprocal patterns in the developing sclerotome. Longitudinal sections through the trunk of 2.5-day chick embryos stained with ephrin–Fc (A) or EphB–Fc (B) reveal the presence of Eph receptors in the rostral half sclerotome (A) and ephrin ligands in the caudal half sclerotome (B). DM, Dermomyotome; r, rostral; c, caudal. (Modified from Krull *et al.*, 1997.)

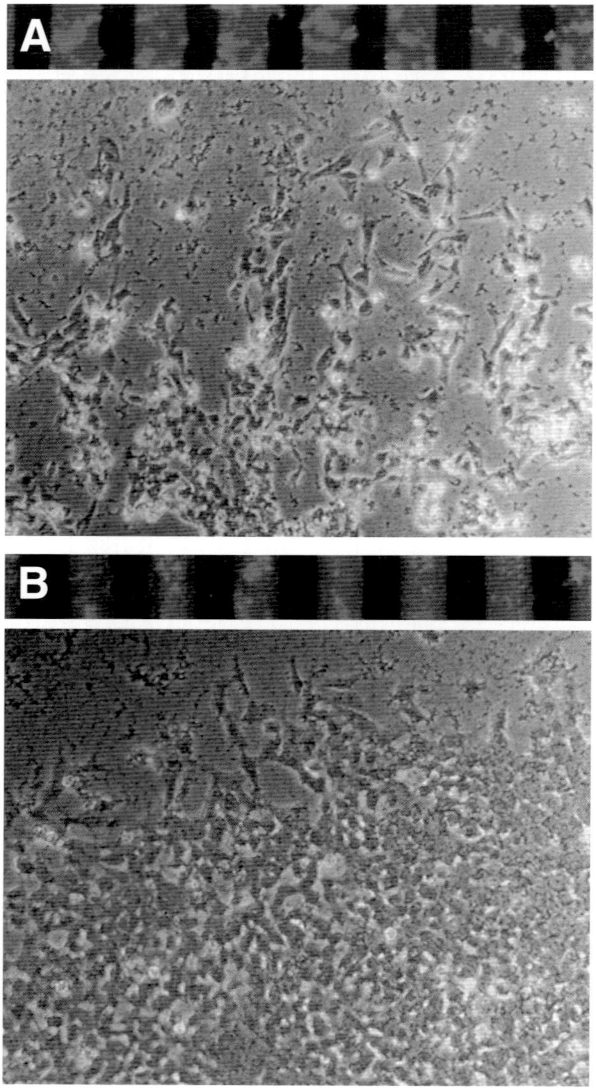

Figure 4 Alternating lanes of ephrin-B1–Fc *in vitro* inhibit neural crest migration. (A) When migrating neural crest cells are exposed to a patterned two-dimensional substrate composed of alternating lanes of fibronectin versus antibody-aggregated ephrin-B1–Fc/fibronectin, they migrated selectively on the fibronectin lanes, avoiding the lanes containing multimeric ephrin-B1–Fc. (B) When soluble ephrin-B1–Fc or –myc was added to the culture medium of neural crest migrating on alternating lanes of fibronectin versus antibody-aggregated ephrin-B1–Fc/fibronectin, they migrated equally well on all lanes. This demonstrates that soluble ligand functions as a competitive inhibitor of multimeric ephrin-B1. (Modified from Krull *et al.,* 1997.)

T-cadherin and ephrins in the caudal half sclerotome and Eph receptors in the rostral half sclerotome (Krull *et al.,* 1995, 1997).

The trajectories of migrating neural crest cells can be followed in explants either by injecting the lipophilic dye, DiI, into the neural tube lumen prior to emigration or using immunocytochemistry with the HNK-1 antibody. In normal explants, neural crest cells migrate through the somites in their typical segmental pattern. Using low-light-level videomicroscopy to follow individual cell movements, the mean rates of migration were 10–14 μm per hour, similar to those inferred from analysis of static images. Frequently, the neural crest cells migrated in close-knit groups of two to four cells. The migratory trajectories in individual neural crest cells are often complex, with cells migrating in an episodic fashion encompassing forward, backward, and lateral movements (Krull *et al.,* 1995). The explant system offers several advantages for studying the role of receptor–ligand interactions in neural crest patterning, ranging from the ability to visualize cells as they migrate through their normal three-dimensional embryonic environment, to the ease with which blocking reagents can be applied.

VI. Inhibitory Interactions between Eph Receptors and Their Ligands Contribute to the Segmental Pattern of Trunk Neural Crest Migration

Both the complementary distribution of Eph receptors and ligands and the ability of the ligand to directly inhibit neural crest migration in tissue culture indicate that Eph receptor–ligand interactions are strong candidates for mediating the segmental migration of neural crest cells *in vivo.* The next important step was to test the function of these interactions in living explant tissue using time-lapse videomicroscopy with and without exogenous soluble ligand (Krull *et al.,* 1997). Exposure to ephrin-B1 ligands disrupted the segmental pattern of neural crest migration (Fig. 5). Neural crest cells were present within both rostral and caudal halves of each somitic sclerotome, suggesting that the normal inhibitory action of the caudal sclerotome on neural crest migration had been blocked. In contrast, control explants never contained cells in the caudal somite.

Time-lapse cinematography permitted the migration pathways of the cells to be analyzed in detail and revealed major defects in the trajectories of the treated neural crest cells. Neural crest cells migrated through both the rostral and caudal halves of the somites after treatment with ephrin-B1–Fc or ephrin-B1–myc, invading the normal neural crest-free zone in the caudal somite. Cells were able to cross what would normally serve as a boundary, migrating both rostrocaudally and caudorostrally across the somite midline. In addition, both rostral and caudal cells became more erratic in their migration trajectories in the presence of ephrin-B1–Fc, moving in circles or migrating backward from the periphery to the neural

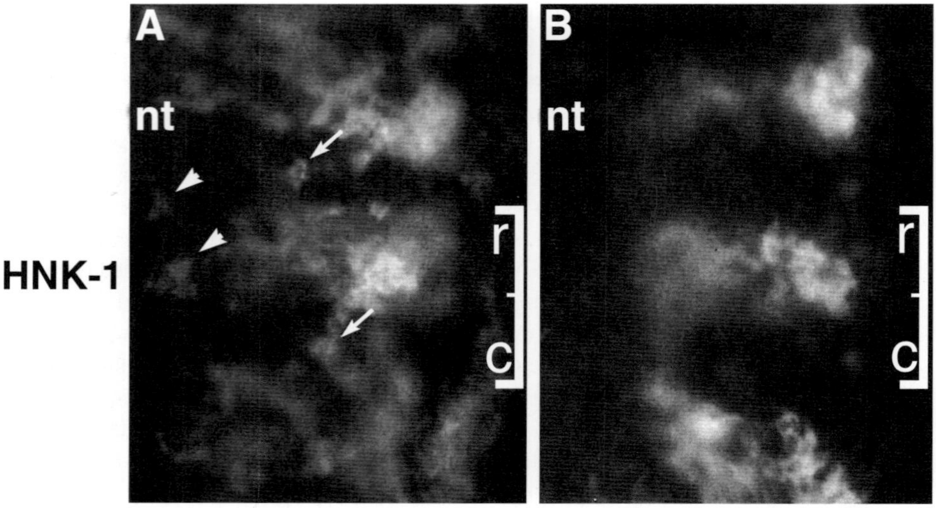

Figure 5 Eprhin-B1–Fc disrupts the segmental pattern of neural crest migration in living explants. (A) Ephrin-B1–Fc or –myc was added to trunk explants and static patterns of neural crest migration were visualized via HNK-1 antibody staining. Neural crest cells enter both rostral and caudal half somites in the presence of exogenous ephrin-B1. (B) In Fc-treated control explants, neural crest migrate in their typical segmental pattern through the rostral half of the somite. Brackets indicate the rostrocaudal extent of a single somite, with the midline representing the intrasomitic border. r, Rostral; c, caudal; nt, neural tube. (Modified from Krull *et al.,* 1997.)

tube in many cases. However, exogenous ligand treatment clearly did not inhibit the ability of the neural crest cells to migrate. In fact, cells within ligand–Fc-treated explants migrated at rates similar to cells in Fc-treated control explants. Furthermore, deficits in the organization of migratory behavior by both rostral and caudal cells were apparent, suggesting that ephrin-B1–Fc disrupted the coordinated movement of neural crest cells. These findings stress the utility of observing cells as they migrate *in situ.*

These results suggest that Eph family receptor tyrosine kinases and their transmembrane ligands are involved in interactions between neural crest and somite cells, mediating an inhibitory activity necessary to constrain neural precursors to specific territories in the developing nervous system. Our functional studies, performed on neural crest cells migrating *in situ,* are in good agreement with complementary studies using different assay systems. Wang and Anderson (1997) have demonstrated an inhibitory influence of receptor tyrosine kinase ligands ephrin-B1 and ephrin-B2 on both trunk neural crest and motor axons *in vitro.* Whereas we have used nonclustered ligands as competitive inhibitors, Wang and Anderson used preclustered ligand–Fc fusion proteins as receptor agonists. Our loss-of-function experiments *in situ* and their gain-of-function manipulations *in vitro* yield complementary results, with the former demonstrating a disruption of the metameric pattern of neural crest migration in the presence of excess ligand and the latter showing a restriction of migration in response to an inhomogeneous distribution of ligand. Furthermore, related studies in the developing hindbrain of *Xenopus* demonstrate that the receptor tyrosine kinases Eph A4 and Eph B2 and the ligand ephrin-B2 influence the segmental organization of branchial arch neural crest migration (Smith *et al.,* 1997). An emerging theme is that interactions between Eph receptors and ligands inhibit the migration of neural crest cells *in vivo* and *in vitro.*

An interesting question is whether the disruption in the pattern of neural crest migration is direct or indirect. We noted that Eph receptors are present both on neural crest and rostral sclerotomal cells. Neural crest cells that emerge from the neural tube apposing the caudal half of the sclerotome turn either rostrally or caudally on contacting the adjacent somite, as if being inhibited from this region. Our data raise the intriguing possibility that direct interactions between Eph receptors on neural crest cells and ligands on caudal half sclerotomal cells restrict the immediate entrance of neural crest cells. Once neural crest cells enter the rostral half sclerotome, they mingle with somite cells that express Eph receptors. The inability of neural crest cells to cross from the rostral to caudal half territory could be caused directly by repulsion of neural crest cells at the intrasomitic border. This idea is supported by our *in vitro* experiments demonstrating that Eph ligands can directly inhibit migrating neural crest cells (Krull *et al.,* 1997).

VII. Perturbation of Peanut Lectin Binding Molecules Alters the Segmental Pattern of Neural Crest Migration

The lectin peanut agglutinin (PNA) stains cells within the caudal half sclerotome (Norris *et al.,* 1989; Oakley and Tosney, 1991). A functional role for a peanut lectin-binding glycoprotein fraction derived from caudal sclerotome is suggested by experiments in which liposomes containing these molecules inhibit sensory growth cones *in vitro* (Davis *et al.,* 1990) and the finding that addition of exogenous PNA disrupts the segmental migration of neural crest cells (Krull *et al.,* 1995). Treatment of trunk explants with the lectin peanut agglutinin altered the patterning of neural crest cell migration. In the presence of PNA, neural crest cells invaded both rostral and caudal sclerotomes; rostrally located neural crest moved in a generally unperturbed fashion, while caudally located neural crest cells traveled in a highly disorganized manner. These results suggest that peanut agglutinin-binding molecules are required for the segmental patterning of trunk neural crest migration.

Time-lapse recordings reveal different migratory behaviors in the presence of PNA compared to ephrin-B1–Fc. In soluble ephrin-B1–Fc-treated explants, both rostrally and caudally migrating neural crest cells traveled in a nonlinear, disoriented manner. In PNA-treated explants, by contrast, rostrally located neural crest cells moved normally, while caudally located neural crest cells traveled in a highly disorganized manner (Krull *et al.*, 1995), resembling neural crest cells in the presence of dimeric ephrin-B1–Fc (Fig. 5). This similarity suggests that neural crest cells may be receiving repulsive signals in both cases, either from the exogenous ephrin-B1–Fc or from the ephrin-B1 expressing caudal sclerotome cells. The migratory trajectories observed in the presence of PNA are very similar to those in the presence of ephrin-B1–myc, which does not stimulate receptor phosphorylation and signaling (Krull *et al.*, 1997), altering the segmental migration without eliciting abnormal cell behaviors. It is as yet unclear if the PNA binding molecule is ephrin-B1. However, we cannot rule out the possibility that there may be multiple inhibitory mechanisms in the caudal sclerotome.

VIII. Cell–Matrix Interactions Are Important for Neural Crest Emigration but Not Segmental Migration

After their emigration from the neural tube, neural crest cells enter a cell-free zone filled with extracellular matrix molecules. Many ECM molecules also line neural crest migratory pathways. Integrins are the primary receptors for extracellular matrix molecules on the surface of neural crest cells. Microinjection of integrin antibodies (Bronner-Fraser, 1985, 1986) or antisense oligonucleotides (Kil *et al.*, 1996) into embryos during cranial neural crest migration causes severe perturbations in neural crest development in the head.

Although blocking integrin function clearly affects the emigration of cranial neural crest cells (Bronner-Fraser, 1986), little has been known about their role in the trunk region. We have used the three-dimensional trunk explant system to examine the role of cell–matrix interactions in truck neural crest migration. By adding function-blocking antibodies that interfere with the $\alpha4$ subunit of integrin to whole trunk explants, we observed a number of interesting defects in neural crest cell movement. Neural crest emigration was completely blocked in about half of the cases, with a profound reduction in the overall number of neural crest cells that entered the rostral sclerotome in the other half (Kil *et al.*, 1998). Many neural crest cells appear to remain clustered at the intersomitic border rather than entering the sclerotome, as if the balance of adhesion was shifted away from cell–matrix adhesion and diverted toward cell–cell adhesion. However, the segmental pattern of migration was unperturbed in these explants. These results suggest that perturbing integrin function alters the properties of migratory cells, but does not alter their overall metameric pattern of migration. Thus, cell–matrix interactions may play a permissive role in the segmental migration of neural crest cells through the somites, but cannot play an instructive role.

IX. Other Inhibitory Cues in the Trunk

In addition to being absent from the caudal half of each sclerotome, avian neural crest cells are never observed in a space around the notochord, raising the possibility that the notochord may inhibit neural crest migration. When neural crest cells are cocultured with notochord, they avoid the region surrounding notochords (Newgreen *et al.,* 1986). Furthermore, grafting a notochord within the somites results in neural crest cells avoiding the implanted notochord (Pettway *et al.,* 1990). The inhibitory action of the notochord is age-dependent, being optimal in notochords derived from 2-day-old embryos and declining after 3 days (Pettway *et al.,* 1996). In addition to inhibiting neural crest cells, the notochord is thought to be inhibitory for motor axons (Tosney and Oakley, 1990). The notochord-derived substance appears to be a chondroitin sulfate proteoglycan of the aggregan family (Pettway *et al.,* 1996). Together with the data demonstrating that Eph ligands mediate inhibitory interactions in the caudal half of the sclerotome, the finding that the notochord inhibits neural crest cells suggest that the dominant "guidance" cues for trunk neural crest migration are inhibitory. Thus, the patterning of cell movement in the trunk may largely occur by exclusion from nonpermissive areas.

X. Conclusions

Neural crest migration is intimately linked to both the formation and segmentation of the somites. After differentiation of somites into dermomyotome and sclerotome, the sclerotomal compartment is subdivided into rostral and caudal halves. Although overtly similar, the caudal half sclerotome has a higher cell density (Ranchst and Bronner-Fraser, 1991) and different molecular markers. Notably, ephrins are selectively expressed in the caudal half sclerotome whereas their cognate Eph receptors are expressed on neural crest cells and rostral sclerotomal cells. Functional interactions between Eph receptors and ligands appear to restrict neural crest cells to the rostral half sclerotomal domain. This in turn leads to their segmental migration and the subsequent metameric distribution of neural crest-derived sensory and sympathetic ganglia.

References

Becker, N., Seitanidou, T., Murphy, P., Mattei, M.-G., Tipilko, P., Nieto, M. A., Wilkinson, D. G., Charnay, P., and Gilardi-Hebenstreit, P. (1994). Several receptor tyrosine kinase genes of the Eph family are segmentally expressed in the developing hindbrain. *Mech. Dev.* **47,** 3–17.

Bergemann, A. D., Cheng, H. J., Brambilla, R., Klein, R., and Flanagan, J. G. (1995). ELF-2, a new member of the Eph ligand family, is segmentally expressed in mouse embryos in the regions of the hindbrain and newly forming somites. *Mol. Cell. Biol.* **15,** 4921–4929.

Brambilla, R., and Klein, R. (1995). Telling axons where to grow: A role for Eph receptor tyrosine kinases in guidance. *Mol. Cell. Neurosci.* **6,** 487–495.

Bronner-Fraser, M. E. (1985). Alterations in neural crest migration by an antibody that affects cell adhesion. *J. Cell Biol.* **101,** 610–617.

Bronner-Fraser, M. (1986). Analysis of the early stages of trunk neural crest migration in avian embryos using the monoclonal antibody HNK-1. *Dev. Biol.* **115,** 44–55.

Bronner-Fraser, M. (1986). An antibody to a receptor for fibronectin and laminin perturbs cranial neural crest development. *Dev. Biol.* **117,** 528–536.

Bronner-Fraser, M., and Stern, C. (1991). Effects of mesodermal tissues on avian neural crest cell migration. *Dev. Biol.* **143,** 213–217.

Cheng, H.-J., Nakamoto, N., Bergemann, A. D., and Flanagan, J. G. (1995). Complementary gradients in expression and binding of ELF-1 and Mek4 in development of the topographic retinotectal projection map. *Cell* **82,** 371–381.

Davis, J., Cook, M., Stern, C. D., and Keynes, R. J. (1990). Isolation from chick somites of a glycoprotein that causes collapse of dorsal root ganglion growth cones. *Neuron* **4,** 11–20.

Davis, S., Gale, N. W., Aldrich, T. H., Maisonpierre, P. C., Lhotak, V., Pawson, T., Goldfarb, M., and Yancopoulos, G. D. (1994). Ligands for the Eph-related receptor tyrosine kinases that require membrane attachment or clustering for activity. *Science* **266,** 816–819.

Drescher, U., Kremoser, C., Handwerker, C., Loschinger, J., Noda, M., and Bonhoeffer, F. (1995). *In vitro* guidance of retinal ganglion cell axons by RAGS, a 25kDa tectal protein related to ligands for Eph receptor tyrosine kinases. *Cell* **82,** 359–370.

Duband, J. L., and Thiery, J. P. (1987). Distribution of laminin and collagens during avian neural crest cell development. *Development* **101,** 461–478.

Fan, C. M., and Tessier-Lavigne, M. (1994). Patterning of mammalian somites by surface ectoderm and notochord evidence for sclerotome induction by a hedgehog homolog. *Cell* **79,** 1175–1186.

Flenniken, A. M., Gale, N. W., Yancopoulos, G. D., and Wilkinson, D. G. (1996). Distinct and overlapping expression patterns of ligands for Eph-related receptor tyrosine kinases during mouse embryogenesis. *Dev. Biol.* **179,** 382–401.

Gale, N. W., Holland, S., Valenzuela, D., Flenniken, A., Pan, L., Henkemeyer, M., Strebhardt, K., Hirai, H., Wilkinson, D. G., Pawson, T., Davis, S., and Yancopoulos, G. D. (1996). Eph receptors and ligands comprise two major specificity subclasses, and are reciprocally compartmentalized during embryogenesis. *Neuron* **17,** 9–19.

Henkemeyer, M., Orioli, D., Henderson, J. T., Saxton, T. M., Roder, J., Pawson, T., and Klein, R. (1996). Nuk controls pathfinding of commissural axons in the mammalian central nervous system, *Cell* **86,** 35–46.

Holash, J. A., and Pasquale, E. B. (1995). Polarized expression of the receptor protein-tyrosine kinase CEK5 in the developing avian visual system. *Dev. Biol.* **172,** 683–693.

Irving, C., Nieto, M. A., DasGupta, R., Charnay, P., and Wilkinson, D. G. (1996). Progressive spatial restriction of Sek-1 and Krox-20 gene expression during hindbrain segmentation. *Dev. Biol.* **173,** 26–38.

Kenny, D., Bronner-Fraser, M., and Marcelle, C. (1995). The receptor tyrosine kinase QEK5 m-RNA is expressed in a gradient within the neural retina and the tectum. *Dev. Biol.* **172,** 708–716.

Keynes, R. J., and Stern, C. D. (1984). Segmentation in the vertebrate nervous system. *Nature* **310,** 786–789.

Kil, S., Krull, C., Cann, G., Clegg, D., and Bronner-Fraser, M. (1998). The alpha4 subunit of integrin is essential for neural crest migration. *Devel. Biol.* **202,** 29–42.

Kil, S. H., Lallier, T., and Bronner-Fraser, M. (1996). Inhibition of cranial neural crest adhesion *in vitro* and migration *in vivo* using antisense oligonucleotides. *Devel. Biol.* **179,** 91–101.

Krotoski, D., Domingo, C., and Bronner-Fraser, M. (1986). Distribution of a putative cell surface receptor for fibronectin and laminin in the avian embryo. *J. Cell Biol.* **103,** 1061–1072.

Krull, C. E., Collazo, A., Fraser, S. E., and Bronner-Fraser, M. (1995). Segmental migration of trunk neural crest: Time-lapse analysis reveals a role for PNA-binding molecules. *Development* **121,** 3733–3743.

Krull, C. E., Lansford, R., Gale, N. W., Marcelle, C., Collazo, A., Yancopoulos, G., Fraser, S. E., and Bronner-Fraser, M. (1997). Interactions between Eph-related receptors and ligands confer rostrocaudal polarity to trunk neural crest migration. *Curr. Biol.* **7,** 571–580.

Lallier, T., and Bronner-Fraser, M. (1991). Avian neural crest cell attachment to laminin: Involvement of divalent cation dependent and independent integrins. *Development* **113,** 1069–1084.

Layer, P. G., Alber, R., and Rathjen, F. G. (1988). Sequential activation of butyrlcholinesterase in rostral half somites and acetylcholinesterase in motoneurones and myotomes preceding growth of motor axons. *Development* **102,** 387–396.

Macdonald, R., Xu, Q., Barth, K. A., Mikkola, I., Holder, N., Fjose, A., Krauss, S., and Wilson, S. W. (1994). Regulatory gene expression boundaries demarcate sites of neuronal differentiation in the embryonic zebrafish forebrain. *Neuron* **13,** 1039–1053.

Munsterberg, A., and Lassar, A. (1995). Combinatorial signals from the neural-tube, floor plate and notochord induce myogenic bHLH gene-expression in the somite. *Development* **121,** 651–660.

Munsterberg, A., Kitajewski, J., Bumcrot, D., McMahon, A., and Lassar, A. (1995). Combinatorial signaling by Sonic hedgehog and Wnt family members induces myogenic bHLH gene-expression in the somite. *Genes Dev.* **9,** 2911–2922.

Nakagawa, S., and Takeichi, M. (1995). Neural crest cell–cell adhesion controlled by sequential and subpopulation-specific expression of novel cadherins. *Development* **121,** 1321–1332.

Newgreen, D. F., and Thiery, J. P. (1980). Fibronectin in the early avian embryo: Synthesis and distribution along the migration pathways of neural crest cells. *Cell Tissue Res.* **211,** 269–291.

Newgreen, D. F., Scheel, M., and Kastner, V. (1986). Morphogenesis of sclerotome and neural crest in avian embryos: *In vivo* and *in vitro* studies on the role of notochordal extracellular matrix. *Cell Tissue Res.* **244,** 299–313.

Norris, W. E., Stern, C. D., and Keynes, R. J. (1989). Molecular differences between the rostral and caudal halves of the sclerotome in the chick embryo. *Development* **105,** 541–548.

Oakley, R. A., and Tosney, K. W. (1991). Peanut agglutinin and chondroitin-6-sulfate are molecular markers for tissues that act as barriers to axon advance in the avian embryo. *Dev. Biol.* **147,** 187–206.

Perris, R., Krotoski, D., and Bronner-Fraser, M. (1991a). Collagens in avian neural crest cell development: Distribution *in vivo* and migration-promoting ability *in vitro. Development* **113,** 969–984.

Perris, R., Krotoski, D., Domingo, C., Lallier, T., Sorrell, J. M., and Bronner-Fraser, M. (1991b). Spatial and temporal changes in the distribution of proteoglycans during avian neural crest development. *Development* **111,** 583–599.

Pettway, Z., Guillory, G., and Bronner-Fraser, M. (1990). Absence of neural crest cells from the region surrounding implanted notochords *in situ. Dev. Biol.* **142,** 335–345.

Pettway, Z., Domowicz, M., Schwartz, N. B., and Bronner-Fraser, M. (1996). Age-dependent inhibition of neural crest migration by the notochord correlates with alterations in the S103L chondroitin sulfate proteoglycan. *Exp. Cell Res.* **225,** 195–206.

Ranscht, B., and Bronner-Fraser, M. (1991). T-cadherin expression alternates with migrating neural crest cells in the trunk of the avian embryo. *Development* **111,** 15–22.

Rickmann, M., Fawcett, J. W., and Keynes, R. J. (1985). The migration of neural crest cells and growth cones of motor axons through the rostral half of the chick somite. *J. Embryo. Exp. Morphol.* **90,** 437–455.

Smith, A., Robinson, V., Patel, K., and Wilkinson, D. G. (1997). The Eph A4 and Eph B1 receptor tyrosine kinases and ephrin B2 ligand regulate targetted migration of branchial neural crest cells. *Curr. Biol.* **7,** 561–570.

Stern, C. D., and Keynes, R. J. (1987). Interactions between somite cells: The formation and maintenance of segment boundaries in the chick embryo. *Development* **99,** 261–272.

Stern, C. D., Norris, W. E., Bronner-Fraser, M., Carlson, G. J., Faissner, A., Keynes, R. J., and Schachner, M. (1989). J1/tenascin-related molecules are not responsible for the segmented pattern of neural crest cells or motor axons in chick embryo. *Development* **107,** 309–320.

Tan, S. S., Crossin, K. L., Hoffman, S., and Edelman, G. M. (1987). Asymmetric expression in somites of cytotactin and its proteoglycan ligand is correlated with neural crest cell distribution. *Proc. Natl. Acad. Sci. U.S.A.* **84,** 7977–7981.

Tanaka, H., Agata, A., and Obata, K. (1989). A new membrane antigen revealed by monoclonal antibodies is associated with axonal pathways. *Dev. Biol.* **132,** 419–435.

Tosney, K. W., and Oakley, R. A. (1990). The perinotochordal mesenchyme acts as a barrier to axon advance in the chick embryo: Implications for a general mechanism of axon guidance. *Exp. Neurol.* **109,** 75–89.

Wang, H. U., and Anderson, D. J. (1997). Roles of Eph family transmembrane ligands in repulsive guidance of truck neural crest migration and motor axon outgrowth. *Neuron* **18,** 383–396.

Xu, Q., Alldus, G., Holder, N., and Wilkinson, D. G. (1995). Expression of truncated Sek-1 receptor tyrosine kinase disrupts the segmental restriction of gene expression in the xenopus and zebrafish hindbrain. *Development* **121,** 4005–4016.

Xu, Q., Alldus, G., MacDonald, R., Wilkinson, D. G., and Holder, N. (1996). Function of the Eph-related kinase RTK1 in patterning of the zebrafish forebrain. *Nature* **381,** 319–322.

Index

Please note: Index includes entries for volumes 47 and 48. Volume numbers are in **bold.**

F

Q

R

Contents of Previous Volumes

Volume 44

Volume 45

Volume 46